S0-CKS-238

ALLE · ZEIT · WACH
1842

L. Håkanson · M. Jansson

Principles of Lake Sedimentology

With 187 Figures

Springer-Verlag
Berlin Heidelberg New York Tokyo 1983

Dr. Lars Håkanson
National Swedish Environment Protection Board
Water Quality Laboratory Uppsala
S-750 08 Uppsala, Sweden

Dr. Mats Jansson
Institute of Limnology
University of Uppsala
S-751 22 Uppsala, Sweden

ISBN 3-540-12645-7 Springer-Verlag Berlin Heidelberg New York Tokyo
ISBN 0-387-12645-7 Springer-Verlag New York Heidelberg Berlin Tokyo

Library of Congress Cataloging in Publication Data. Main entry under title: Principles of lake sedimentology. 1. Sedimentation and deposition. 2. Lakes. I. Hakanson, Lars. II. Jansson, M. (Mats), 1947-. QE571.P74. 1983. 551.3'5. 83-16791.

Offsetprinting: Beltz Offsetdruck, Hemsbach/Bergstr.
Bookbinding: J. Schäffer OHG, Grünstadt.
2132/3130-543210

To create order in nature is the ultimate goal of most scientific endeavours. With this book, we cannot hope to achieve that goal, only a higher level of disorder.

Lake Sedimentology is a science with tradition:

"For example, in the area where the Danube flows out of Europe into the Pontus through several mouths a bank has been formed opposite the river by the mud which has been discharged from the mouths; it is nearly 120 miles in length, and a day's voyage out to sea, so that ships which are navigating the Pontus and are far away from the shore may easily, if they are sailing unwarily, run aground on certain parts of these shoals, which sailors call 'the Breasts'. The fact that these deposits are not formed close inshore but are driven so far away from the land must be accounted for as follows. In so far as the currents of the rivers are stronger than those of the sea and force a way through it, the earth and other matter washed down by the stream must continue to be thrust forward, and cannot be allowed to come to rest or subside. But when the impetus of the current has become spent amid the increasing depth and volume of the sea, then the earth sinks by reason of its natural weight and settles. This is why in the case of large and swift rivers where the sea near the coast is deep, the deposits build up at some distance away, but in the case of small and sluggish streams, the sand-banks are formed close to their mouths. This fact is clearly demonstrated when heavy rains occur, because at such times even quite small streams, when they have overpowered the waves at their mouths, force their mud out to sea to a distance which corresponds to the strength of their currents. We must certainly not refuse to believe in the extent of the sand-bank formed by the Danube or in the quantity of the stones, timber and earth which are washed down by the other rivers; indeed it would be foolish to do so when we often see with our own eyes some insignificant winter torrent quickly swell into a flood, scoop out a bed, and force its way through high ground, sweeping down with it every variety of wood, soil and stones, and forming deposits of such a size that in a short time the area may so change its appearance as to become unrecognizable."

From Polybius (ca. 200–113 B.C.), Book IV, Chapter 41,
The Rise of the Roman Empire (Penguin Classics 1979)

Contents

1 Prologue

Looking at the present textbooks in General Sedimentology, e.g., Leeder's (1982) *Sedimentology*, Friedman and Sanders' (1978) *Principles of Sedimentology*, Reineck and Singh's (1975) *Depositional Sedimentary Environments*, Allen's (1971) *Physical Processes of Sedimentation*, or Krumbein and Sloss' (1963) *Stratigraphy and Sedimentation*, it is evident that sedimentological processes in lakes, if mentioned at all, occupy only very few pages. From a strict geological perspective, that is not entirely unreasonable. But from a limnological, hydrological and "environmental" point of view, we believe that the present situation is unsatisfactory and that a textbook on the principles of sedimentation in lakes would fill a niche. The overall aim of this book is to present a comprehensive outline on the basic sedimentological principles for lakes, to focus on environmental aspects and matters related to lake management and control – on lake ecology rather than lake geology.

To achieve this, we have tried to create not a state-of-the-art publication or a catalogue on "who did what", but a "how and why" book, which in one volume comprises the fundamentals of lake sedimentology. Our hope is that the book may be used as a guide to those who plan, perform, and evaluate lake sedimentological investigations.

Our purpose is to give a multi-disciplinary perspective. This philosophy is shown in Fig. 1.1, where Lake Sedimentology is depicted as the hub of a wheel of Limnology, Ecology, Hydrology/Hydraulics, Geosciences, Pollution Control, Ecotoxicology, Chemistry/Physics, Mathematics/Statistics, and Methodology/Techniques. We do not,

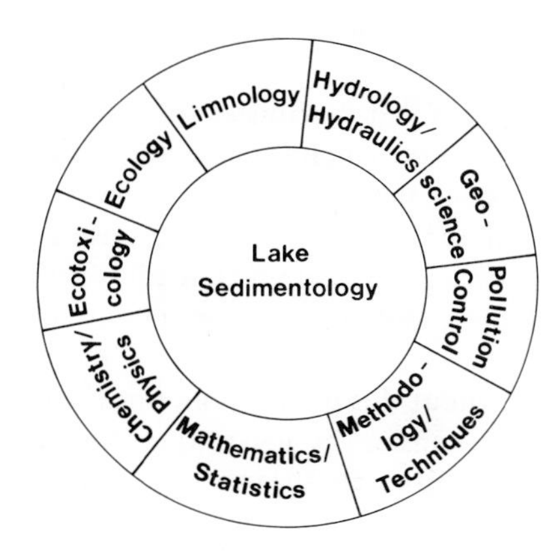

Fig. 1.1. Lake Sedimentology as a multi-disciplinary subject

however, advocate that Lake Sedimentology is the heart of all these sciences, only that Lake Sedimentology cannot be properly evaluated without the help of other related disciplines.

It is evident that no single textbook could or should encompass all these disciplines in an adequate manner, and that is not the aim in this context. Good textbooks and papers have already been published on:

- General sedimentology (as already mentioned).
- Early diagenesis (see, e.g., Kukal 1971, Berner 1980).
- Flocculation and formation of aggregates (see, e.g., Krone 1978, Kranck 1979, 1981).
- Entrainment of deposited particles and bottom dynamics (see, e.g., Fukuda 1978, Sheng and Lick 1979, McCall and Fisher 1980, Fukuda and Lick 1980, Håkanson 1982b).
- Sediment structures, varves and lamination (see, e.g., Reineck and Singh 1973, Edmondson 1975, Sturm 1979, Renberg 1981a).
- Turbidity currents and mass movements (see, e.g., Johnson 1964, Sturm and Matter 1978, Lüthi 1980).
- Bioturbation and sediment mixing (see, e.g., Davis 1974, Benninger et al. 1979, Brinkhurst and Cook 1980).
- Sediment pollution (see, e.g., Förstner and Müller 1974, Förstner and Wittmann 1979, Postma 1981).
- Sediment dating (see, e.g., Robbins 1978, Krishnaswami and Lal 1978).
- Deltaic sedimentation (see, e.g., Axelsson 1967, Gilbert 1975).
- Shore and coastal hydraulics (see, e.g., Muir Wood 1969, Swift et al. 1972).
- Estuarine sedimentation (see, e.g., Dyer 1979, Olausson and Cato 1980).
- Lake hydrodynamics (see, e.g., Lerman 1978, Fischer et al. 1979, Simons 1980).
- Limnology (see, e.g., Hutchinson 1957, 1967, 1975, Golterman 1975, Wetzel 1975).

All these aspects of sedimentology, and several more, are of interest in the present context. We have included this list here to emphasize the multi-disciplinary aim of this work *and* the impossibility of grasping everything in detail. So, from the broad perspective illustrated in Fig. 1.1, we will focus on some specific themes (Fig. 1.2), which have received less attention in the literature, in spite of the impressive list of publications just given.

This book is thus fairly detailed on bottom and sediment dynamics, on the geographical aspects of sedimentation in lakes, on methods of sampling and equipment, and on the fundamental principles and sediment properties necessary for a proper understanding of physical, chemical and biological sedimentological mechanisms. We will not discuss laboratory analyses of sediment parameters in great detail, since several textbooks are already available on that subject (see, e.g., Shoughstad et al. 1979).

Figures and diagrams play a prominent role throughout this publication, since our intention has been to increase the degree of comprehension by illustrating almost everything brought forward. The mathematics and statistics are meant to be "basic level", although we know in advance that many a "biologist" will regard some of

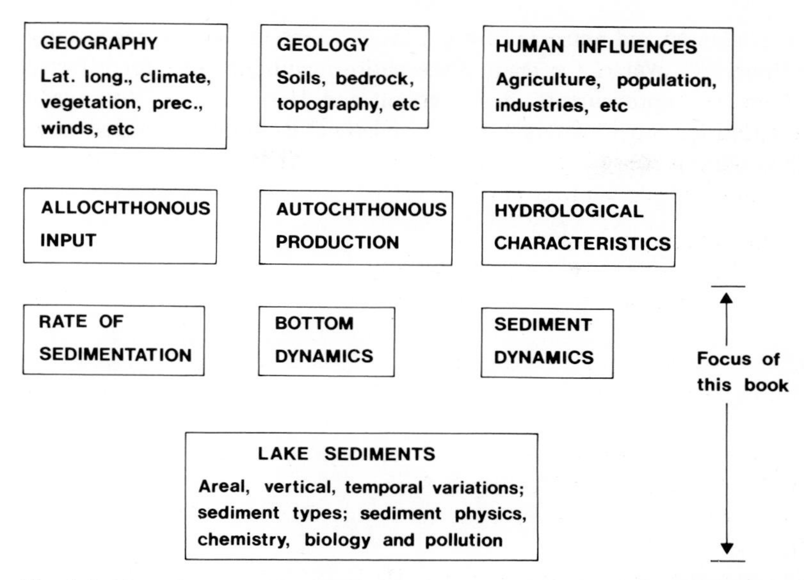

Fig. 1.2. Many factors influence the quality and quantity of particles settling in lakes, the processes of sedimentation, the processes in the sediments and the information value of sediment samples. The sedimentological principles in special focus in this book are indicated in this figure

the formulas with distrust and many "engineers" will look at the same formulas with mild contempt.

References are kept short because our aim has been to concentrate on facts rather than scientists. This philosophy entails, however, in this context that Swedes and Scandinavians have been somewhat overrepresented in the reference list in relation to their real influence on lake sedimentology. Another reason for this discrepancy is that to obtain a comprehensive framework for this book, we have used only a limited number of lakes as type lakes to illustrate the various principles. In that context, it was natural to us to use results from environments with which we are most familiar.

This book is not primarily directed towards professional sedimentologists, but more towards students at different levels, people at consultant bureaus without this specialization, and officials at central and regional offices responsible for lake management and control.

The birth of the work can partly be linked to two very successful conferences entitled *Interactions between sediments and freshwater,* held in Amsterdam in 1976 (see Golterman 1977) and in Hamilton in 1981 (see Sly 1982), and to the subsequent formation of the International Association for Sediment Water Science (I.A.S.W.S.), the objective of which is to support research and scientific exchange in the topic of *Environmental Sedimentology.* We believe that this is an interesting field of the natural sciences with great future potential.

We would also like to express our appreciation to all those who have influenced and assisted us in the arduous task of making this book. The infrastructure to the subject has been planted and nourished by numerous scientists. We are particularly indebted to Robert G. Wetzel for reading the entire manuscript and for offering many suggestions to improvements. Bert Karsson and Hans Kvarnäs have made substantial contributions to Chapters 6 and 7, and Nigel Rollison has helped us to minimize all Swenglish pecularities.

2 Lake Types and Sediment Types

In this chapter, we will first give a short outline of classification systems for lakes and lake sediments, and then demonstrate that a relationship exists between lake type and sediment type. "Lake sediments are the product of lake life. Consequently, they reflect the lake type". This quotation emanates from the old maestro of lake sedimentological studies in Sweden, Gösta Lundqvist, who wrote several major theses on the relationship between sediments, lake types and lake surroundings during the 1930's and 1940's (see, e.g., Lundqvist 1938, 1942). Thus, the central role of the bottom sediments was recognized more than 40 years ago. It was not, however, until during the last 15 years that lake sedimentology as an integrated, multi-disciplinary subject attracted a wider range of interest.

2.1 Lake Classifications

Lakes and lake sediments can be classified and named according to different principles and for different reasons. The aim of the subsequent section is to provide a brief background to the main sedimentological theme of this book and present three commonly used ways to describe and classify lakes.

2.1.1 Genetic Lake Types

In a geological perspective, lakes may be regarded as temporary objects on the face of the earth. As for geomorphological or biological species, lakes exist in three successive stages: youth, maturity and old age. Lakes generally originate from some drastic geological event, like volcanism, earthquake, or glaciation. The relief of the drainage area and the lake basin is under constant alteration by exogenic forces: weathering, erosion, transport and deposition. Practically all lakes in, for example, northern Europe and America have been influenced by the enormous form-creating power of the last glacial era, and if no tectonic catastrophe happens to these areas, all lakes will, in due course, be filled up by sediments and be transformed into land. These Nordic lakes are in focus in this book, but it should be noted that such lakes make up only a part of all existing types of lake. This is eminently clear from Hutchinson (1957). In the first chapter, entitled *The origin of lake basins,* of his large volume, he differentiates between 11 major lake types, which are divided in

76 sub-types (on 163 pages). In this context, where the emphasis is on sedimentology and not on the genetics, geology or morphology of lakes, we will only briefly summarize the 11 major lake types:

1. Tectonic Lakes

This group includes all lakes formed by movements of the deeper part of the earth's crust, e.g.:

- Lakes formed by epeirogenetic movements, for example, the Caspian Sea, the Sea of Aral and other lakes from the Ponto-Caspian region, which have been separated from the sea by several events of land elevation, e.g., the mountain-building that created the Alps.
- Lakes formed by tilting, folding or warping, e.g., the Central African Rift lakes. Some of the largest lakes on earth belong to this class of lakes (see Table 2.1).

Table 2.1. Morphometric data of some large tectonic lakes (Golterman 1975)

	Baikal	Tanganyika	Victoria	Dead Sea	Black Sea	Caspian Sea
Max. depth (m)	1,600	1,435	80	400	2,245	1,000
Surface area (km^2)	31,500	31,900	66,000	1,000	412,000	440,000
Max. width/max. depth	45	20	2,500	40	114	300

2. Volcanic Lakes

This rather diversified group of lakes includes, e.g.:

- Maars, calderas and crater lakes, as illustrated in Fig. 2.1A.
- Lakes formed by volcanic damming, as illustrated in Fig. 2.1B for Snag Lake, California.

3. Landslide Lakes

For example, lakes held by rockslides, mudflows and screes. Such lakes are often transitory, since the debris of the slide may be more or less easily eroded away by subsequent high waters (or similar events).

4. Glacial Lakes

"No lake-producing agency can compare with the effects of the Pleistocene glaciations. During most of the earth's history, there must always have been some tectonic, volcanic, and solution basins, and some lakes due to the action of wind and to the building activity in mature river valleys. The immense number of small lakes produced by glacial activity which now exist is, however, a quite exceptional phenomenon, presenting to the limnologist many times the number of individual basins which would have been available for study during most of the Mesozoic Tertiary eras." This quotation from Hutchinson (1957) highlights the reason why we focus on lakes from Nordic environments: they are abundant, they even dominate the landscape in northern Europe and America (see Fig. 2.2), they are of great ecological and practical importance.

Very many different types of glacial lakes exist and it may be difficult, not only to give a clear-cut distinction between various types of glacial lakes, but also between

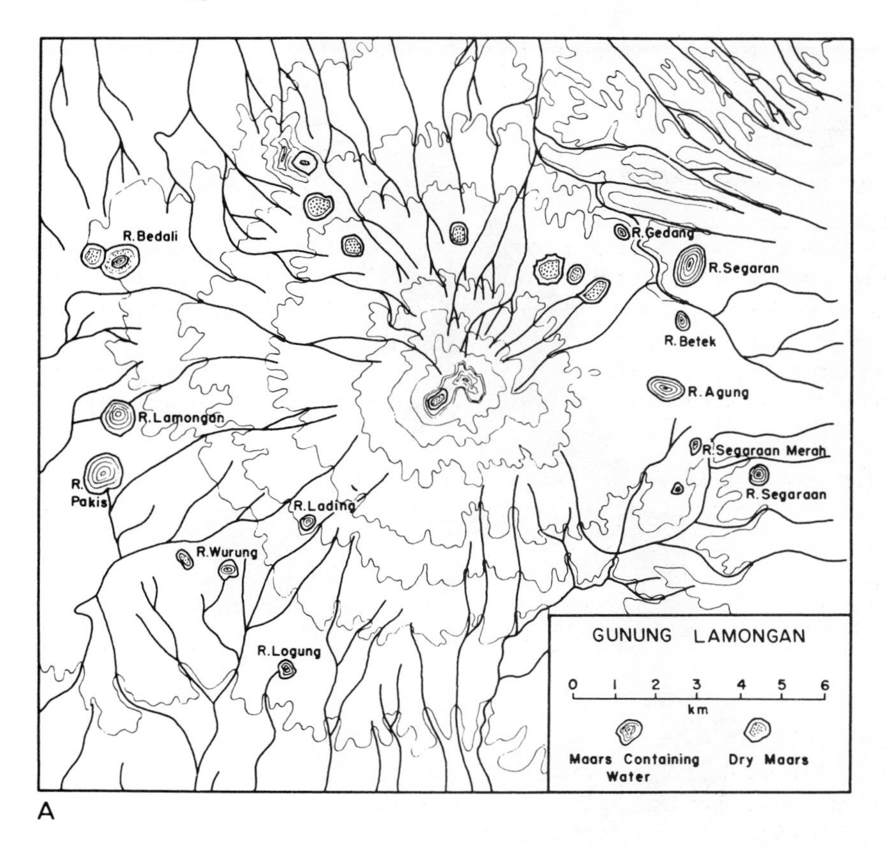

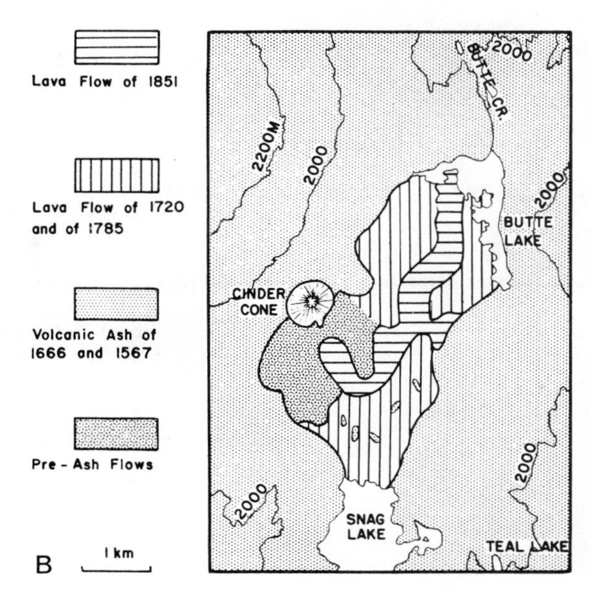

Fig. 2.1A, B. Examples of volcanic lakes. **A** Parasitic maars on the slopes of the volcano Gunung Lamongan, Java (Ruttner 1931). **B** Snag Lake, Lassen National Park, California, created by damming from lava flows from the volcano Cinder Cone (Finch 1937). Both figures from Hutchinson (1957)

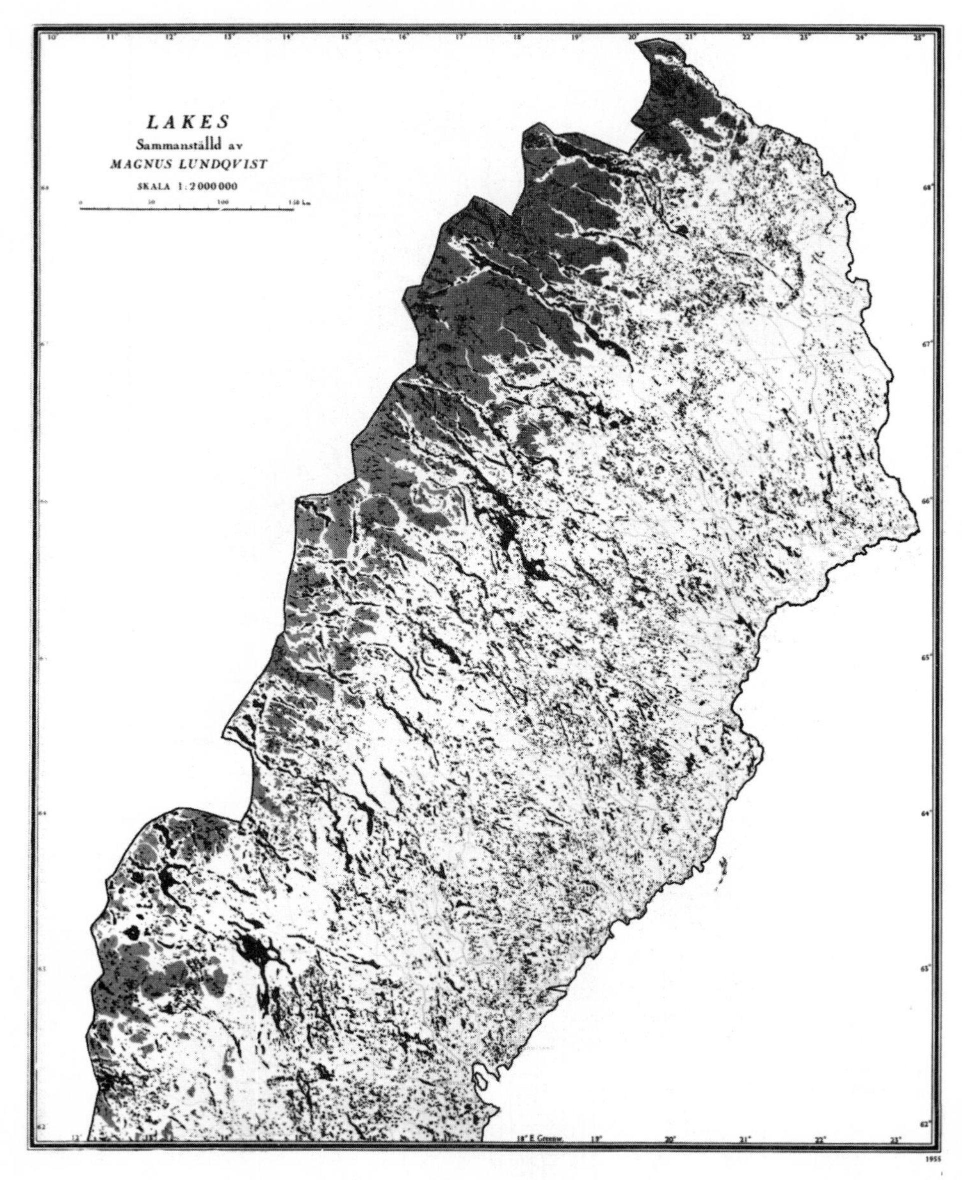

Fig. 2.2. Lakes constitute an important part of the Swedish landscape (Atlas över Sverige). Note, this reproduction does not clearly distinguish between the lakes and the fjell terrain in NW Sweden

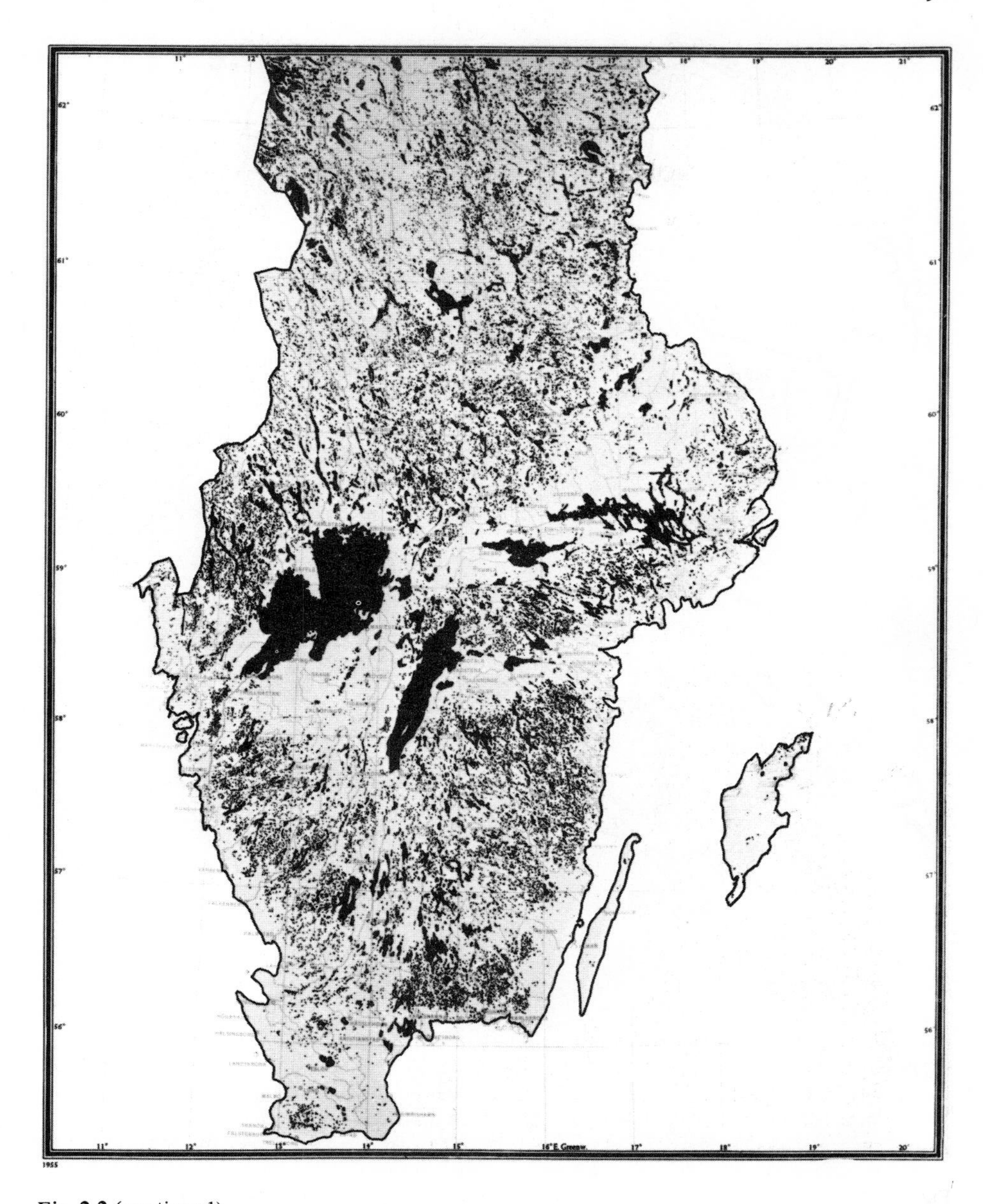

Fig. 2.2 (continued)

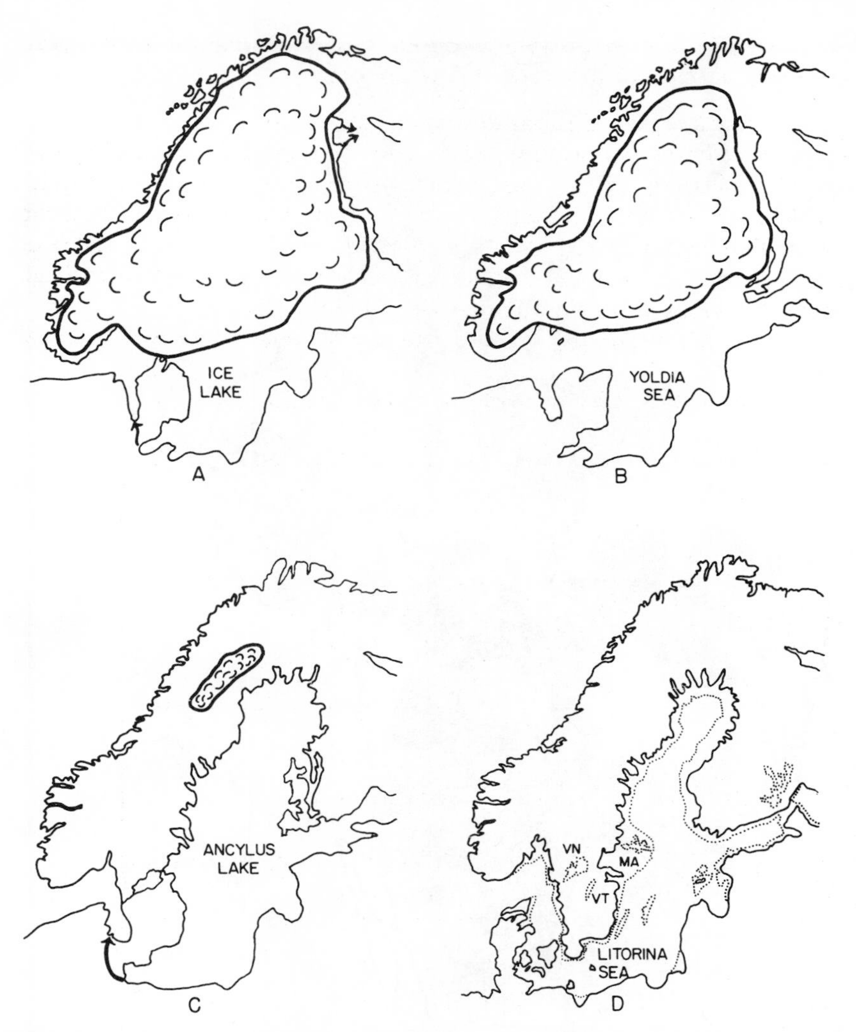

Fig. 2.3 A–D. The postglacial development of the Baltic. **A** The Baltic Ice Lake at about 8000 B.C. **B** The Yoldia Sea about 7900 B.C. **C** The Ancylus Lake, 6500 to 6000 B.C. **D** The Litorina Sea at about 5000 B.C., with the modern coast line and the positions of certain lakes indicated by *dotted lines*. *VT* Lake Vättern; *VN* Lake Vänern; *MA* Lake Mälaren (Hutchinson 1957)

glacial lakes and, for example, tectonic lakes. This is evident from Fig. 2.3, which illustrates the step-wise evolution of the Baltic region since the time of the last, or rather latest, glaciation. The whole landscape is formed by the glacial activity *and* by the isostatic land uplift (i.e., the land mass tends to rise after being depressed by the immense ice shield). Figure 2.3 illustrates the successive periods of freshwater (the Ice Lake and the Ancylus Lake) and seawater (the Yoldia Sea and the Litorina Sea) that have characterized the Baltic region from about 8000 B.C. to the present(dotted line in Fig. 2.3D). This history determines to a great extent the present limnological and sedimentological status of all lakes in the given region.

The following four main types of glacial lake may be distinguished:

- Lakes in direct contact with ice, e.g., lakes on or in ice and lakes dammed by ice.
- Glacial rock basins, e.g., cirque lakes, piedmont (or sub-alpine) lakes and fjord lakes. This important group includes the lakes of the English lake district, the lakes of the European Alps (e.g., Geneva, Constance and Brienz), the large lakes of sub-arctic Canada (e.g., the Great Slave Lake), the fjord lakes of Norway (e.g., Loenvatn and Mjoesa), and the Laurentian Great Lakes.
- Morainic and outwash lakes, e.g., lakes created by terminal, recessional or lateral moraines.
- Drift basins, e.g., kettle lakes and thermokarst lakes (see Fig. 2.4).

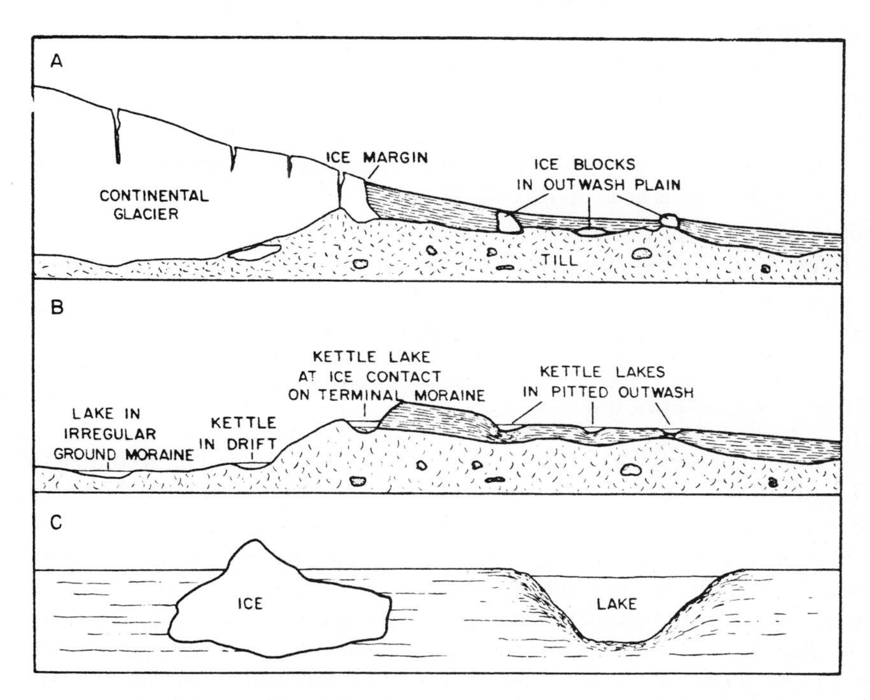

Fig. 2.4 A–C. Schematic illustration of formation of various types of kettle lakes. **A** Retreating continental ice with outwash plain containing stagnant ice blocks. **B** Lakes formed in outwash and in till by melting of ice blocks and in irregularity in ground moraine. **C** Formation of kettle lake in hole from melted ice block (Hutchinson 1957)

5. Solution Lakes

Water percolating any deposit of soluble material, like limestone, gypsum or rock salt, may produce cavities. A famous region for solution lakes is the Karst area at the Dalmatian coast. But solution lakes also exist in Florida, where the older metamorphic rocks are covered with Tertiary limestone sediments, and in the calcareous parts of the Alps.

6. Fluvial Lakes
Several types of lakes can be produced by the activity of running water:
- Plunge pool lakes or erosion lakes are created by cataracts or by the corrosive power of (present and past) water falls or rapids.
- Fluviatile dams are formed when rivers are dammed by sediments deposited across their courses from tributaries, e.g., delta lakes and meres.
- Meander lakes are associated with mature flood plains, e.g., oxbow lakes or levee lakes.

7. Aeolian Lakes
Lake basins formed by wind action are called aeolian. Such basins occur primarily in arid regions. Usual types are:
- Deflation basins formed by wind erosion.
- Lakes dammed by wind-blown sand (loess).

8. Shoreline Lakes
The general mode for the formation of shoreline lakes is by damming of material transported by longshore currents, e.g., tombolo lakes and spits lakes.

9. Organic Lakes
This is a small category of lakes which includes phytogenic dams, formed by blocking of vegetation, beaver dams and coral lakes.

10. Anthropogenic Lakes
That is, lakes, dams and excavations formed by man.

11. Meteorite Lakes
These lakes created in a very drastic way through the impact of a meteorite.

2.1.2 Trophic Level Classifications

The previous section dealt with a genetic approach to classify lakes, i.e., according to the geological history. In this section, we will use another approach and very briefly summarize a system based on the trophic concept, introduced in limnology by Naumann (1932) and Thienemann (1925, 1931) and later elaborated by numerous limnologists (see, e.g., Rodhe 1958, 1969, Wetzel 1975). The classical scheme, illustrating the trophic lake type (oligotrophic, eutrophic, dystrophic) in relation to the rate of supply of organic matter from allochthonous and autochthous sources, is given in Fig. 2.5. In this context, we will focus on these three lake types and not discuss more specific types of lakes (see, e.g., Wetzel 1975 or Eugster and Hardie 1978).

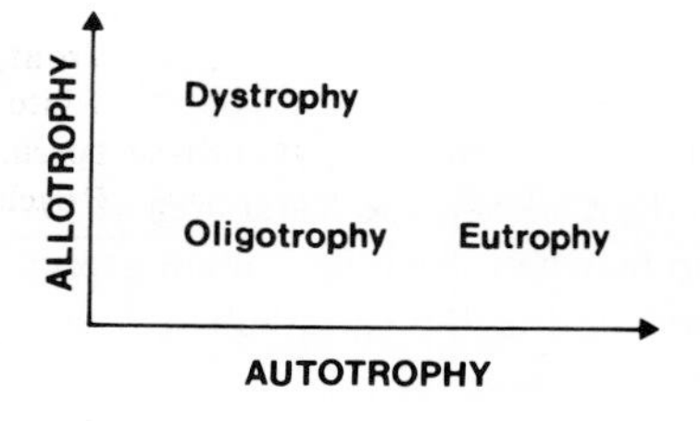

Fig. 2.5. Trophic types of lakes; from the relationship between the supply of organic matter from autotrophic (inside) and allotrophic (outside) sources (Wetzel 1975, Rodhe 1969)

Oligotrophic Lakes. The bioproductivity of lakes can be defined in different ways and is linked to a very complex pattern of causes and effects (see Fig. 2.6). Oligotrophic lakes are characterized by a low primary productivity, a high Secci-disk transparency, a low algal volume, comparatively low concentrations of the nutrient elements (P and N), and a fish fauna dominated by species like trout and whitefish. This is illustrated by numerical data in Table 2.2. It should be stressed that the numerical values in Table 2.2 separating the given classes *cannot* be applied uncritically. To emphasize that a continuous rather than discontinuous change exists between oligotrophy and eutrophy, the term *mesotrophy* is frequently used to define this area of transition.

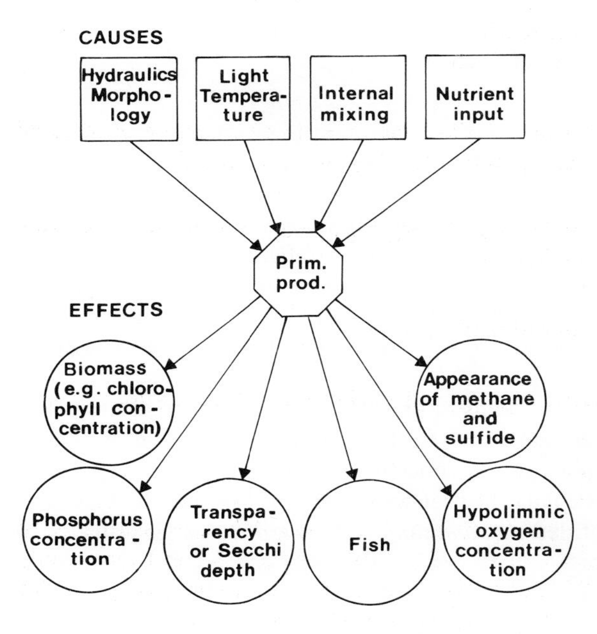

Fig. 2.6. Schematical illustration of the relationship between various trophic state indicators (biomass, P-conc., Secci disk transp., etc.), primary production and driving forces (or causes) (Imboden and Gächter 1979)

Table 2.2. Characteristic features in lakes of various trophic level (Håkanson 1980a)

Trophic level	Primary production ($g\,C\,m^{-2}\cdot yr^{-1}$)	Secci disk transparency (m)	Chlorophyll-a [a] ($mg\,m^{-3}$)	Algal volume [a] ($g\,m^{-3}$)	Total P [b] ($mg\,m^{-3}$)	Total N [b] ($mg\,m^{-3}$)	Dominating fish
Oligotrophic	< 30	> 5	< 2	< 0.8	< 5	< 300	Trout, White fish
Mesotrophic	25–60	3–6	2–8	0.3–1.9	5–20	300–500	White fish, Perch
Eutrophic	40–200	1–4	6–35	1.2–2.5	20–100	350–600	Perch, Roach
Hypertrophic	130–600	0–2	30–400	2.1–20.0	> 100	> 1000	Roach, Bream

[a] Mean value for the vegetation period (May–October)
[b] Mean value for the spring circulation

Eutrophic Lakes. Lakes with a high bioproductivity are called eutrophic. Such lakes generally have a fish community dominated by perch, roach and bream, high concentrations of nitrogen and phosphorus, high values of chlorophyll a, and a low transparency.

Dystrophic Lakes. Lakes that receive major amounts of their organic input from allochthonous sources are often called "brown water lakes", "cognac lakes", or, more scientifically, dystrophic or polyhumous, in reference to their high content of humic materials. Most of the organic content of dystrophic lakes is thus imported from surrounding terrestrial areas and not, like in the eutrophic lake type, produced in the lake itself. The productivity is generally low in dystrophic lakes.

2.1.3 Thermal Lake Types

From a sedimentological perspective, it is often important to scrutinize the prevailing type of stratification in the lake water, since this has bearings on the fate and distribution of the material dissolved and suspended in the water mass. F.A. Forel, who is often referred to as the father of limnology, is also the originator of the useful terminology to classify lakes according to their thermal/stratigraphic properties. Hutchinson and Löffler (1956) and Hutchinson (1957) have modified and improved earlier stages of that system into the one given here as Fig. 2.7. Before penetrating this figure, however, it is necessary to introduce the terms in general use to describe thermal stratifications in lakes.

Since the density varies with the temperature in a most characteristic way for water (see Fig. 2.8), with a maximum at +4°C, the thermal stratification in lakes is an interesting act of balance linked to the concepts of variation in time and space. In the "typical" case, one finds colder, heavier water near the bottom, the *hypolimnion* layer, a transition zone entitled *metalimnion,* and a warmer, lighter water mass

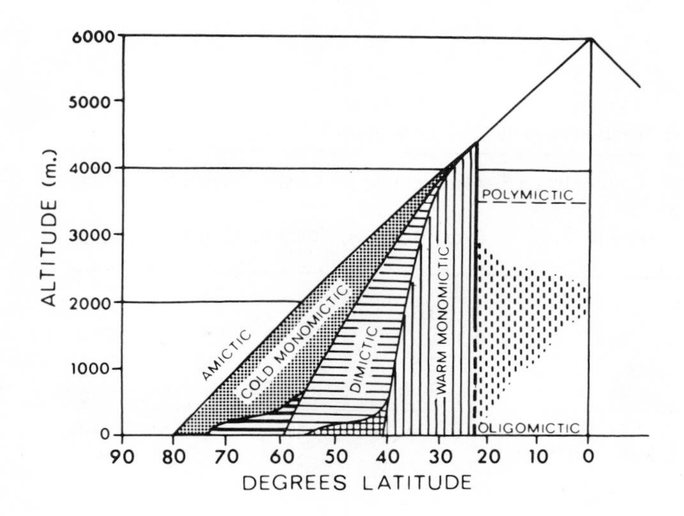

Fig. 2.7. Classification of thermal lake types with latitude and altitude. (Modified from Hutchinson and Löffler 1956, by Wetzel 1975)

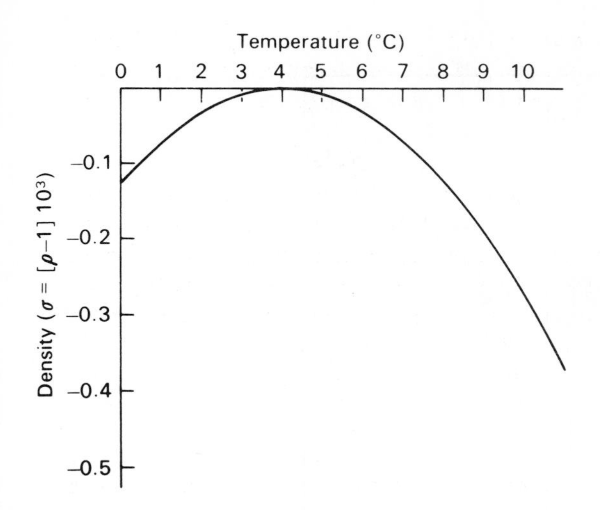

Fig. 2.8. The relationship between density and temperature of water (Lindell 1980)

at the surface, the *epilimnion* (see Fig. 2.9). This type of stratification can, however, be upset and altered in many different ways. Figure 2.10 illustrates the yearly variation in thermal stratification in a typical Swedish lake. During summer, from May 31 to October 18 in this particular year, the lake was thermally stratified with warmer water overlying colder water. The lake circulates, is homothermal, during spring and autumn and is inversely stratified during the winter, when the epilimnic water is cooler than the hypolimnic water.

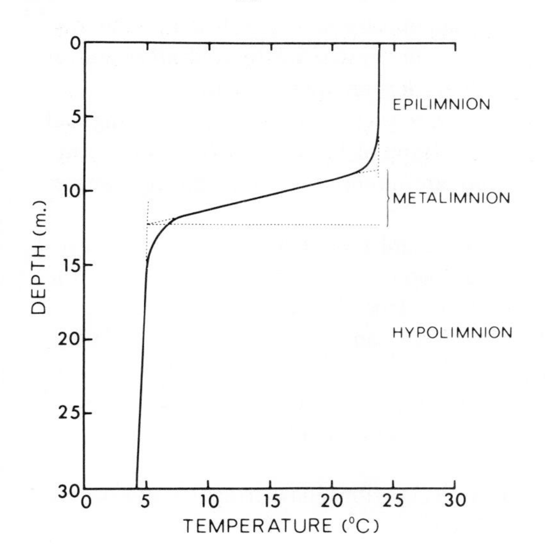

Fig. 2.9. Typical thermal stratification of a lake with epi-, meta-, and hypolimnic strata (Wetzel 1975)

The following thermal lake types may be differentiated:

- *Amictic lakes* never circulate and are always frozen. They predominate in Antarctica or the Arctic. Most heating to these lakes is by means of direct insolation through the ice and from the bottom through the sediments. These lakes have a specific inverse stratification (see Fig. 2.11).

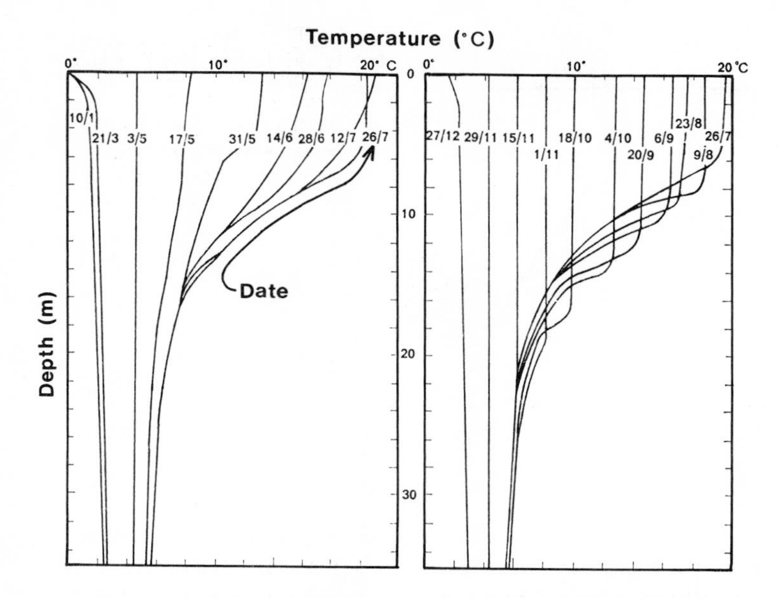

Fig. 2.10. The yearly variation in thermal stratification in a dimictic lake (Falkenmark and Forsman 1966)

- *Cold monomictic lakes* never have water temperatures above + 4°C and only one period of circulation (in the summer). They are restricted to cold areas and/or high altitudes and are often connected with glaciers and permafrost.
- *Dimictic lakes* circulate, by definition, twice a year, generally in the spring and fall. These lakes represent the most common type of thermal stratification within cool temperate regions of the world. They are directly stratified during summer and inversely stratified during winter.
- *Warm monomictic lakes* circulate once a year and their temperature never drops below + 4°C. They are characteristic of temperate zones, mountain areas in the subtropical latitudes and areas influenced by oceanic climates.
- *Polymictic lakes* have frequent circulation. They can be separated into *cold* polymictic lakes, which circulate at temperatures close to + 4°C, and *warm* polymictic lakes, which circulate at higher temperatures. Polymictic lakes are characteristic of regions with rapidly alternating winds, great daily temperature variations and small seasonal temperature variations.
- *Oligomictic lakes* are generally found in tropical regions, have irregular and/or rare circulations and temperatures well above + 4°C.

It should be stressed that this classification presupposes lakes with sufficient depth to form a hypolimnion. Thus, very shallow lakes are excluded.

It should also be noted that this classification is focussed on lakes in which circulation occurs throughout the entire water column, i.e., *holomictic* lakes. Many lakes are not holomictic but *meromictic,* i.e., do not undergo complete mixing.

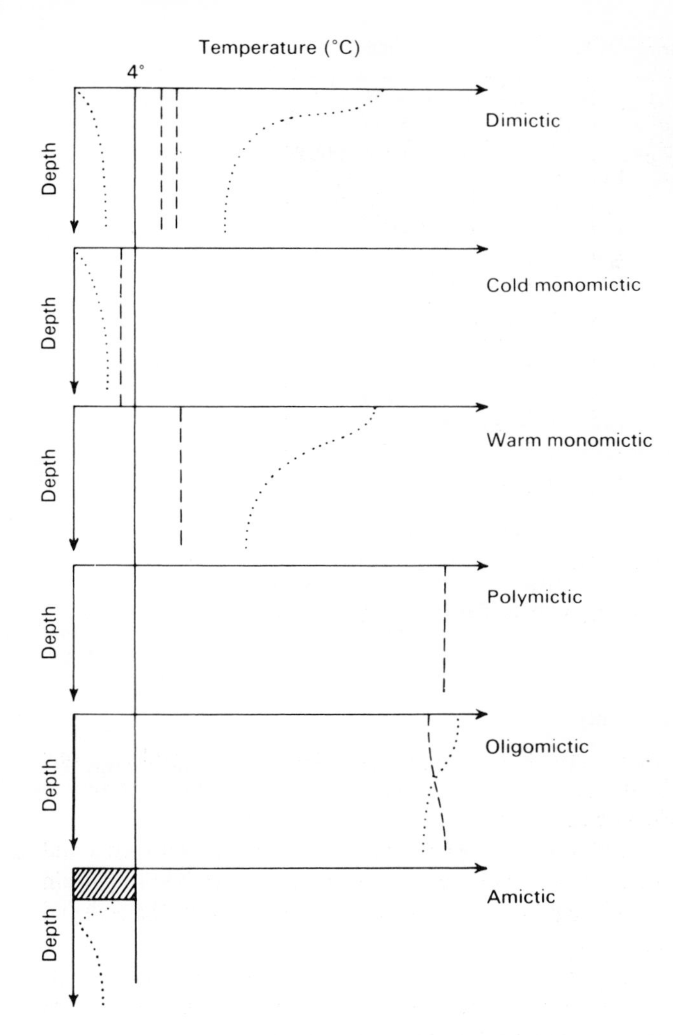

Fig. 2.11. Thermal lake types. *Dotted lines* winter and/or summer stratification. *Dashed lines* circulation temperatures and *shaded area* ice cover (Lindell 1980)

If the stratification in a lake is not due to thermal gradients but to salinity, the two strata are separated not by a *thermocline* but by a *halocline,* which is sometimes referred to as a *chemocline.*

2.2 Sediment Classifications

Sediments, like lakes, can be classified according to several principles and by means of many sediment parameters. At the outset, it should be emphasized that some confusion presently exists concerning classifications and nomenclature on lake sediments; no generally accepted classification system is available. The situation is significantly better for marine sedimentologists (see Seibold and Berger 1982). A systematical and elegant approach to classify marine deposits is given in Table 2.3. While waiting for a trenchant and practically useful general classification system for limnic sediments,

Table 2.3. Classification of marine sediment types (Seibold and Berger 1982)

Lithogenous Sediments. Detrital products of disintegration of pre-existing rocks (igneous, metamorphic, or sedimentary) and of volcanic ejecta (ash, pumice). Transport by rivers, glaciers, winds. Redistribution through waves and currents. Nomenclature based on *grain size* (gravel, sand, silt, clay). Additional qualifiers are derived from the *lithologic components* (terrigenous, bioclastic, calcareous, volcanogenic etc.), and from the *structure* and *color* of the deposits. Typical examples are as follows (environment in parentheses):

- Organic-rich *clayey silt,* with root fragments (marsh)
- Finely laminated *sandy silt,* with small shells (delta-top)
- Laminated quartzose *sand,* well sorted (beach)
- Olive-green homogeneous *mud* rich in diatom debris (upper cont. slope) (*mud* is the same as terrigenous clayey silt or silty clay)

Fine-grained lithogenous sediments (which become *shale* upon aging and hardening) are the most abundant by volume of all marine sediments (about 70%). This is largely due to the great thickness of continental slope sediments.

Biogenous Sediments. Remains of organisms, mainly carbonate (calcite, aragonite), opal (hydrated silica), and calcium phosphate (teeth, bones, crustacean carapaces). *Organic sediments,* while strictly speaking biogenous, are commonly treated separately. Arrival at the site of deposition by in situ precipitation (benthic organisms living there), or through settling via water column (pelagic organisms; coarse shells fall singly, small ones commonly arrive as aggregates). Redeposition by waves or currents. Redissolution common, either on sea floor or within sediment. Nomenclature is based on *type of organism* and also on *chemical composition.* Additional qualifiers from *structure, color, size, accessory matter.* Typical examples:

- Oyster bank (lagoon or embayment)
- Shell sand (tropical beach)
- Coral reef breccia (slope below coral reef) (*breccia* is coarse broken-up material, in this case reefal debris)
- Oolite sand, well sorted (strand zone, Bahamas)
- Light grey calcareous ooze, bioturbated (deep-sea floor)
- Greenish grey siliceous ooze (deep-sea floor)

Biogenous sediments are widespread on the sea floor, covering about one half of the shelves and more than one half of the deep ocean bottom, for a total of 55%. About 30% of the volume of marine sediments being deposited at the present time may be labelled biogenous, although they may have considerable lithogenous admixture.

Hydrogenous Sediments. Precipitates from seawater or from interstitial water. Also products of alteration during early chemical reactions within freshly deposited sediment. Redissolution common. Nomenclature based on *origin* ("evaporites") and on *chemical composition.* Additional qualifiers from *structure, color, accessories.* Typical examples:

- Laminated translucent halite (salt flat)
- Finely bedded anhydrite (Mediterranean basin, subsurface)
- Nodular greyish white anhydrite (ditto)
- Manganese nodule, black, mammilated, 5 cm diam. (deep Pacific)
- Phosphatic concretion, irregular slab, 15 cm diam., 5 cm thick, light brown to greenish, granular (upwelling area)

Hydrogenous sediments, while widespread (ferromanganese!) are not important by volume at present. At times in the past, when thick salt deposits were laid down in the newly opening Atlantic, in the Mesozoic, and much later in a dried-up Mediterranean (end of Miocene) the volume of hydrogenous sediments produced was considerable. The salinity of the ocean may have been appreciably lowered during those times.

we will discuss the two major approaches used to classify sediments: the genetic mode and the descriptive way.

2.2.1 Genetic Sediment Types

Seibold and Berger (1982) base their classification on the following gentic concepts:

- *Lithogenous* sediments, i.e., deposits that enter the oceans as particles and aggregates and settle onto the sea floor. In limnological context, the term allochthonous would be better for this type of sediment.
- *Hydrogenous* sediments, i.e., deposits emanating from materials precipitated out of solution.
- *Biogenous* sediments, i.e., materials connnected to organisms. In freshwater terminology, it is usual to name these two latter types of deposits *endogenic,* since they are the result of processes taking place within the water mass. But it should be noted that these genetic terms could be used in all aquatic environments, independent of salinity.

Another genetic approach is to focus on the geological rather than the "geographical" origin and to distinguish between, e.g., glacial, fluvial, glaciofluvial, postglacial or aeolian deposits, or to focus on the processes forming the sediments, e.g., shore sediments, slope sediments or turbidites. Norrman and Königsson (1972) used a genetic approach based on both geology and processes; their result on Lake Vättern, Sweden, is given in Fig. 2.12A.

A geochemical approach to the classification of recent sediments has been made by Berner (1981). His system is based on the presence or absence of dissolved oxygen and sulfide in the sediment. Sediments are first divided into *oxic* and *anoxic* categories. Anoxic deposits are then separated into *sulfidic* and *non-sulfidic,* depending upon the presence or absence of measurable amounts of dissolved sulfide. Anoxic non-sulfidic environments are further divided into *post-oxic,* resulting from oxygen removal without sulfate reduction, and *methanic,* resulting from complete sulfate reduction and methane gas formation. These classes often succeed one another during the course of early diagenesis in the order: oxic, post-oxic, sulfidic and methanic (see Chap. 5.4). These geochemical classes are intimately linked to the pH, Eh (redox potential), and iron and manganese status of the sediments. The system is given in Table 2.4.

Table 2.4. A geochemical classification of sedimentary environments. C refers to concentration in moles per liter. H_2S signifies total sulfide (Berner 1981)

Environment	Characteristic phases
I. Oxic ($C_{O_2} \geqslant 10^{-6}$)	Hematite, goethite, MnO_2-type minerals; no organic matter
II. Anoxic ($C_{O_2} < 10^{-6}$)	
A. Sulfidic ($C_{H_2S} \geqslant 10^{-6}$)	Pyrite, marcasite, rhodochrosite, alabandite; organic matter
B. Non-sulfidic ($C_{H_2S} < 10^{-6}$)	
1. Post-oxic	Glauconite and other Fe^{2+}-Fe^{3+} silicates (also siderite, vivianite, rhodochrosite); no sulfide minerals; minor organic matter
2. Methanic	Siderite, vivianite, rhodochrosite; earlier formed sulfide minerals; organic matter

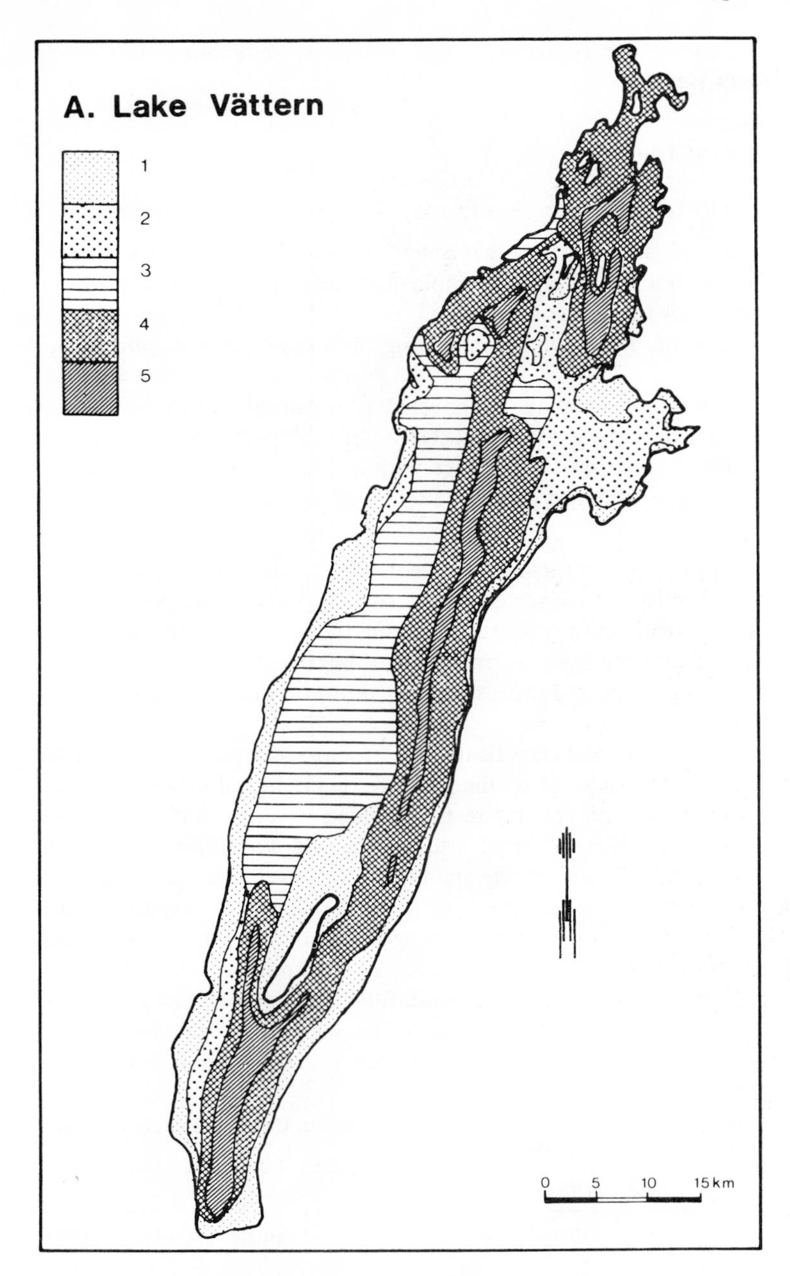

Fig. 2.12. A The sediment distribution in Lake Vättern according to Norrman and Königsson (1972). *1* Shore sediments (> 0.2 mm); *2* Glaciofluvial ice-margin deposits (> 0.2 mm); *3* Glaciofluvial suspended sediments (varved clay); *4* Postglacial to recent subaquatic slope deposits (sand to clay); *5* Silty sediments with a significant organic content (silty mud-gyttja)

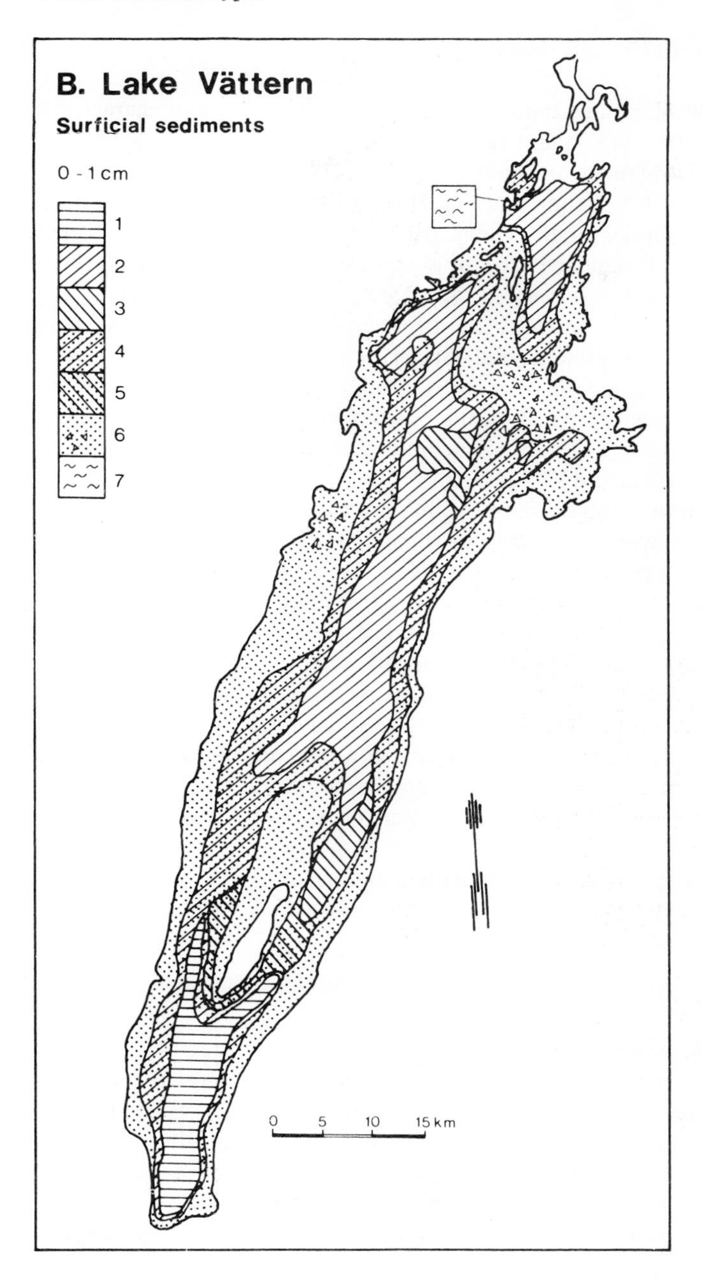

Fig. 2.12. B Distribution of surficial (0–1 cm) sediment types in Lake Vättern (Håkanson and Ahl 1976). *1* Grey to dark grey fine, recent sediments; *2* Brown to dark brown fine, recent sediments; *3* Light brown fine, old sediments; *4* Brown to dark brown mixed sediments; *5* Light brown mixed sediments; *6* Coarse sediments (sand, pebbles); *7* Black sediments with fibre

2.2.2 Descriptive Sediment Classifications

Descriptive sediment classifications can be based on numerous sediment characteristics, like color, texture, structure, grain size, organic content, algal content or benthic community. Certain schools focus on the organic part of the sediments (e.g., Wesenberg-Lund 1901, Potonié 1908, Naumann 1931, Hansen 1959a,b), others concentrate on the mineralogy, the physics or the chemistry of the sediments (see e.g., Kukal 1971, Thomas et al. 1972, Bågander 1976), or on any combination between all these possible alternatives. One alternative based on color (simple labels like grey – dark grey – light grey, brown – dark brown – light brown, black), texture (coarse, fine or mixed), age (recent or old = preindustrial), specific characteristics (fibre) and sediment depth (only surficial sediments 0–1 cm) was used for Lake Vättern by Håkanson and Ahl (1976). The areal distribution map of different sediment types as given by this system is shown in Fig. 2.12B to facilitate a comparison with the genetic approach adopted by Norrman and Königsson (1972).

A very elaborate, somewhat outdated, but still useful classification system focussed on the organic materials was introduced by Lundqvist (1938). He differentiated between various forms of *gyttja* and *dy* (see also v. Post 1862, Hansen 1959a) by means of determinations of the content of detritus, mineral grain size, limonite, and dominant plankton groups (diatoms and blue-green algae). In the present context, we will not enter the nomenclature jungle of organic sediments, and we will avoid using the terms gyttja and dy to describe lake sediments. Instead, and until a thorough systematic review has been presented, we would like to advocate the use of the loss on ignition (IG) as a rough estimate of the organic content and concentrations (and ratios) of carbon, nitrogen, phosphorus, silicon and carbonate, as a simple, practical and non-confusing means of describing the sediment type, in combination with the physical sediment parameters (see Chap. 4.1).

Another descriptive way to classify sediments is by means of the cone apparatus (see Chap. 3.5). The penetration depths of the three cones (L_1, L_2, L_3) will reveal the character of the sediments, i.e., the sediment type.

- Hard bottoms of, e.g., rocks, boulders, pebbles or shells yields low and/or irregular penetration depths.
- Sandy bottoms yield low and more regular penetration depths.
- Sandy silty bottoms or bottoms dominated by vegetation, typical for areas of transportation in lakes, give greater penetration depths, e.g., $L_1 = 1.5$, $L_2 = 5$, $L_3 = 8$ (cm). Hard horizons of layers of discontinuity will be revealed by the cones; e.g., a hard layer at 4 cm would yield (as compared to the 1.5–5–8 values above); $L_1 = 1.5$, $L_2 = 4$, $L_3 = 4.5$ (cm).
- Fine sediments (mud, gyttja, dy, clay, silt, etc.) yield comparatively high penetration depths.

The cone method is rapid, objective and can be used on all bottoms to obtain numerical values of the sediment type. It is, however, rather crude and may not distinguish between, for example, shells and sand. The method is summarized in Table 2.5.

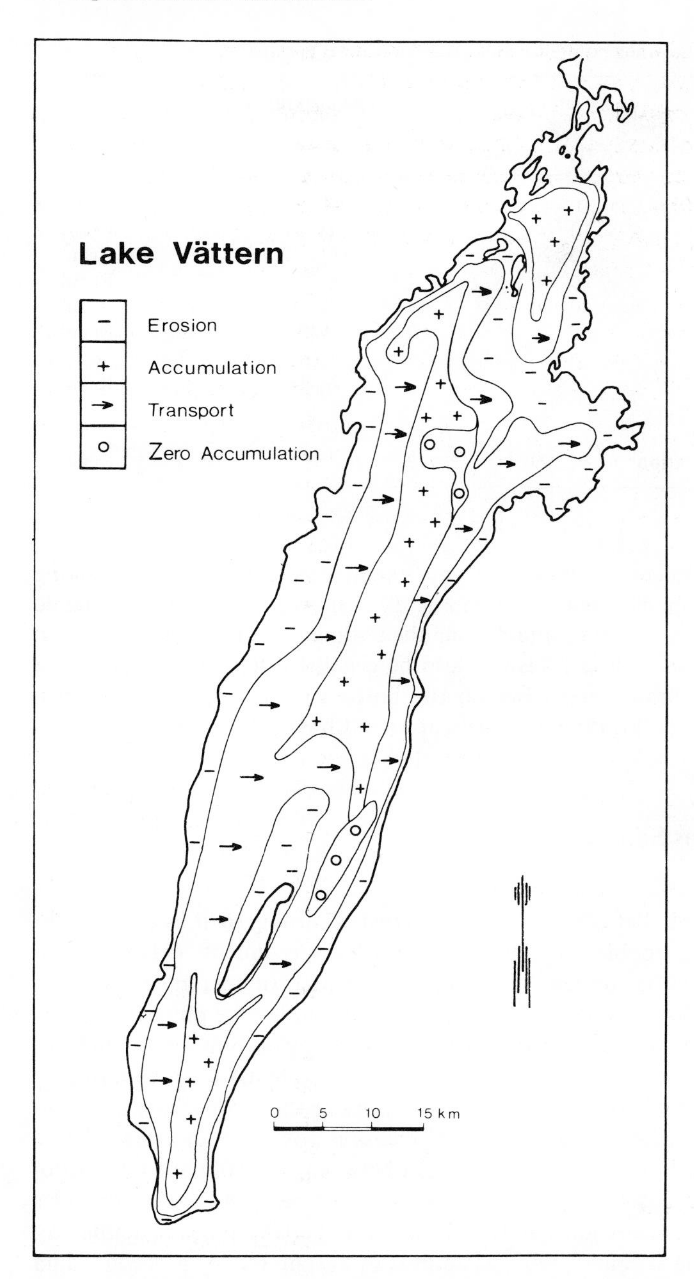

Fig. 2.13. Bottom dynamic types in Lake Vättern. Zero accumulation of fine particles in areas of low supply of material and high water energy (Håkanson and Ahl 1976)

Table 2.5. Classification of sediment types by means of the cone apparatus

Sediment type	Bottom dynamic type	Water content (in %) of superficial sediments	Cone penetration depths (cm) L_1	L_2	L_3	Notes
Rock, boulder	Erosion	0	5	0	2	Irregular
Pebbles, mussles	Erosion	⩽ 10	1	2.5	0	Irregular
Gravel, sand	Erosion	10–20	0–1	0.5–2.5	1–5	Regular
Sand	Erosion	20–50	< 1.5	1–4	1.5–7	Regular
Sandy silt	Transportation	50–80	< 3	3–10	5–15	Regular
Hard clay	Transportation	50–60	< 2	3–5	5–10	Regular
Fine sediments (silty)	Accumulation	75–85	1.5–4	6–15	8–20	Regular
Very fine sediments (muddy)	Accumulation	85–99	> 2	> 15	> 20	Regular

When discussing sediment types, it is always essential to consider the prevailing botton dynamic situation. This is clear from Fig. 2.13, showing the bottom dynamic map of Lake Vättern – the key to interpreting the sediment distribution maps given in Fig. 2.12. Sediments as well as lakes can also be classified descriptively from the type of benthic invertebrates prevailing on the bottom (see, e.g., Saether 1979). However, we will return to this type of classification in Chapter 5.3.

2.3 Lake Type versus Sediment Type

It has been demonstrated that lakes may be classified according to at least three different principles (origin, trophic level, stratification). Lake sediments may be distinguished by means of their origin and by numerous sediment parameters. Thus, when comparing lake type and sediment type, it is important to define the presuppositions. It should, at the outset, be noted that this is an interesting and important field of scientific endeavour, which is not very thoroughly or systematically explored at present. The aim of this section is not to try to synthesize the incongruous and scattered data from as many lake types and their sediments as possible into a new scheme that highlights probable but hitherto hidden links between the water and its "tachometer", as the sediments may be called. Instead, the aim is to emphasize the principles for *one* such search for correlations, and to do this for the lake types as defined by the trophic level and the sediment type, as defined by certain readily accessible data on the nutrient elements (P, N and C, see Hansen 1959a,b, 1961, Tadajewski 1966, Horie 1969, Kemp 1971, Premazzi and Ravera 1977, Ravera and Parise 1978). It should be stressed that no biological sediment parameters, like diatom frustules (Stockner 1972) or bottom fauna (Wiederholm 1979) will be discussed in this context.

It is probable that the best result would be obtained if several standard chemical sediment variables (inorganic-N, organic-N, total-N, apatite-P, NAI-P, organic-P, total-P, sulfides, carbonates, humic and fulvic substances, Fe/Mn-oxides and hydroxides, detritus, inorganic biogenic matter, silicates, chlorophyll, loss on ignition, etc.) were used, thereby obtaining a more balanced index of general validity. The objective here, however, is not to scrutinize these possibilities, but rather to demonstrate the information that could be gained from a *minimal* number of parameters.

We will start from the following facts and arguments (see Håkanson 1983c):

1. The trophic level of a lake can be expressed in terms of several more or less interrelated measurements. e.g., primary productivity, water transparency, chlorophyll a content, algal volume, concentrations of nutrients (N and P) and type of community of fish and bottom fauna (Table 2.2). These individual means of expressing trophic level only tell us part of the entire "trophic story", and they all show different types of relationships with the sediments (Hutchinson 1957, 1967, 1973, Wetzel 1975).

2. The average composition of planktonic materials is $C_{106}N_{16}P$, which gives a "natural" C/N ratio of 5.6 for sediments from very eutrophic waters dominated by plankton.

3. Phosphorus is generally distributed in a very complex manner in lake sediments, since P may appear in various chemical forms, which may be differently mobile and dependent on the sediment "climate" (e.g., pH and Eh; see J.D.M. Williams et al. 1971, 1976, Boström et al. 1982). Thus, a priori, one may assume that the P-content of lake sediments should not provide the same resolution of the trophic level of the lake as the N-content and/or the C-content.

4. The elemental composition of humic substances varies roughly accordingly (see Gjessing 1975, Golterman 1975):

C: 45%–60%, N: 3%–5%, P: $\approx$ 0.5% (of the dry weight).

This gives a C/N ratio for humus in the range of 10–20, which is significantly higher than for plankton. Thus, the C/N ratio of lake sediments may be used as a criterion to distinguish humosity of lakes (Hansen 1961).

5. The predominantly minerogenic deposits, like sand, varved (glacial) clay, igneous rock and soils, have a low organic content and low concentrations of N and P, but generally high or very high C/N ratios. This is illustrated in Table 2.6, and means that many oligotrophic lakes, which are dominated by minerogenic materials, would have sediments with high C/N ratio *and* low organic content. Lakes dominated by humus would also have sediments with high C/N ratio, but here the organic content would be high.

6. The sediments contain the following three major (in this context) components: the *organic* matter, which can be expressed in a simple but crude way by the loss on ignition (IG, the *minerogenic* matter (Si), and the *inorganic biogenic material,* which is made up by diatome frustules and biogenic precipitated calcium carbonate (the D-group). From these premises Hansen (1961) used the difference in minerogenic matter (Si) and the inorganic biogenic matter (D) as a measure of the level of oligotrophy to eutrophy; in oligotrophic lakes the minerogenic component (Si) is large and the content of inorganic biogenic matter low; in eutrophic lakes the opposite is valid. He also used the C/N ratio as a rough measure of humosity; dystropic (or polyhumic)

Table 2.6. Data on nitrogen (N), carbon (C) and loss on ignition (IG) in various media

	N (%, ds)	C (%, ds)	C/N	Reference
Varved clay	0.08	0.55	6.9	Kögler and Larsen (1979)
Igneous rock	0.02	0.20	10	Bowen (1966)
Soils	0.1	2	20	Bowen (1966)
Fresh water	$0.23 \cdot 10^{-4}$	$11 \cdot 10^{-4}$	48	Bowen (1966)
Land animals	10	46.5	4.7	Bowen (1966)
Land plants	3	45.4	15	Bowen (1966)
Water vegetation (Fontinalis)	3.2	29.8	9.3	Hansen (1956)
Lake sediments	N	IG	C/N	
Sand, Lake Vänern	0.3	1.8	> 20	Håkanson (1977b)
Sand, Lake Vättern	0.4	2.2	> 18	Håkanson (1977b)
Silt [a], Lake Vänern	0.6	3.2	> 18	Håkanson (1977b)
Silt, Lake Vättern	0.7	3.3	> 16	Håkanson (1977b)

[a] Post glacial and preindustrial silt

lakes should have a C/N ratio larger than 10–15 (by definition dystrophic means rich in humus); oligohumic lakes should have C/N ratios lower than 10. The diagram is illustrated in Fig. 2.14. The Si-content can be determined as the difference between total SiO_2 (quartz) and alkali soluble SiO_2.

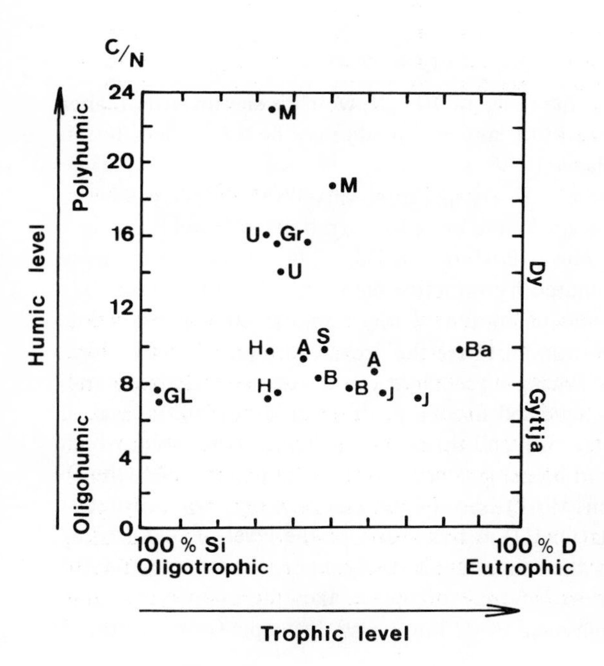

Fig. 2.14. Relationship between sediments and lake type in Danish lakes. *A* Almind Sø ; *B* Borresø; *Ba* Bavelse Sø; *GL* Grane Langsø; *Gr* Gribsø; *H* Hampen Sø; *M* Mørkesø; *S* Slauensø; *U* Uglesø. (After Hansen 1961)

7. It is well established that a strong positive relationship exists between nitrogen content and organic content (or loss on ignition) in surficial sediment samples from different sites in a lake (Mackereth 1966, Kemp 1971). This is also demonstrated in Table 2.7 with data from nine Swedish lakes/basins. The correlation between N-content and loss on ignition (IG) is very important in this approach, and to describe this relationship we will define two measures describing two characteristics of the regression line: (1) the slope coefficient will be called BPN (bioproduction number), and (2) the N-content on the regression line for the IG-value of 10% will be called BPI (bioproduction index). It should be stressed that this nomenclature is *only* utilized for reasons of simplicity to describe these two properties of the regression line. The concepts are illustrated in Fig. 2.15. It should be noted that the BPN-value includes data from many sample sites distributed over the entire lake area, from shallow waters with low IG-values to the deep-hole, which generally has the highest IG-value. The BPI-value, on the other hand, gives a normalized measure related to one specific organic content, namely IG = 10%.

Table 2.7. The relationship between N-content and IG-content of surficial (0–1 cm) sediment samples from eight Swedish lakes/basins (Håkanson 1981b)

Lake/basin	n	r	Regression line	BPI	BPN
Hjälmaren [a]	35	0.73	N = 0.70 · IG – 0.16	6.8	0.70
Hjälmaren [b]	21	0.96	N = 0.59 · IG + 0.34	6.3	0.59
Mälaren	18	0.97	N = 0.40 · IG + 0.36	4.4	0.40
Vänern	95	0.86	N = 0.29 · IG – 0.02	2.9	0.29
Vättern	90	0.78	N = 0.24 · IG + 0.42	2.9	0.24
Blacken	24	0.73	N = 0.38 · IG + 1.23	5.0	0.38
Västerås bay	18	0.81	N = 0.57 · IG – 0.64	5.1	0.57
Dalbosjön	32	0.87	N = 0.32 · IG – 0.06	3.1	0.32
Värmlandssjön	63	0.84	N = 0.30 · IG – 0.09	2.9	0.30

[a] Data from the 1977 sampling programme
[b] Data from the 1971 sampling programme
n = number of analysis; r = correlation coefficient, BPI = bioproduction index, BPN = bioproduction number

8. In this context, we will not consider the vertical variation in sediment cores and the factors influencing N, C and IG in this "historical" dimension (Mackereth 1966, Bloesch 1977, Nikaido 1978). The focus here is entirely on surficial sediments and the areal dimension. It is evident that the information value of suficial sediments is influenced by several factors, like rate of sedimentation, resuspension, bioturbation, degradation, mobilization, etc., and that the sediments from a defined layer (0–1 cm) yield time-integrated information. This time-resolution is, generally, primarily governed by the rate of sedimentation, which varies areally and temporally in a lake, implying that the 0–1 cm layer would represent a different time-span at different sites. Consequently, the sediments would only provide a long-term average of the trophic level of a lake.

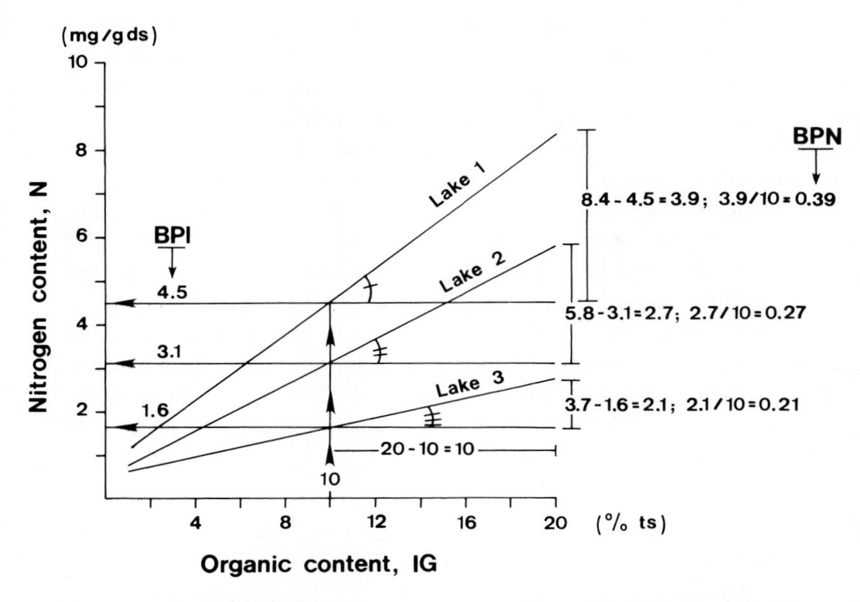

Fig. 2.15. Determination of bioproduction number *(BPN)* and bioproduction index *(BPI)* for three hypothetical lakes. The BPN-value is obtained as the slope coefficient of the regression line between N- and IG-content of surficial (0–1 cm) sediment samples which provide an even area cover of the bottom area, and the BPI-value is given as the N-content on the regression line for the IG-content 10% (Håkanson 1981b)

These are the key factors underlying the diagram illustrated in Fig. 2.16. We may summarize the information in this diagram accordingly:

I. The relationship between N-content (total-N) and loss on ignition (IG) of surficial sediments can be used to classify lake trophic level if more than 50% of the bottom area has surficial sediments with IG-contents lower than 20%. Then the slope coefficient, defined as the BPN-value, will yield best resolution of the trophic character of the lake. This is illustrated in Table 2.8 with empirical data from 12 Swedish lakes. The sediment indicators of trophic level (BPN, BPI, C/N ratio) are tested against a weighted measure of lake trophic level (TL, based on data on transparency, chlorophyll a and the ratios between total-N of lake water to mean water depth, $\bar{D}$, and total-P to $\bar{D}$). The main reason for this conclusion is that the slope coefficient (BPN) accounts for the areal spread and the prevalent bottom dynamics to yield a *lake* specific measure of lake trophic level, whereas any point on the regression line, as given, for example, by the C/N ratio only yields *site* specific information.

II. Oligotrophic lakes have BPN-values smaller than 0.33, mesotrophic lakes have BPN-values between 0.33 and 0.45, eutrophic lakes have BPN-values in the range of 0.45 to 0.65, and very eutrophic lakes have BPN-values larger than 0.65.

III. For lakes dominated by IG-contents in the range from 20% to 30%, the BPN-approach may be used to estimate lake trophic level, but with reduced validity.

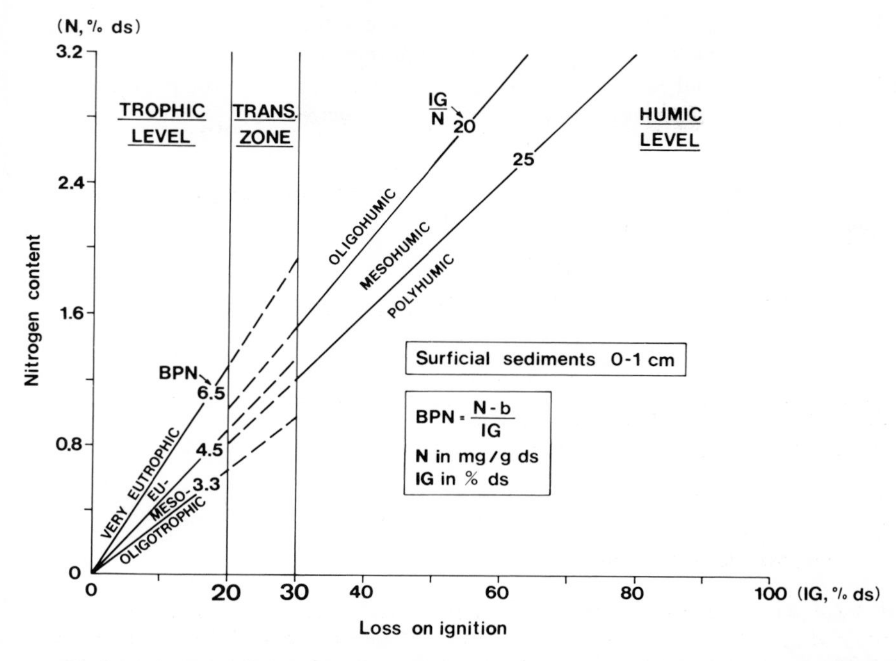

Fig. 2.16. Diagram illustrating the relationship between nitrogen concentration and loss on ignition of surficial sediments (0–1 cm) relative to lake trophic level and lake humic level (Håkanson 1983c)

Table 2.8. Regression lines and correlation coefficients for some of the most important relationships between water indicators (TL and P/D̄) and sediment indicators (BPN, BPI, C/N) of trophic level. Calculations based on data from 12 Swedish lakes/basins (Håkanson 1981b)

Regression line	Corr. coeff.
TL = 180.28 · BPN – 33.59	0.86
TL = 18.07 · BPI – 32.89	0.72
TL = – 6.81 · C/N + 128.45	– 0.76
TL = 1.16 · P/D̄ + 10.70	0.98
P/D̄ = 159.43 · BPN – 40.01	0.90
P/D̄ = 17.05 · BPI – 44.56	0.81
P/D̄ = – 6.06 · C/N + 103.75	– 0.80
BPN = 0.099 · BPI + 0.012	0.83
BPN = – 0.04 · C/N + 0.88	– 0.84

IV. The BPN-method *cannot* be used to express trophic level when more than 50% of the lake surface has higher IG-contents than 30%. This is illustrated in Fig. 2.17 with empirical data from 71 lakes covering most types of trophic and humic levels. For such lakes the relationship between N-content and IG-content can be used (see

Hansen 1961) to express lake humic level. Oligohumic lakes will have IG/N ratios lower than 20, mesohumic lakes, which constitute the transition zone between oligo- and polyhumic lakes, would have IG/N ratios between 20 and 25. And true polyhumic lakes, would have IG/N ratios higher than 25.

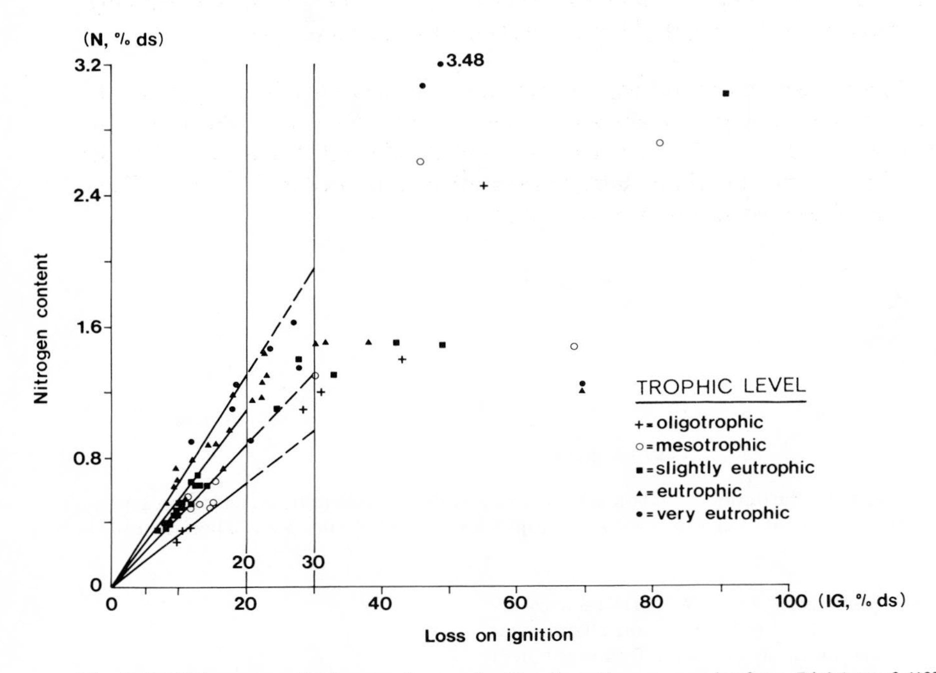

Fig. 2.17. Nitrogen content versus loss on ignition in sediment samples from 71 lakes of different trophic level (Håkanson 1983c)

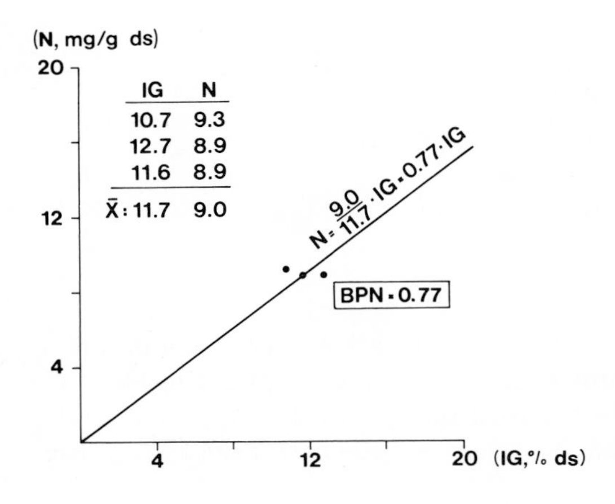

Fig. 2.18. Example illustrating the determination of the BPN-value from only three empirical data from surficial lake sediments (0–1 cm)

It should be stressed that:

- The dystrophic lake type is not fundamentally different from the oligotrophic-eutrophic series of lake (Hansen 1956).
- In nature there are only smooth and continuous transitions between various lake types.
- Lakes may, for example, be both eutrophic (have a high bioproduction) *and* poly-humic (have a high content of humic substances) at the same time.

If only a limited number of sediment data are available from a lake, e.g., from the deepest part, then a regression analysis cannot be made and a BPN-value cannot be determined. In that case, the most adequate estimate of the BPN-value would be the slope coefficient determined according to the method illustrated in Fig. 2.18 utilizing three empirical pairs of data from Lake Hjälmaren, Sweden.

3 Methods of Sampling

The aim of this chapter is to give a brief account of theoretical and practical aspects, and advantages and disadvantages of various types of sediment samplers, to discuss sampling nets and sample preparation and, finally, to give a short outline on the use and limitations of the sediment trap technique. Reviews of samplers and sampling techniques have been given by, e.g., Hopkins (1964), H.E. Wright et al. (1965), Bouma (1969) and Sly (1969).

3.1 General Requirements on Sampling Equipment

A large number of samplers have been designed for specific purposes and for sampling in different environments. Most sediment sampler can be classified as *grab* samplers (see Fig. 3.1) or *core* (or gravity or punch, see Fig. 3.2) samplers. The lack of uniform methods of sampling and sample preparation has impeded interpretations and comparisons of results from lake sedimentological studies. Since the benefits of comparability are generally accepted, the conditions for sampling in relatively remote and unaccessible lakes shoud determine the technical limitations of equipment constructed for general use. The main problem, however, is to take undisturbed samples, and in order to achieve this certain requirements must be met:

1. To eliminate an undesirable pressure wave under the sampler during lowering, the equipment should permit a free water passage. This is particularly important in bottom faunistic studies and when the sediments are loose. Most of the grab samplers in Fig. 3.1 do not meet the requirement for free water passage during lowering.

2. To minimize the frictional resistance and the risks for sediment deformation and compaction during sediment penetration, it is important to use coring tubes with a relatively small wall thickness compared to the sample area, tubes with smooth inside surfaces, with a moderate outside clearance and with a sharp edge and a small edge angle; the valve must allow unrestricted flow during sediment penetration and the mechanism must not close until the sampler is pulled up. Sediment compaction in coring tubes can be revealed by means of a double-stitched tape placed on the outside wall of the tube – sediment intrusion should be at the same level as the "dirt" verge of the tape. Hvorslev (1949) recommended that the cross-section area of the edges should be less than 10% of the sample area (the Hvorslev ratio). This requirement cannot be met for small tubes.

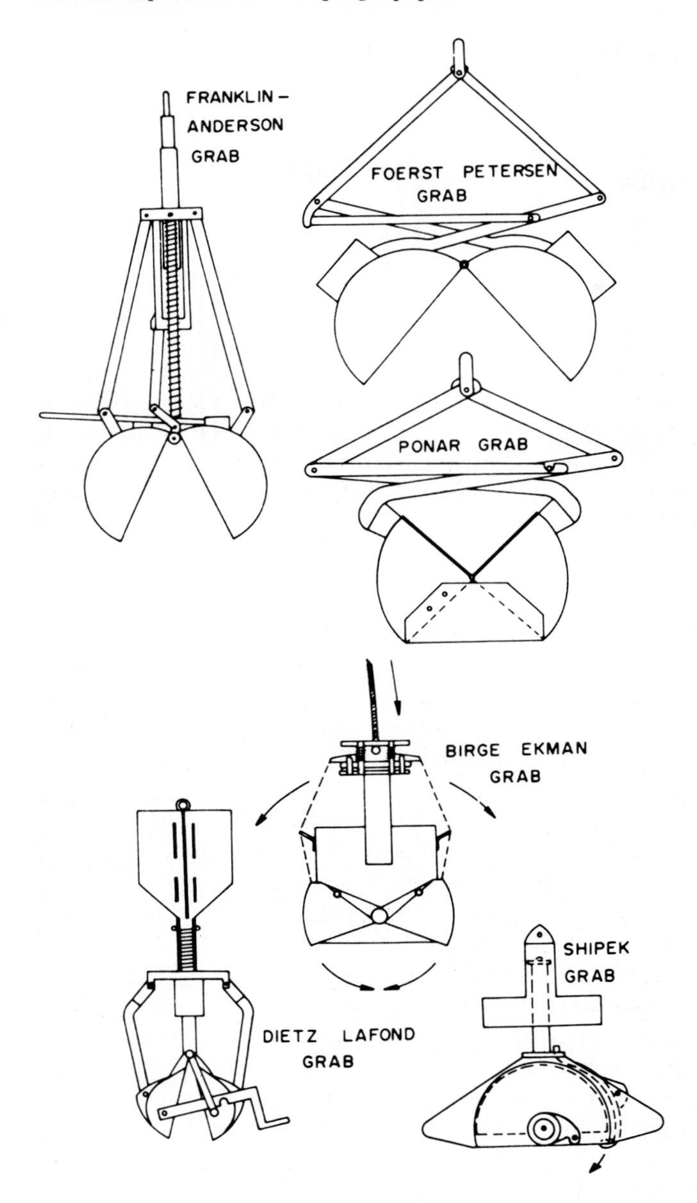

Fig. 3.1. Various types of grab samplers

3. To prevent the material from being lost during retrieval, the sampler must be fitted with an efficient closing mechanism. The ideal closing mechanism would operate in situ and close tight at both ends of the tube. The Jenkin sampler (Fig. 3.2) closes at both ends of the tube, but causes severe outside clearance problems; small tubes do not require closing at the bottom end, but for small tubes it may be difficult to fulfill the recommendation given by Hvorslev. For example, if the edge is 0.1 cm thick, then the inner diameter must be 4.1 cm; if the edge is 0.2 cm, the inner

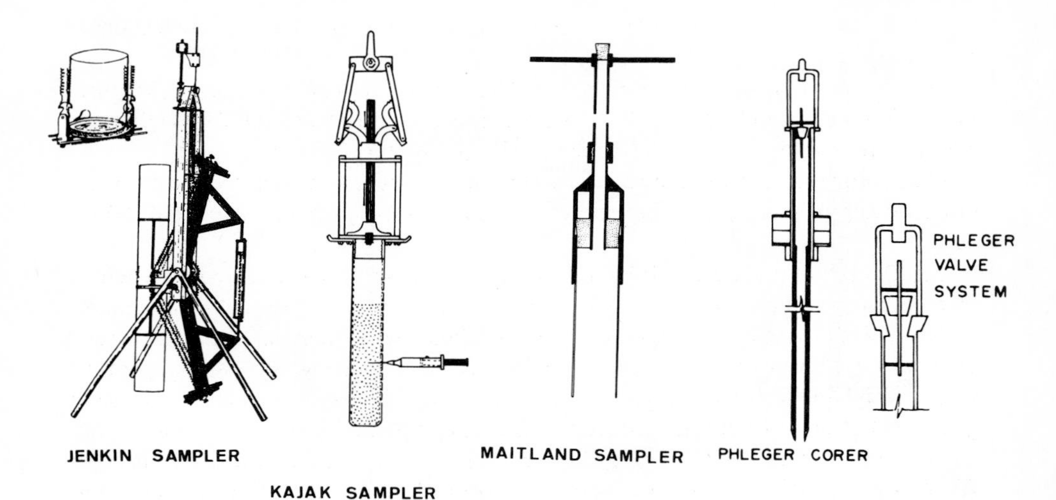

Fig. 3.2. Various types of core samplers

diameter must be at least 8.2 cm. But if the inner diameter of tubes that are open at the bottom exceeds about 5 cm, very loose sediments may be lost during retrieval.

4. To be able to sample certain interesting levels and photograph the sediments before extrusion, it is desirable that the sampler be fitted with at least one transparent side.

5. To prevent especially loose sediments from being distorted during extrusion, the construction of samplers, especially grab samplers, should allow for sub-sampling from the sampler; the sediments should be accessible from the top by, for example, an ordinary ladle or small cores. This is an important requirement, which is not met by most grab samplers in Fig. 3.1. It is practically impossible to obtain undisturbed samples is mostly sub-sampled by means of an extruding rod with a tight-fitted disk, out in a highly unpredictable manner when the jaws are opened. The material in core samples is mostly sub-saumpled by means of an extruding rod with a tight-fitted disk, which should be pulled in the direction in which the core penetrated the tube.

6. To be able to use the sampler on various bottoms (loose, hard, sand, silt, etc.), the construction should permit the attachment of easily exchangeable weights, as well as a stop collar. It may also be useful to have the exchangeable weights on an adjustable frame, so that the center of gravity can be placed as low as possible; but not too low to significantly increase the outer clearance.

7. To be able to use the sampler under different circumstances without any extra aid or equipment, and from a trailer-carried rowing boat, the sampler should be manually operated. The maximum weight should not exceed 20 kg.

8. To be able to make different kinds of analysis (chemical, physical and biological) on material from the same sample site, and to minimize the corresponding field work, it may be an advantage that the sample area and penetration depth should be as large as possible. It is evident that this requirement conflicts with requirements 7 and 2.

9. It is, finally, important that the equipment is easy to operate, also by less experienced personnel.

No single sediment sampler can meet all these, sometimes conflicting, requirements. The core samplers (Fig. 3.2) operate best on rather loose and fine sediments, with a water content of surficial sediments (W_{0-1}) of about 60% or more.

Grab samplers, like the Ponar, can preferably be used on coarse sediments or on bottoms rich in shells etc., when the water content (W_{0-1}) is less than about 50%–60%.

Fig. 3.3 illustrates a modified Ponar grab sampler which has been constructed to meet as many as possible of the given requirements on a hand-operated grab sampler.

Fig. 3.4 shows two photographs of a modified Ekman sampler, which has been designed to improve the performance of an old and thoroughly tested apparatus. It should be noted that this type of sampler is unsuitable on sandy bottoms; the sand grains will effectively prevent the jaws from closing due to increased friction. Brinkhurst (1974), with obvious compassion and personal experience, dedicated his book *The benthos of lakes* to "all those who have hauled an Ekman by hand from 50 m in a small boat on a windy day to find a stone caught between the jaws".

Fig. 3.5 illustrates one of many possible alternatives of core samplers that meet many of the given requirements. This particular example is equipped with transparent rectangular sides (for X-ray studies).

Table 3.1 demonstrates some basic data of these three samples and how well they meet the given requirements.

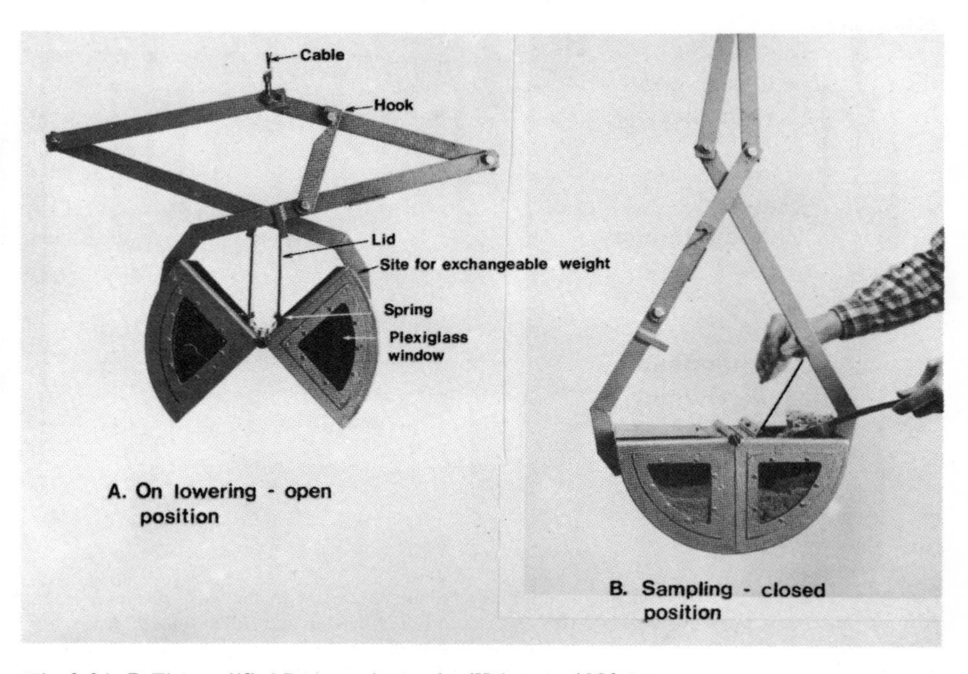

Fig. 3.3A, B. The modified Ponar grab sampler (Håkanson 1982a)

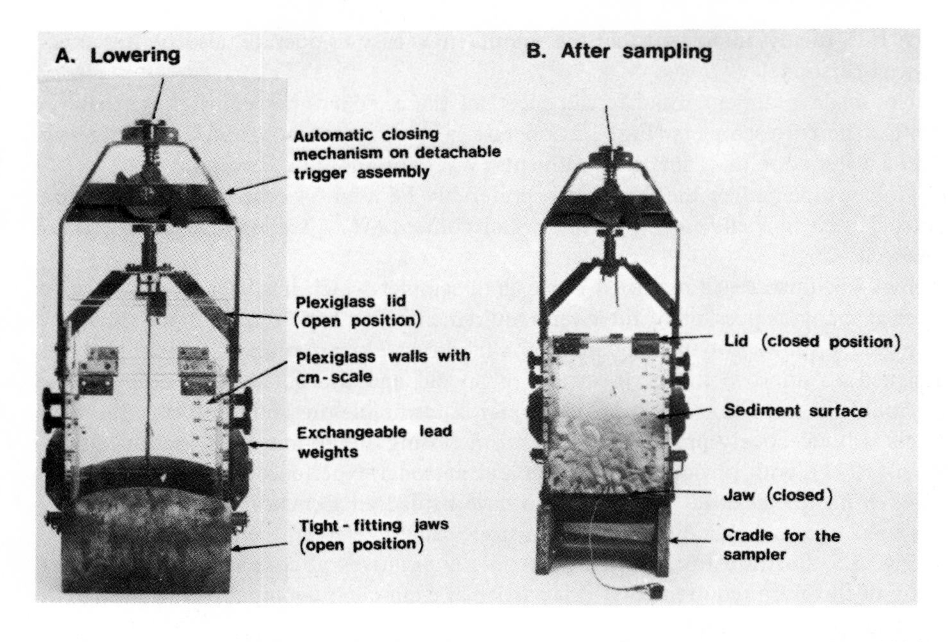

Fig. 3.4A, B. The modified Ekman sampler. **A** When lowering; **B** after sampling (Håkanson 1981b)

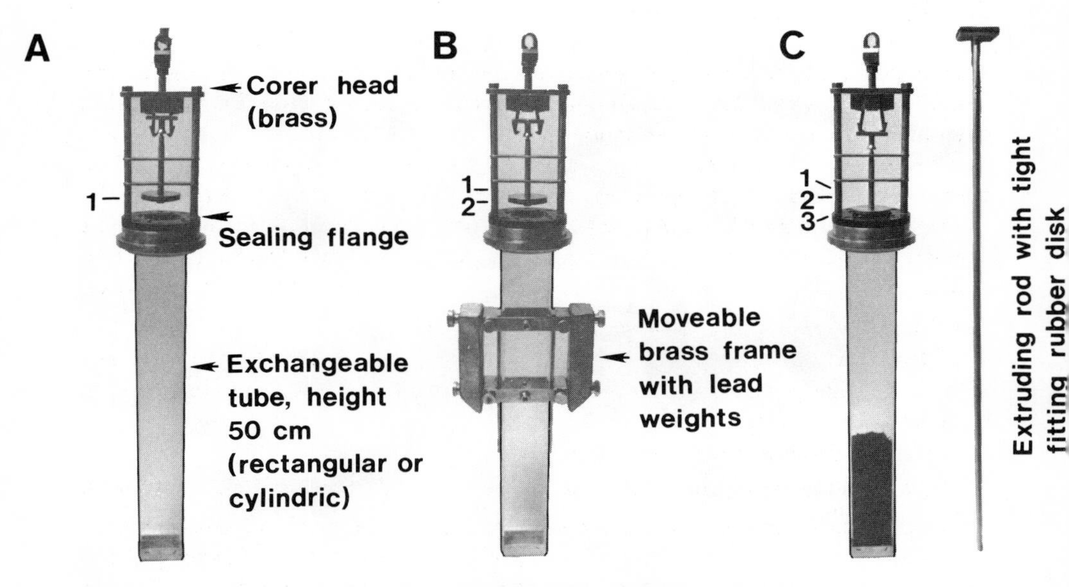

Fig. 3.5A–C. A gravity corer, in this case with rectangular tubes, with the valve system in open position one (**A**), in open position two (**B**), and closed (**C**) (Axelsson and Håkanson 1978)

Table 3.1. Performance of three samplers constructed to meet certain requirements of efficiency

	Grab samplers		Core sampler
	Mod. Ekman	Mod. Ponar	Axelsson-Håkanson
Sample area (cm^2)	300	420	40–80
Penetration depth (cm)	25	15	25–150
Weight (kg)	5	5	5
Free water passage	Yes	Yes	Yes
Closing on retrieval	Yes	Yes	Yes
Inner/outer clearance	Min.	Min.	Min.
Hvorslev ratio	< 5%	< 3%	8%–11% [a]
Closing at top	Yes, lid	Yes, lid	Yes, lid
Closing at bottom	Yes, jaw	Yes, jaw	No
Transparent side	Yes	Yes	Yes
Subsampling from sampler	Yes	Yes	No
Extruding rod	No	No	Yes
Variable weights	Yes, 20 kg	Yes, 20 kg	Yes, 20 kg
Hand operable	Yes	Yes	Yes
Type of closing	Automatic or trigger wright	Automatic	Automatic
Range (W_{0-1})	$W_{0-1} < 60$	$15 < W_{0-1} < 90$	$W_{0-1} > 60$
Not usable on	Sand and coarser	Very loose sediments	Sand or very consolidated sediments

[a] Data for cylindrical tube of 2 mm wall thickness

3.2 Types of Sampling System

This section contains a survey of different types of sampling nets, their benefits and limitations, as well as a discussion on the sedimentological and statistical factors that determine the yield of a given sampling program.

At the outset, it should be emphasized that the choice of sampling net should be governed by the aim of the investigation, and in this respect we will assume that the primary aim of most lake sedimentological studies is to obtain information on: (1) a characteristic, lake-typical, mean or median value, (2) the areal distribution pattern of the investigated variable, and (3) the vertical variation in certain sediment cores.

The following types of sampling system may be distinguished:

- Deterministic systems; based on given presuppositions, information or purposes. Such nets are generally denser in given areas of special interest and attenuated in areas of low priority.
- Stochastic systems; based on random sampling.
- Regular grid systems, which can be put at random or deterministically on the lake.

No thorough investigation has yet been published where the information value from these types of sampling nets has been properly evaluated. The regular grid system has, however, certain advantages and it is also the most utilized type of system (see Sly 1975). It is easy and straight-forward to apply and provides an even area cover of the investigated lake. One such sample system is illustrated in Fig. 3.6. But how many samples would be required and what factors regulate the statistical validity?

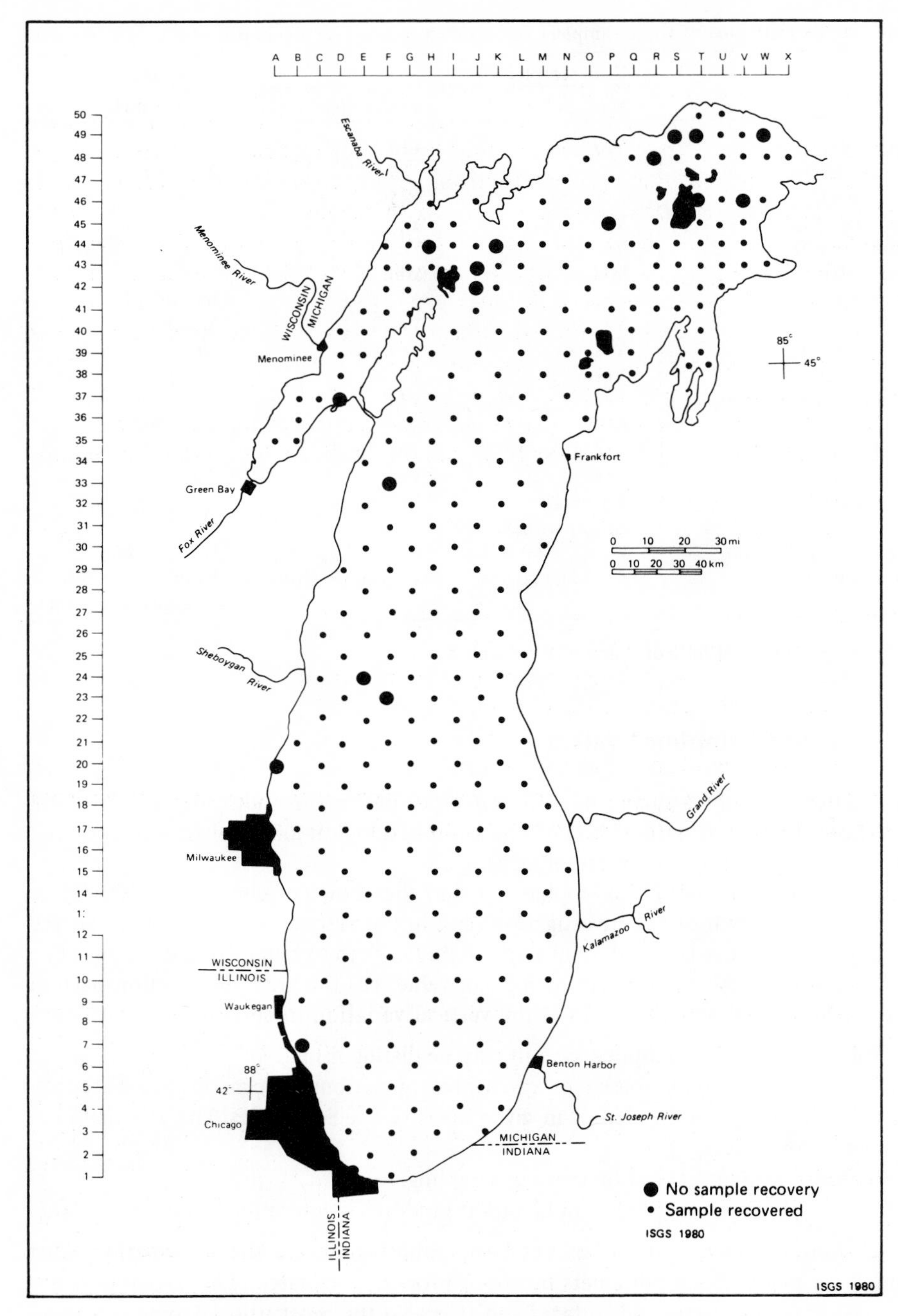

Fig. 3.6. Grid sample system utilized in Lake Michigan (Cahill 1981)

3.2.1 Number of Samples

The following factors influence the information value of sediment samples:

1. The water system; different presuppositions exist in lakes with and without areas of accumulation (i.e., continuous deposition of fine particles), in rivers and bays.

2. The prevailing bottom dynamics (erosion, transportation, accumulation); areas of erosion are characterized by hard or consolidated deposits (like rock, gravel, sand, glacial clays); areas of transportation (i.e., discontinuous deposition of particles finer than medium silt) are often very diverse; areas of accumulation are almost always characterized by loose sediments, sometimes with a high content of pollutants.

3. The lake area; more samples would generally have to be taken in larger lakes than in smaller to obtain the same information value.

4. The lake bottom roughness (R, dimensionless); more samples would be required in lakes with high R-values than in lakes with smooth and undramatic bottom configuration. The R-value is defined for whole lakes accordingly (see Håkanson 1981a):

$$R = \frac{0.165 \cdot (l_c + 2) \cdot \sum_{i=0}^{n} l_i}{D_{50} \cdot \sqrt{a}}, \qquad (3.1)$$

where R = the (normalized) lake bottom roughness;
l_c = the contour line interval in m;
l_i = the length of the given contour line in km;
D_{50} = the median water depth in m;
a = the lake area in km^2;
n = the number of contour lines.

The form roughness (R_f, dimensionless) may be utilized to quantitatively compare the degree of topographical irregularity of various sub-basins in a lake.

$$R_f = \frac{0.165 \cdot (l_c + 2) \cdot \sum_{i=0}^{n} l_i}{a'}, \qquad (3.2)$$

where a' = the sub-area in km^2.

5. Anthropogenic factors/type of pollution; different contaminating substances from the same source often reveal similar distribution patterns (lobes of decreasing concentrations with increasing distance from the source of pollution) in lake sediments (both areally and vertically), which are generally clearly distinguished from the distribution pattern of non-contaminating or conservative variables, which generally covariate with physical sediment parameters, like water content and bulk density.

6. The sediment chemical "climate"; the sediment distributions of certain elements, like P, Fe and Mn, are highly dependent on variable factors like the pH, Eh (= redox potential) and O_2-content in the sediment-water interface. Other elements, like Pb, are known to be much less dependent on the sediment chemical "climate".

7. The sediment physics and the sediment biology; the physical character (e.g., water content, bulk density, grain size, organic content and porosity) depends on the quantity and quality of the deposited materials (geology), but also to a high degree on the sediment biology, e.g., the bioturbation.

8. The number of samples.
9. The type of sampling net (deterministic, stochastic, regular, grid, etc.).
10. The quality of the sampling equipment, i.e., the possibility of taking undisturbed samples.
11. The sub-sampling, sample preparation, and finally,
12. The reliability of the laboratory analysis.

All these factors may, potentially, have an influence on the result. As far as we are aware, no systematic study has yet been introduced which accounts for even half of these 12 factors. Thus, some time may elapse before a tested sedimentological sample formula is introduced, especially since one is often faced with skew populations, as in the case of contaminating substances, and few samples ($n < 30$).

3.2.2 The Sample Formula

While waiting for a scientifically tested sample formula, describing at least the major causal relationships, the following pilot sample formula may be utilized (Håkanson 1981b).

$$n = 2.5 + 0.5 \cdot \sqrt{a \cdot F}. \tag{3.3}$$

This formula is based upon knowledge of only two morphometric standard parameters:

- a, the lake area (km^2); more samples (n) should be taken in larger lakes than in smaller;
- F, the shore development, which is used as an indirect measure of the bottom roughness; there exists a marked positive relationship between the shore development (F) and the bottom roughness (R) (see Håkanson 1974), which is understandable for morphological reasons, and in lakes with high F-values more samples (n) should be taken than in lakes with F-values close to 1.

$$F = \frac{l_0}{2 \cdot \sqrt{\pi \cdot A}}, \tag{3.4}$$

where l_0 = the normalized shoreline length in km;
A = the total lake area in km^2 (i.e., the water surface, a, plus the area of islands).

The F-value illustrates the relationship between the actual length of the shoreline and the length of the circumference of a circle with an area equal to the total lake area. A perfectly circular basin has an F-value of 1. F-values larger than 10 are rare; "normal" values for lakes in the Nordic countries lie in the order of 2–4.

The method presupposes that the samples must provide *an even area cover* of the whole lake area, e.g., according to a regular grid as illustrated in Fig. 3.6 or according to the square-net technique illustrated in Fig. 3.7. In this example, we can assume that $F = 2$ and $A = a = 20$, i.e., $n = 2.5 + 0.5 \cdot \sqrt{20 \cdot 2} = 5.7$; or $n = 6$. These six samples should be distributed evenly over the lake surface. Each site should represent about $20/6 = 3.3\ km^2$, which corresponds to a square-net side of $\sqrt{3.3} = 1.8$ km. In the example given in Fig. 3.7, we have assumed that the map scale and the size of the

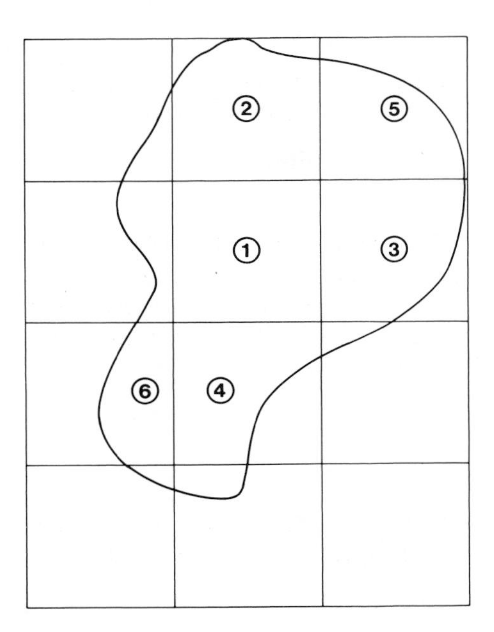

Fig. 3.7. Example of sampling net utilizing a transparent square-net placed upon the map of the lake. In this case six samples are obtained with an even areal distribution

square-net have been adjusted to these presuppositions, then the net can be placed at random on the map of the lake; the first sample site is in the middle of the square which completely falls within the limits of the shoreline; the next site is in the centre of the next largest square within the limits of the shoreline, and so on. It should be noted that this is just one example of a sampling system to get an even area cover in a systematic and objective manner; alternative systems along, e.g., sections, or subjectively without square-nets, may often give just as good results. The sample formula (3.3) means that it is seldom advisable for either statistical or sedimentological reasons to take less than three samples in a lake. It is also obvious that it would not be advisible to use any linear relationship between the number of samples (n) and the given variables (a and F); a linear relationship would imply that, in principle, 300 samples would be required in a lake where $a = 100\ km^2$ if three samples constitute the lower limit in a lake where $a = 1\ km^2$. Hence, the sample formula (3.3) is based on the assumption that n should not increase linearily with a and F, but rather more moderately.

The sample formula (3.3) is graphically illustrated in Fig. 3.8. From this nomogram it is clear that n is always larger than 3. The mathematical character of the sample formula implies that special attention must be given to the actual definition of the limitations of the investigated area, as demonstrated in Table 3.2. For Lake Hjälmaren, which is a Swedish lake with four topographically distinguished sub-basins, Hem Bay, Mellan Bay, Great-Hjälmaren, and East Hjälmaren, the sample formula yields different results depending on the definitions:

- if the four sub-basins are treated separately, seven samples would be required in Hem Bay and Mellan Bay, eight in East Hjälmaren and 22 in Great-Hjälmaren, i.e., a total of 44 samples;
- if, on the other hand, Lake Hjälmaren is treated as a single unit, then only 28 samples would be required according to the formula.

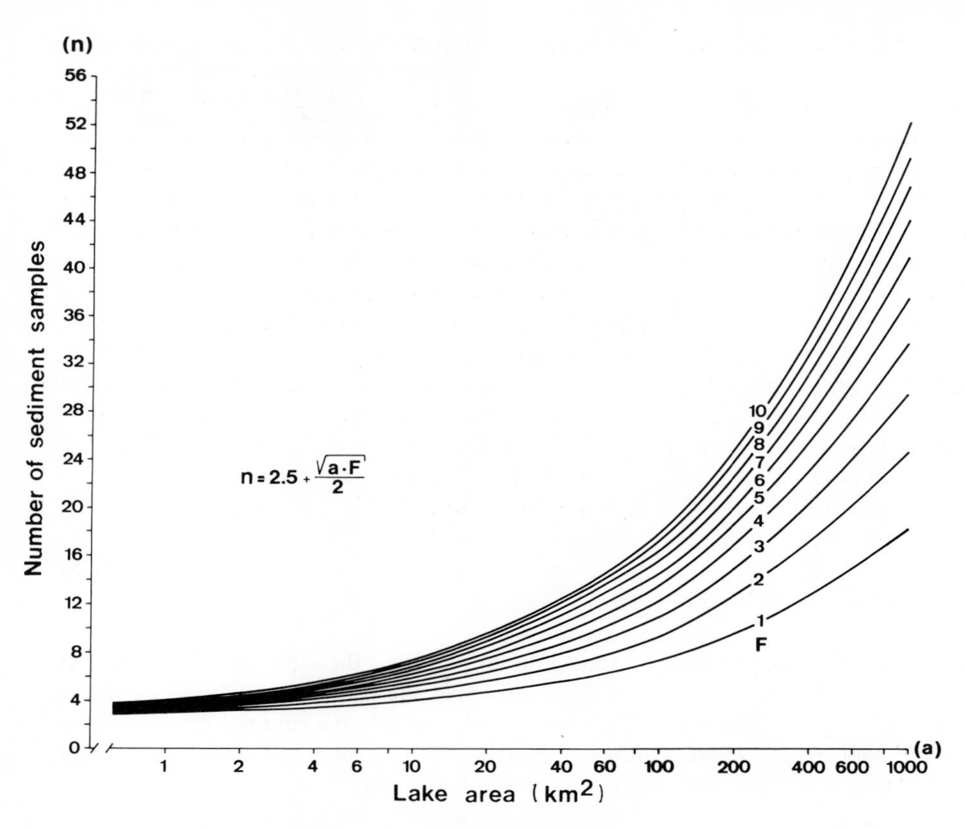

Fig. 3.8. Nomogram – the sampling formula (Håkanson 1981b)

Table 3.2. The results when using the sampling formula in a multi-basin lake (Hjälmaren) when (1) each sub-basin ins treated separately, and (2) the entire lake is the sampling area (Håkanson 1981b)

	Area km^2 a	Shore development F	Number of sediment samples n
Hem Bay	25.41	2.80	6.72 ∿ 7
Mellan Bay	40.13	1.88	6.84 ∿ 7
Great-Hjälmaren	376.58	4.11	22.17 ∿ 22
E. Hjälmaren	35.72	3.32	7.94 ∿ 8
Σ Sub-basins			43.67 ∿ 44
The whole of Lake Hjälmaren	477.8	5.43	27.97 ∿ 28
	Difference: 28 = 44 – 28 =		16

3.2.3 Sampling in Different Environments – Statistical Aspects

Any given sample, or site, is meant to represent not only the given location but also a surrounding area. If the variation in physical, chemical, contaminational or biological sediment character within that area is small, then few samples would be required to obtain a representative picture; if, on the other hand, the variability is large, many samples are required to achieve the same statistical representation. In the first case, any given sample site or sample has a high information value.

In this section, we will assume that the analytical technique yields adequate data, and instead focus on the statistical side of the problem of getting representative results. We would like to answer the following questions: How many samples (N) would be required to determine a representative value (in this case the mean value, $\bar{x}$) for a given area with a given statistical certainty? We will assume that, with a 90% certainty, the error in the determination of the mean should be less than 10% of the mean. And secondly: What factors regulate the information value?

Utilizing basic statistical concepts (see e.g., Spiegel 1972), we have:

$$\bar{x} = \frac{1}{n} \cdot \sum_{i=1}^{n} x_i \tag{3.5}$$

$$s_x^2 = \frac{1}{n-1} \cdot \sum_{i=1}^{n} (x_i - \bar{x})^2 = \frac{1}{n-1} \cdot \left[\sum_{i=1}^{n} x_i^2 - n \cdot \bar{x}^2 \right] \tag{3.6}$$

$$V = \frac{s_x}{\bar{x}} \cdot 100 \tag{3.7}$$

$$y \cdot \bar{x} = t_c \cdot \frac{s_x}{\sqrt{n-1}} \tag{3.8}$$

where $\bar{x}$ = the mean value of the given empirical data (x_i, i = 1, 2, 3 . . . . ., n);
n = the number of data;
s_x = the standard deviation (s_x^2 = the variance);
V = the coefficient of variation (in %);
y = the accepted error in percent of the mean value; in this case we will use $y = 10\%$ of $\bar{x}$, i.e., $y = 0.1$;
t_c = the confidence coefficient (see Fig. 3.9 for illustration). Here we will use a 90% confidence interval and a two-tailed test, which means that the the requested confidence coefficient can be obtained directly from standard statistical tables (Student's t-distribution, see Appendix 1) for $t_{0.95}$. For example, if 8 (n = 8) samples have been taken within the given sample area, then the number of degrees of freedom (ν) is equal to n – 1 = 7 and the requested $t_{0.95}$-value is 1.90.

It should be stressed that this standard procedure is only valid for normal (or approximately normal) populations, and that we have used the traditional small sampling theory ($n < 30$), since this ought to be appropriate in this sedimentological context. From these premises, we may rewrite Eq. (3.8) as:

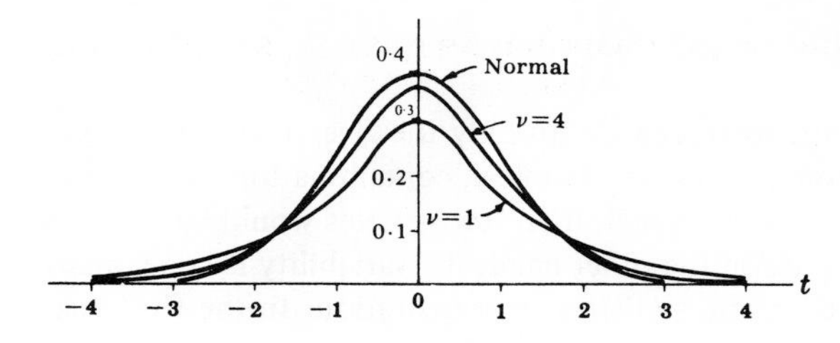

Fig. 3.9. Student's t-distribution for various degrees of freedom (ν)

$$\sqrt{N-1} = \frac{t_c \cdot s_x}{0.1 \cdot \bar{x}} = \frac{t_c \cdot V}{10} \tag{3.9}$$

or

$$N = \left(\frac{t_c \cdot V}{10}\right)^2 + 1. \tag{3.10}$$

N in Eq. (3.10) can be considered as an information value; if N is small, then the variability is small, and vice versa. Subsequently, we will give some examples from three different sedimentological environments and ask: How many samples (N) are required to obtain the same information? Can this number be reduced in a simple and standardized manner? These questions have bearings on the problem to define concentrations, which will be discussed in Section 3.6. To illustrate the basic statistical/sedimentological principle, we will utilize a set of data on surficial sediments (0–1 cm) sampled by means of the modified Ponar grab sampler and utilizing a regular grid with nine sample sites 15 m apart.

- A *river* area; a typical area where processes of transportation dominate the bottom dynamics and where the character of the sediments vary a great deal (sand, twigs, silt, leaves, etc.).
- A *river mouth* area; a typical area where great temporal and areal sedimentological variations predominate.
- A *lake* area where continuous deposition of fine particles prevail (i.e., an area of accumulation).

The data used in this example emanate from River Fyris and Lake Ekoln, Sweden (see Fig. 3.10).

These particular sediments were analysed for water content (W), loss on ignition (or organic content, IG) and three metals (Cu, Pb and Cd). The mean value ($\bar{x}$), standard deviation (s_x), coefficient of variation (V), and the number of necessary samples (N) to obtain the given statistical certainty (i.e., that the 90% confidence interval around the mean is less than 10% of the mean) have been determined for these five sediment parameters and for the following steps or corrections:

- Step 0, i.e., direct analysis of the sediment samples;
- Step 0 with correction for water content (compare grain size, porosity or bulk density, all of which may be expressed in terms of the water content);
- Step 0 with correction for the organic content (IG);
- Step 1, i.e., after wet sieving of the sediments through a 63 μm mesh and subsequent analysis of the silt-clay fraction;

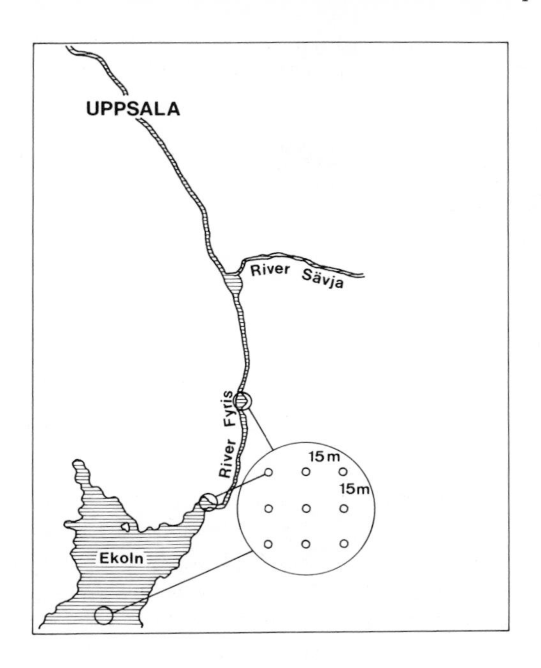

Fig. 3.10. Map showing River Fyris, Lake Ekoln and sites of sampling

- Step 1 with correction for the water content (W);
- Step 1 with IG-correction;
- Step 2, i.e., after centrifugation (5000 r.p.m. for 30 min) and analysis of the top (fine) fraction;
- Step 2 with W-correction;
- Step 2 with IG-correction.

The question posed is: Which of these steps or corrections would yield the least number of necessary samples N?

The theory behind this test is schematically illustrated in Fig. 3.11. For 9 samples and 8 degrees of freedom, $t_{0.95}$ is equal to 1.86. That is, Eq. (3.10) is:

$$N = (0.186 \cdot V)^2 + 1. \tag{3.11}$$

From Table 3.3, we can see that the mean water content, $\overline{W}$, is 57.5% in the river area, only 32.0% in the river mouth area and 82.2% in the central part of Lake Ekoln. The spread is enormous in the river mouth area, where 904 samples would be required to get less than 10% error in the determination of the mean organic content (IG). The corresponding number is only three in the lake area. Table 3.4 gives the results in terms of N [from Eq. (3.11)] for lead and mean values for the three metals studied (Cu, Pb and Cd). It should be noted that Table 3.4 does not provide information on concentrations – just the requested number of samples (N) for a given statistical confidence.

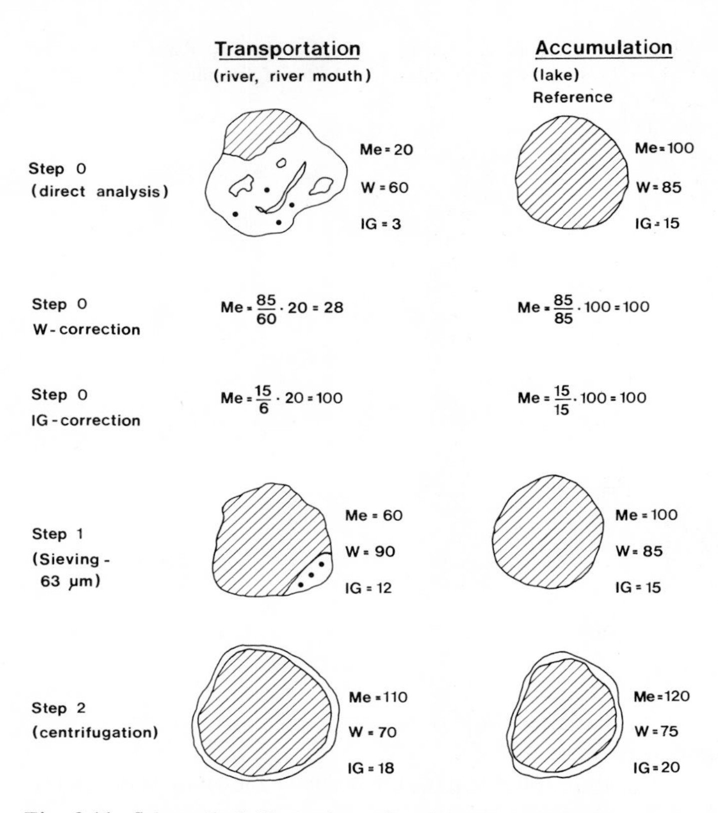

Fig. 3.11. Schematical illustration of various correction measures and their sedimentological implications. *Me* metal, *W* water content, *IG* loss on ignition

Table 3.3. Mean values, standard deviations, coefficients of variation and number of necessary samples for a given level of statistical certainty for data on water content (W) and organic content (IG) from nine sampling sites within three different sedimentological environments (river, river mouth and lake)

	River (R)	River mouth (M)	Lake (L)
n:	9	9	9
$\overline{W}$:	57.5	32.0	82.2
s_W:	9.7	24.3	0.29
V_W:	16.9	76.0	0.35
N_W:	10.9	201.0	1.0
$\overline{IG}$:	6.5	3.6	10.2
s_{IG}:	1.7	5.9	0.78
V_{IG}:	26.6	161.5	7.6
N_{IG}:	25.5	903.8	3.0

Table 3.4. The number of necessary samples (N), for a given level of statistical certainty, in three different sedimentological environments for lead and (Cu + Pb + Cd) after defined steps and corrections

Step/correction	Parameter: Pb		
	River	River mouth	Lake
0	38.3	779.0	4.0
0, W	21.5	175.5	4.1
0, IG	23.2	72.0	7.8
1	10.2	9.5	3.2
1, W	9.9	9.9	3.1
1, IG	17.0	18.1	4.0
2	23.2	13.6	36.4
2, W	17.2	7.8	34.6
2, IG	18.5	14.2	26.3
Step/correction	**Parameters: Cu, Pb and Cd (mean values)**		
	River	River mouth	Lake
0	105.5	788.9	6.0
0, W	72.6	202.1	6.1
0, IG	52.4	76.3	8.3
1	49.1	8.1	3.8
1, W	48.2	7.6	3.8
1, IG	39.9	11.0	3.9
2	72.2	22.1	20.6
2, W	52.6	11.3	19.0
2, IG	50.5	12.4	15.5

The table illustrates clearly that:

– Direct analysis (step 0) of sediment samples from rivers, river mouth areas or transportation areas would yield low information, i.e., many samples (N) would be required to obtain a given statistical certainty. For example, N = 789 for the river mouth area, N = 106 for the river area and N = 6 for the lake area for the mean values for Cu plus Pb plus Cd;
– Correction with the water content (or similar parameter) and the organic content would improve the information value but still not yield optimal results;
– Wet sieving through a 63 μm mesh does not generally alter the deposits from areas of accumulation to a great extent, since the fine, loose materials characterizing such areas pass this mesh size.
– Simple wet sieving through a 63 μm mesh, as indicated by these empirical data, seems to be the best method, yielding the lowest number of necessary analyses (N) for the least amount of work;
– Fractionated centrifugation (or similar method) would not improve the information value, the main reason probably being that the centrifugation also implies fractionation of the different types of organic (authochthonous, allochthonous,

detrital, algal, etc.) and inorganic (e.g., carbonates and minerals) materials, all of which have different abilities to sorb metals, and hence also have different metal concentrations. Thus, the use of centrifugation or other alternative techniques like ultrafiltration that may differentiate the "carrier particles", might imply negative effects from this statistical perspective.

Therefore, the conclusion is that maximum information for minimum work is obtained if all sediment samples from all limnic environments are wet-sieved through a 63 μm mesh before analysis for metals or contaminants. This is in very good agreement with the conclusion reached on concentrations (see Sect. 3.6, and Förstner and Salomons 1981).

3.2.4 Sub-Sampling

The sample preparation must, evidently, be adjusted to the subsequent analysis of the sediments, i.e., the aim of the study. A general rule of thumb is to minimize time of storage and environmental changes in, for example, temperature. A standardized procedure of sample preparation has been introduced by Förstner and Wittmann (1979) (see Fig. 3.12). The aim of this section is to discuss various principles and standard techniques for sediment sub-sampling and not different analytical techniques (see Skougstad et al. 1979). The following alternative principles for sub-sampling exist:

1. Uniform sub-sampling, i.e., the samples are taken at certain given levels determined before the survey, e.g., every second centimeter (0–1, 2–3, 4–5 cm, etc.) or in layers of 3 cm (0–3, 3–6, 6–9 cm, etc.). This type of sub-sampling ignores sedimentary structures, which implies that important information may be lost. Uniform sub-sampling may, however, be practised in environmental control programmes and in larger survey operations, where the drawbacks of any reduced resolution is balanced by advantages in simplicity (time and money).

2. Structural sub-sampling, i.e., the samples are taken according to observed sedimentological structures, laminations, strata or varves. Existing structures may not be visible until the sediments are dried or X-rayed.

The sub-sampling of thin layers (< 1 cm) requires special techniques, especially if the sediments are loose. Various methods to cut thin slices have been successfully used, e.g.:

- The electro-osmotic knife and/or guillotine (see Chmelik 1967), which is based on the electro-osmotic behaviour of clays, i.e., if an electric field is applied to a clay, the interstitial water will flow to the negative electrode. The electro-osmotic cutting knife is a very simple instrument (Fig. 3.13D), the guillotine somewhat more advanced (Fig. 3.13). The curring blade serves as negative electrode and a spatula put into the core as a positive electrode. The electro-osmotic effect creates a thin lubricating water film on the cutting blade, which prevents adhesion of the sediments. The cutter works best on homogeneous clays and loose sediments, which can be quickly cut using 20–30 V (1–3 Amp). Sand requires a higher voltage (40–80 V). This method should not be used for chemical studies of interstitial waters since the positive electrode releases metal ions.

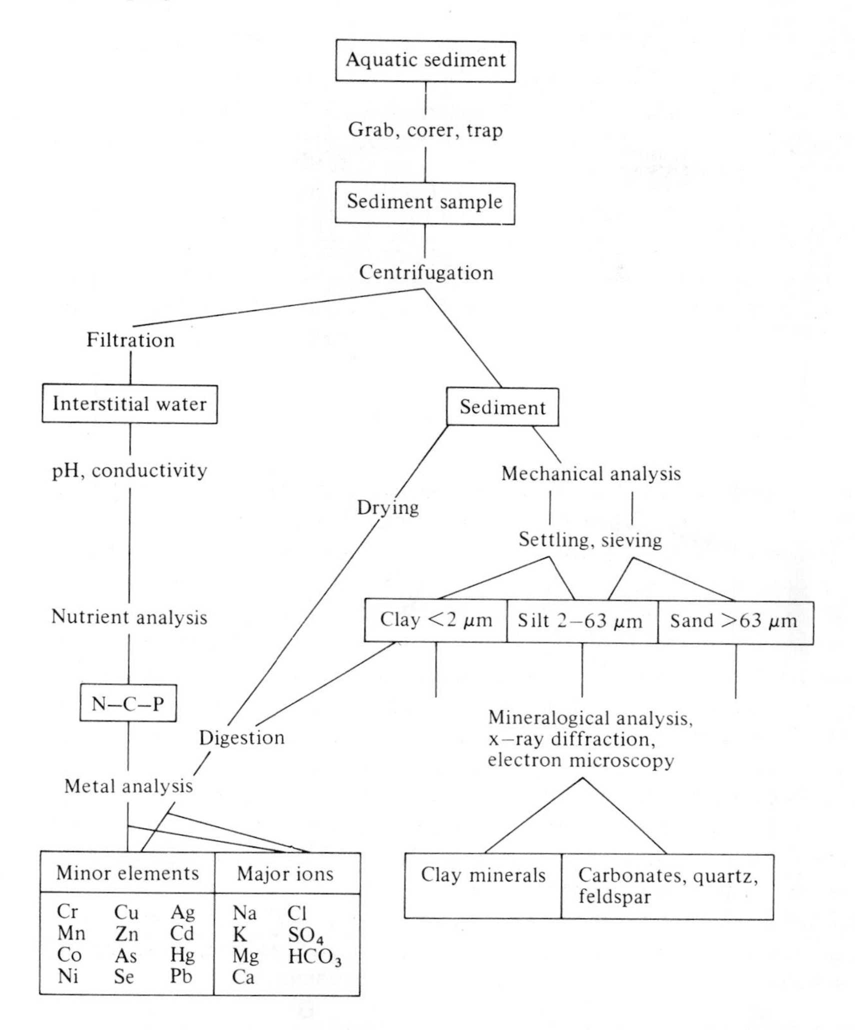

Fig. 3.12. Schematic sequence for analytical procedures of sediment samples for metal determinations (Förstner and Wittmann 1979)

- Close-interval fractionators (see Niemistö 1974, or Fast and Wetzel 1974). The method is illustrated in Fig. 3.14. The sub-sample slide consists, in this case, of a Plexiglas sheet of desired thickness with a hole in the middle for the sediment core. Two gliding sheets protect the sample from contamination by the air.
- Dry-ice freezing ("cold finger") methods (see Shapiro 1958, Hottonen and Meriläinen 1978, Renberg 1981a). The technique illustrated in Fig. 3.15 utilizes liquid nitrogen. The box freezer is lowered into position in situ (in the sediments) by a rod (if th depth is less than about 10–20 m) or by wire. The sediment freezes on the front of the box in about half an hour (2–4 cm thick depending on the character of the sediments). The differences in colour are best observed in fresh samples;

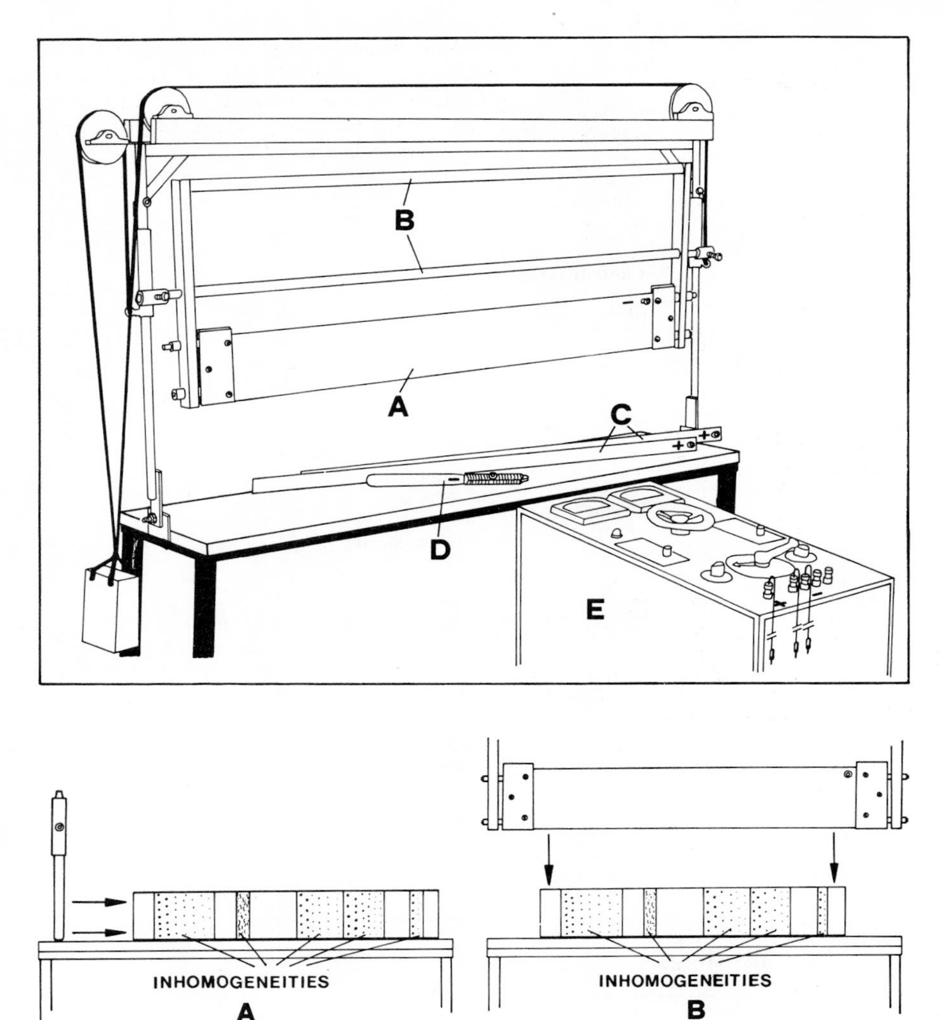

Fig. 3.13. *Top:* Electro-osmotic guillotine; *A* cutting blade; *B* inner frame; *C* positive electrodes; *D* electro-osmotic cutting knife; *E* rectifier. *Bottom: A* core cut by means of electro-osmotic knife starting from one end; *B* parallel core cut by the electro-osmotic guillotine (Sturm and Matter 1972)

varve structures become more visible after drying for 1–2 weeks. Then slices can most easily be taken with a knife when the sample is beginning to melt.

Layers thicker than 1 cm can often be sub-sampled without any special technique or apparatus. If the layers used are too thick, valuable information may be lost. This is illustrated in Table 3.5, where the age distributions of five uniformly sub-sampled layers (0–1, 0–2, 0–3, 0–5 and 0–10 cm) are shown. When these deposits, from Lake Hjälmaren, Sweden, were cut into 1-cm layers, then 21.3% of the material

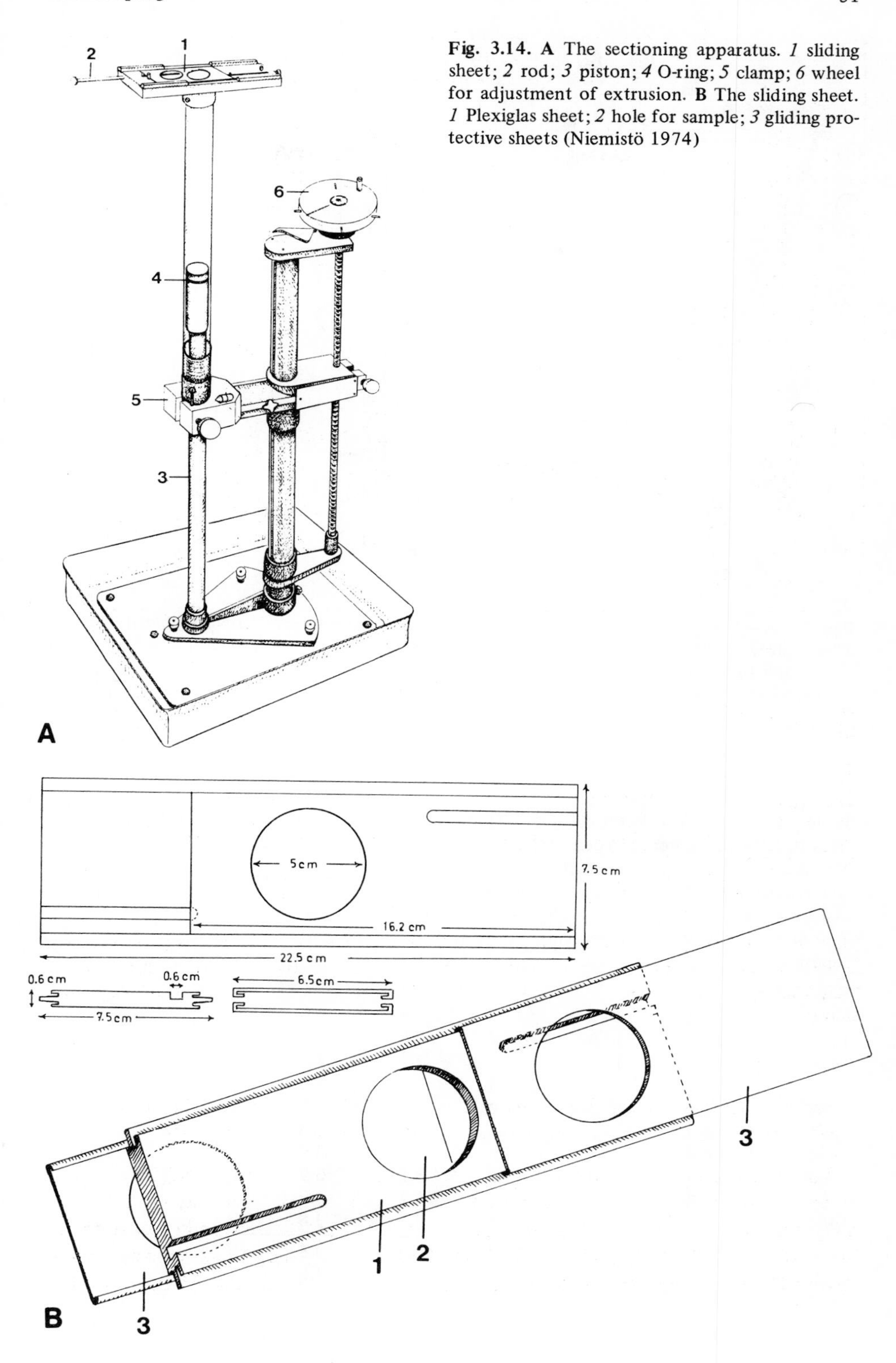

Fig. 3.14. A The sectioning apparatus. *1* sliding sheet; *2* rod; *3* piston; *4* O-ring; *5* clamp; *6* wheel for adjustment of extrusion. **B** The sliding sheet. *1* Plexiglas sheet; *2* hole for sample; *3* gliding protective sheets (Niemistö 1974)

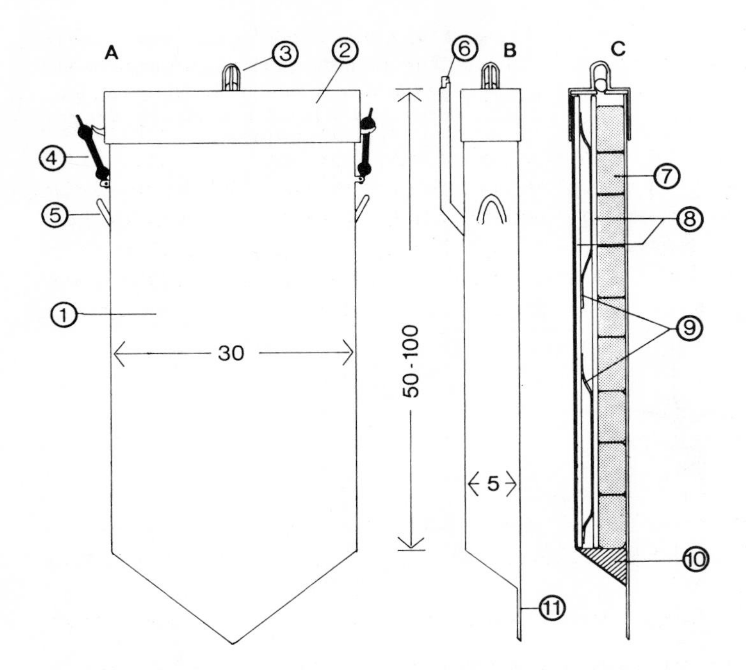

Fig. 3.15A–C. Schematical sketch of the box freezer. **A** Front view. *1* freezing panel; *2* cover; *3* ball valve; *4* rubber fastener; *5* wire handle. **B** Side view. *6* socket. **C** Lateral transsection. *7* dry ice cube; *8* tray of double plywood; *9* spring made of steel plate; *10* lead weight; *11* keel. Dimensions are given in centimeters (Huttunen and Meriläinen 1978)

Table 3.5. The age distribution of surficial sediments, from accumulation areas in Lake Hjälmaren, Sweden, if cut into layers of 0–1 cm, 0–2 cm, 0–3 cm, 0–5 cm and 0–10 cm. The values represent the percentage of material deposited during given years before 1977 (Håkanson 1981b)

Year of deposition	Sediment sample (cm)				
	0–1	0–2	0–3	0–5	0–10
1976	21.3	18.3	12.9	7.8	4.0
1975	15.1	14.3	12.2	7.8	4.0
1974	11.4	11.4	10.8	7.7	4.0
1973	9.0	9.2	9.4	7.6	4.0
1972	7.2	7.5	8.0	7.2	4.0
1971	5.8	6.2	6.8	6.8	4.0
1970	4.8	5.1	5.8	6.3	3.9
1969	4.0	4.3	5.0	5.8	3.9
1968	3.3	3.8	4.2	5.2	3.9
1967	2.8	3.6	3.6	4.7	3.8
Σ	84.7	82.9	78.7	66.9	39.5
Median age	3.2	3.7	4.6	6.8	11.8

emanated from the year 1976, 15.1% from 1975, 11.4% from 1974, etc. If, on the other hand, the top 10-cm layer was utilized, than the compaction and bioturbation would have created a very flat age-frequency curve; each year would then make up about 4% of the sediments. If the 0–1 cm layer is used, 84.7% of the material emanates from the last decade; the median age of the sediments would be 3.2 years. If the 0–10 cm layer is used, only 39.5% would come from the latest decade; the median age would be 11.8 years. The same effect of diminished resolution would be apparent for many sediment variables besides the age. Therefore, in aquatic pollution control programs and general surveys, it is often advisable to utilize uniform 1-cm-thick layers. In specific sedimentological studies, the structural sub-sampling would be preferable.

3.3 Sediment Traps

This section presents a brief outline on:

- The physical principles of sedimentation in vessels; the intention is that the discussion will also reflect important lake sedimentological principles (dealt with more thoroughly in Chap. 6.1).
- Benefits and limitations, construction and use of various types of sediment traps.

Sediment traps are, generally, comparatively simple instruments which may be utilized for very many purposes. Two of the most important fields concern budget/flux calculations and environmental monitoring programmes. It is often practical to distinguish between the following types of sediment traps:

1. Bottom sediment traps, (a) sediment vessel at or close to the sediment surface, and (b) sediment vessel above the sediment surface, in the water phase. Bottom sediment traps placed within zones of accumulation give data of high local representativity, with "legal domicile" on the sample site.

2. Buoy-carried sediment traps, (a) moored sediment traps, and (b) free-drifting sediment traps.

Until quite recently controversial opinions concerning the use of sediment traps, their construction, and the validity of sediment trap data, were widespread, but after some crucial laboratory and field investigations, primarily by Gardner (1980a,b), and critical reviews by Bloesch and Burns (1980) and Blomqvist and Håkanson (1981), the state of the art has been significantly improved.

Subsequently, sediment trap will imply the whole apparatus and sediment vessel the container, e.g., funnel, cylinder or bottle.

3.3.1 Physics of Sedimentation in Vessels

It should be stressed that in lakes there is no steady vertical "rain" of sedimentary particles. Instead, the vertical sinking velocity of settling particles is up to six times lower than the horizontal velocity component (see Table 6.1). Thus, in lakes the small particles settle very slowly and are carried passively in turbulent bodies into a vessel, where the horizontal velocity component is removed and the small vertical component remains to cause the particles to sink. The factors determining the sedimentation are: the flow conditions (laminar, turbulent) around the vessel, the shape of the vessel and the concentration of particles in the vessel.

Bloesch and Burns (1980) demonstrated that turbulence will have no effect on the settling of small (< 500 μm) organic particles, since the flow around such particles is laminar (see Fig. 3.16). Figure 3.17 depicts cylinders, flasks (or rather a vessel with an inverted saucer-shaped plate on top of the cylinder) and funnels under calm (laminar) and flowing (turbulent) conditions.

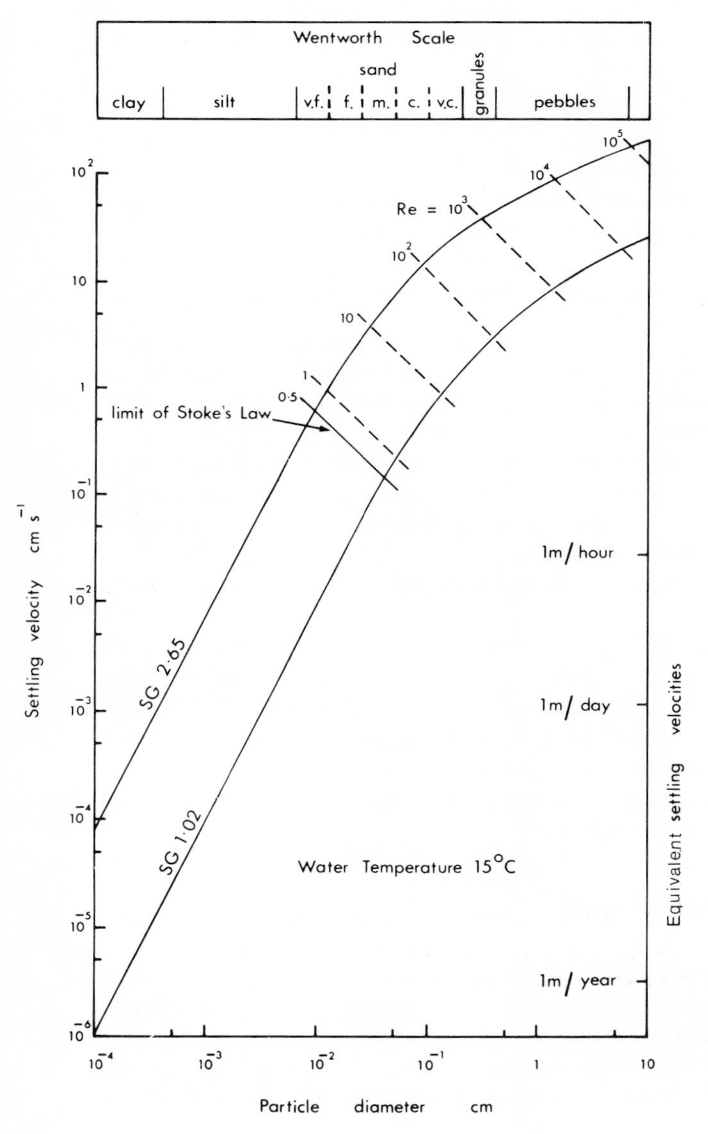

Fig. 3.16. Relationship between settling velocity of spherical particles in water and particle diameter. *SG* specific gravity (Smith 1975)

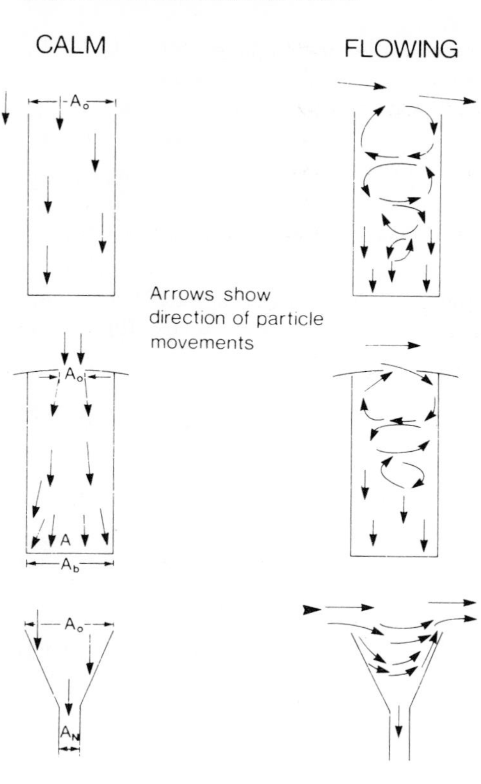

Fig. 3.17. Diagrams illustrating probable paths of particles settling in vessels (Bloesch and Burns 1980)

In calm water (Fig. 3.17), the settling flux (F) of particles will be equivalent around and within the funnel, namely

$$F = V_o \cdot C_o = V_t \cdot C_t, \tag{3.12}$$

where F = settling flux $[ML^{-2}\ T^{-1}]$;
V_o and V_t = mean particle settling velocity $[LT^{-1}]$ outside and inside the vessel;
C_o and C_t = mean particle concentration in water outside and inside the vessel $[ML^{-3}]$.
The total flux, F_t $[MT^{-1}]$ into the trap and onto the bottom is

$$F_t = A_o \cdot F, \tag{3.13}$$

where A_o = area of opening of the vessel $[L^2]$.

Since turbulence does not affect the settling velocity of small particles, also the total flux, F_t, in flowing water (Fig. 3.17) may be given by Eq. (3.13). If the aspect ratio (i.e., the height/diameter quotient) is large enough, there will, even in flowing water, be a stagnant zone at the bottom of the vessel and no resuspension of particles already settled on the floor of the vessel. But the number of particles entering the vessel must equal the number of particles leaving plus the number of particles deposited on the bottom of the vessel, i.e.:

$$\underbrace{A_o \cdot V_o \cdot C_o + Q \cdot C_o}_{\text{in}} = \underbrace{A_o \cdot V_t \cdot C_t}_{\text{bottom}} + \underbrace{Q \cdot C_t}_{\text{out}}, \tag{3.14}$$

where Q = the volume of water entering and leaving the vessel per time unit $[L^3T^{-1}]$. Since Eq. (3.14) should describe balanced conditions, where the material is not destructed or created, C_o must be equal to C_t. This means that the concentrations of particles ouside and inside the vessel are equal and that the flux of settling particles at the bottom of the trap in calm conditions ($F_t = A_o \cdot V_o \cdot C_o$) and during flowing conditions ($F = A_o \cdot V_t \cdot C_t$) also must be equal.

For a vessel placed in flowing water, it is important that the geometric form is such that the concentration of particles is on the same level outside as inside the vessel. This is highlighted in Fig. 3.17 – the bottle case. In calm water (Q = 0 and $V_o = V_t$), the flux through the opening area (A_o) must be equal to the flux on the bottom area (A_b), i.e.:

$$A_o \cdot V_o \cdot C_o = A_b \cdot V_t \cdot C_t \tag{3.15}$$

or

$$C_t = C_o \cdot \frac{A_o}{A_b}. \tag{3.16}$$

That is, if $A_b > A_o$ then $C_t < C_o$.
In flowing water, Eq. (3.14) can be rewritten as:

$$A_o \cdot V_o \cdot C_o = A_b \cdot V_t \cdot C_{tm} + Q \cdot C_{tm}, \tag{3.17}$$

where C_{tm} = concentration of particles inside the bottle $[MT^{-3}]$.

The difference between ($A_b \cdot V_t \cdot C_{tm}$) and ($A_o \cdot V_o \cdot C_o$) would be excess catch (E) due to flask geometry, i.e.:

$$E = Q \cdot C_o - Q \cdot C_{tm} = Q \cdot (C_o - C_{tm}). \tag{3.18}$$

Since Q may attain large values in turbulent waters, even a small difference in ($C_o - C_{tm}$) will give rise to a large excess catch (E).

For the funnel-shaped vessels the effective collecting area in calm waters is A_o, i.e., the area of the mouth of the funnel. In turbulent waters, when vortices penetrate into the funnel, and the material deposited on the sloping walls may be resuspended, the effective collecting area would be that of the neck of the funnel, A_n. In natural waters, where flow conditions vary, the effective collecting area of funnel-shaped vessels would also vary between A_o and A_n. Funnel-shaped vessels would also give rise to selective trapping; rapidly settling particles would not be as easily resuspended and carried out of the funnel as finer particles. Hence, funnels are known to yield undertrapping.

3.3.2 Geometry of Vessels

From the foregoing, it may be concluded that from purely theoretical viewpoints:
- simple cylinders would be the most favourable form for a sediment vessel in all types of waters (stagnant or turbulent);
- bottle-shaped vessels overtrap sediments;
- funnels generally yield under-deposition.

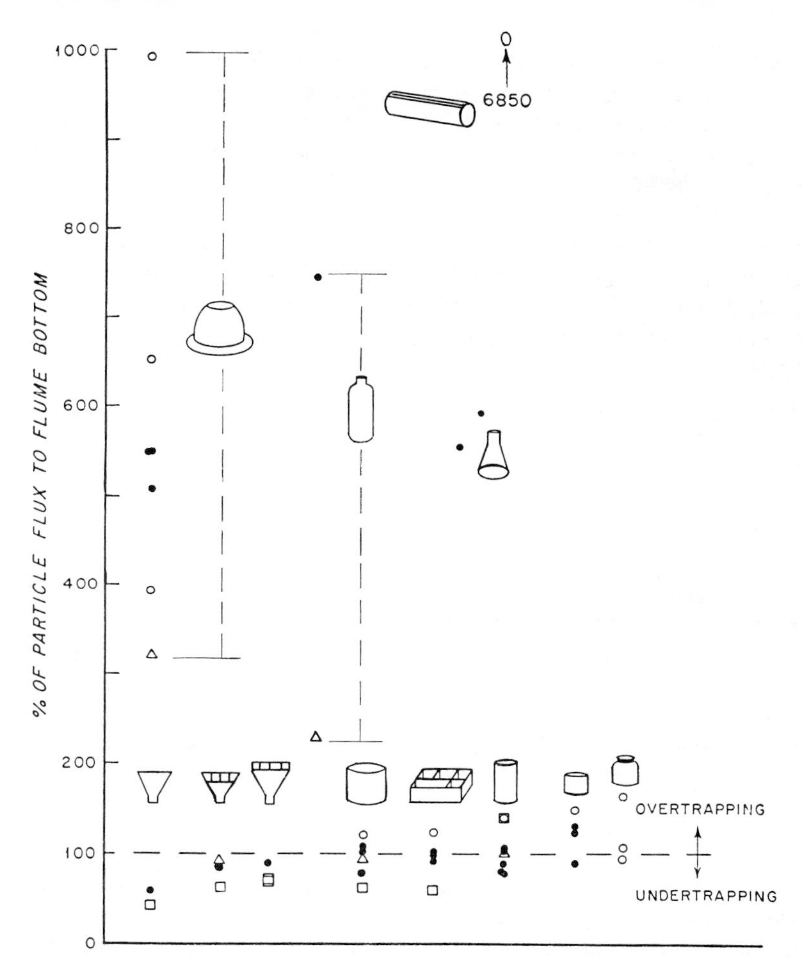

Fig. 3.18. Sediment trapping efficiency of various sediment vessels. Index 100 = no negative effect; < 100 = undertrapping; > 100 = overtrapping; • 4.0–4.5 cm s^{-1}; ○ 9.0–9.5 cm s^{-1}; △ still water; □ 4.3 cm s^{-1} with vessels rotated during experiment (Gardner 1980a)

These theoretical arguments have been empirically verified both by field and laboratory tests. This is exemplified with laboratory data (recirculation flume, varying water velocity, 0–9.5 cm s^{-1}, fluorescein dye, lutite particles with mean diameter 2.6 μm) in Fig. 3.18 for various types of vessels. An ideal sediment vessel would, in this case, get an index of 100. Deviations from this value measure over- or undertrapping; minus for funnels, plus for bottles. Figure 3.19 illustrates comparisons of fluxes from cylinders with different aspect ratios, i.e., height/length ratios. From this figure, it is clear that adequate data are not obtained for H/D-ratios $\leqslant$ 3. In Fig. 3.20, it is demonstrated that variable trapping effects occur in cylinders with diameters < 40 mm. Aspect ratios that are too high (> 15) may, on the other hand,

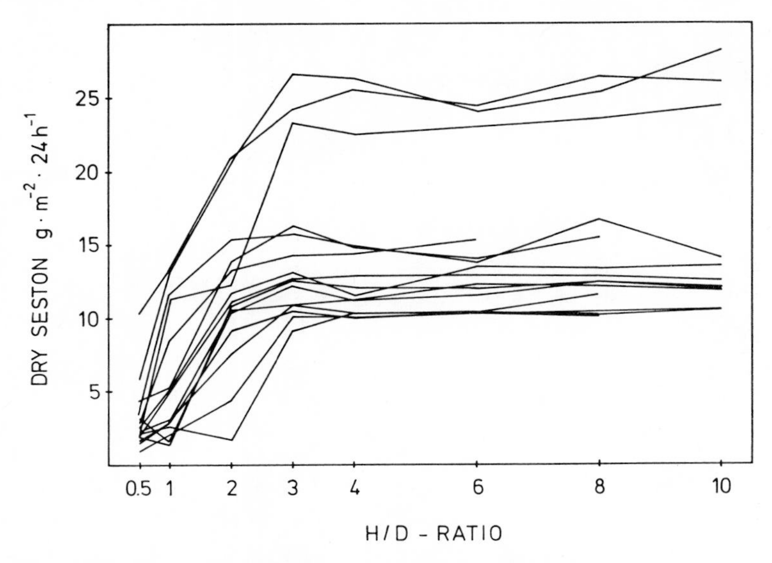

Fig. 3.19. Relationship between amount of deposited seston in cylinders (diameter 21–57 mm) and different H/D ratios (Blomqvist and Kofoed 1981)

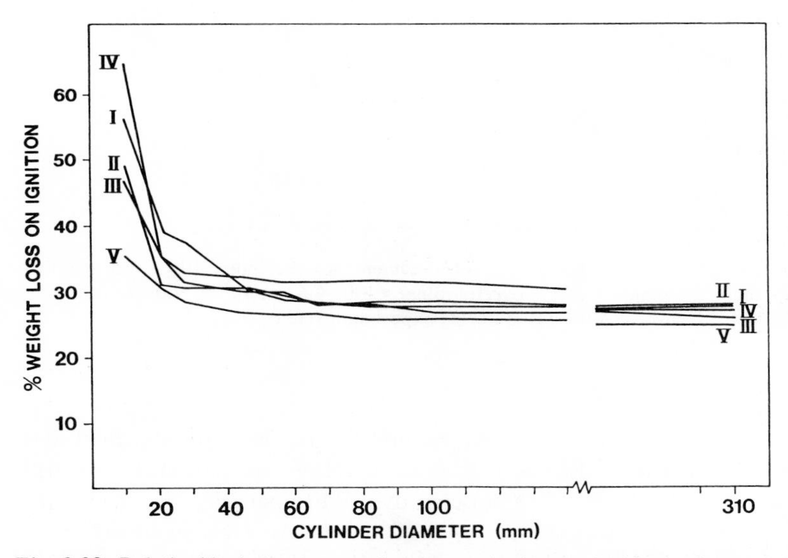

Fig. 3.20. Relationhip between organic content (as loss on ignition) of deposited material and diameter of cylindric sediment vessels. *Roman numerals* denote various periods of registration (Blomqvist and Kofoed 1981)

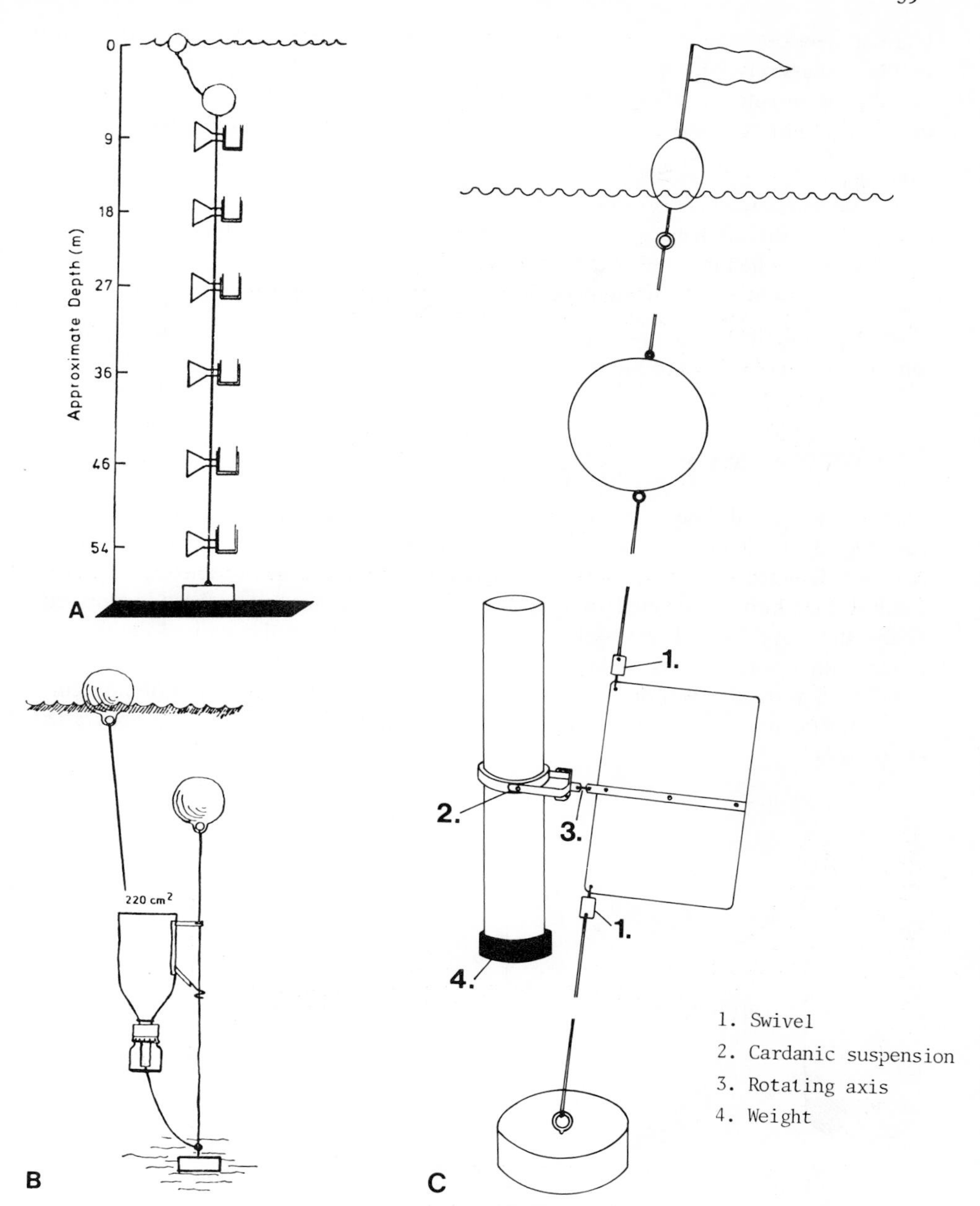

Fig. 3.21A–C. Examples of sediment traps with solitary sediment vessels according to **A** Ansell (1974), **B** Lastein (1976), and **C** a cardanic suspension. (Modified from sketch given by Swedaq, HB, Sweden)

increase development of anoxic conditions, which may change the quality of the sediments. Excessively large vessels may be difficult to handle. Collars, lattices, baffles, radially asymmetrical vessels or reference chambers (for correction of attached growth) are not necessary and would not improve the sediment trap efficiency.

Conclusions:
- Cylinders are the best type of vessel.
- Cylinders should have aspect ratios larger than 3; in very turbulent water this value should be increased to 10.
- Cylinders should have diameters larger than 4 cm.

If those requirements are met, parallel vessels usually give satisfactory results within ± 10% (Bloesch and Burns 1980).

3.3.3 Practical Aspects

The most frequently used types of traps are buoy-suspended anchored sediment traps (see Fig. 3.21). Many different mooring systems have been used, but three typical ones are depicted in Fig. 3.21. They are based on an anchor and a cable stretched by a sub-surface buoy connected to a surface buoy. Sub-surface buoys should be placed below the wave base. The vessels are fixed to the cable at the desired depth in a frame with several vessels or in balanced single fittings. It is essential to keep the vessel in a vertical position (Figs. 3.21 and 3.22) which permits rotation and current adjustment, e.g., by means of a fin on the downstream side of the apparatus (Fig. 3.23).

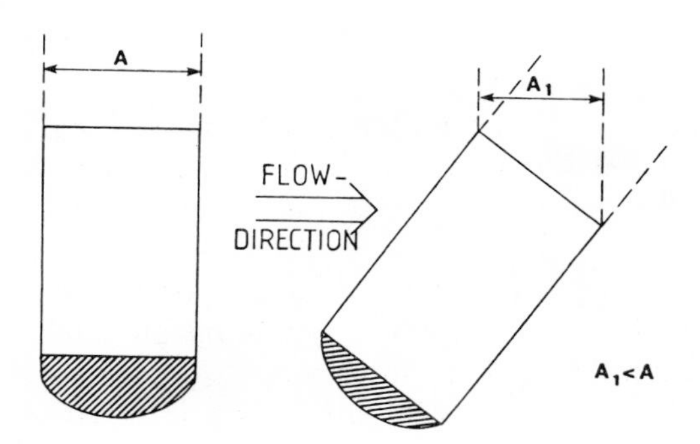

Fig. 3.22. Inclination of sediment vessel will cause a reduction in effective area of deposition (Blomqvist and Håkanson 1981)

Bottom sediment traps (Fig. 3.24) should be placed below the wave base in zones of accumulation (see Chap. 7). On retrieval, the vessel should be carefully examined for sediment disturbances. If no disturbances are visible, then the supernatant water can be siphoned off. If the sediments have been disturbed during retrieval, then the whole suspended water mass must constitute the sample.

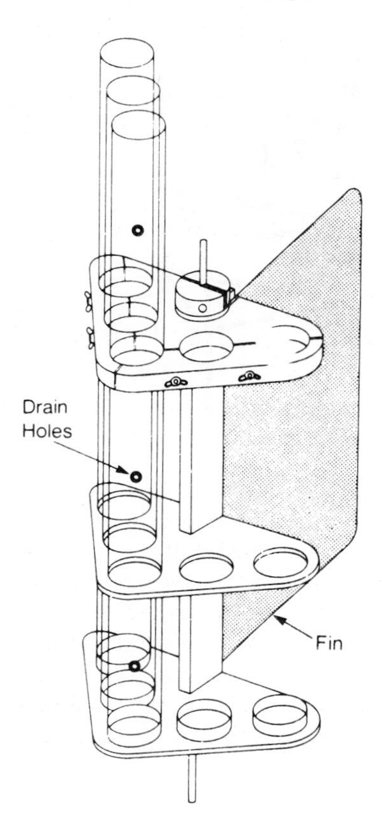

Fig. 3.23. Example of sediment trap with five sediment vessels. (Design of M.N. Charlton, F. Roy and N.M. Burns, from Bloesch and Burns 1980)

3.3.4 Problems with Sediment Traps

Even if good, apparently undisturbed data from parallel sediment vessels have been obtained, one would still face interpretational problems. Sediment trap data must always be treated with great care. Some specific problems still remain concerning the sediment trap technique, e.g.:

1. The problem of mineralization, preservation and exposure time. Bloesch and Burns (1980) have shown that the biweekly loss of material can be in the order of 10% (for parameters like dry weight, particulate carbon, nitrogen and phosphorus) and that the losses increase with exposure time. The question of preservation has not yet been fully answered. Preservatives should not be necessary if the exposure time is short ($<$ 10 days) and might cause more problems than they would solve. The present opinion is that preservatives (such as $HgCl_2$, NaCl, formalin, iodine solution, etc.) should be avoided.

2. The lid problem. The disadvantages of lids concern complications of construction, use and interference. If the aspect ratio is higher than 5, then present experience indicates that no significant resuspension appears during hauling up. If lids are to be used, Fig. 3.24A illustrates one type of system. A cable from the centre tube of the vessel goes through a subsurface plastic (fishing) ring, which keeps the cable taut,

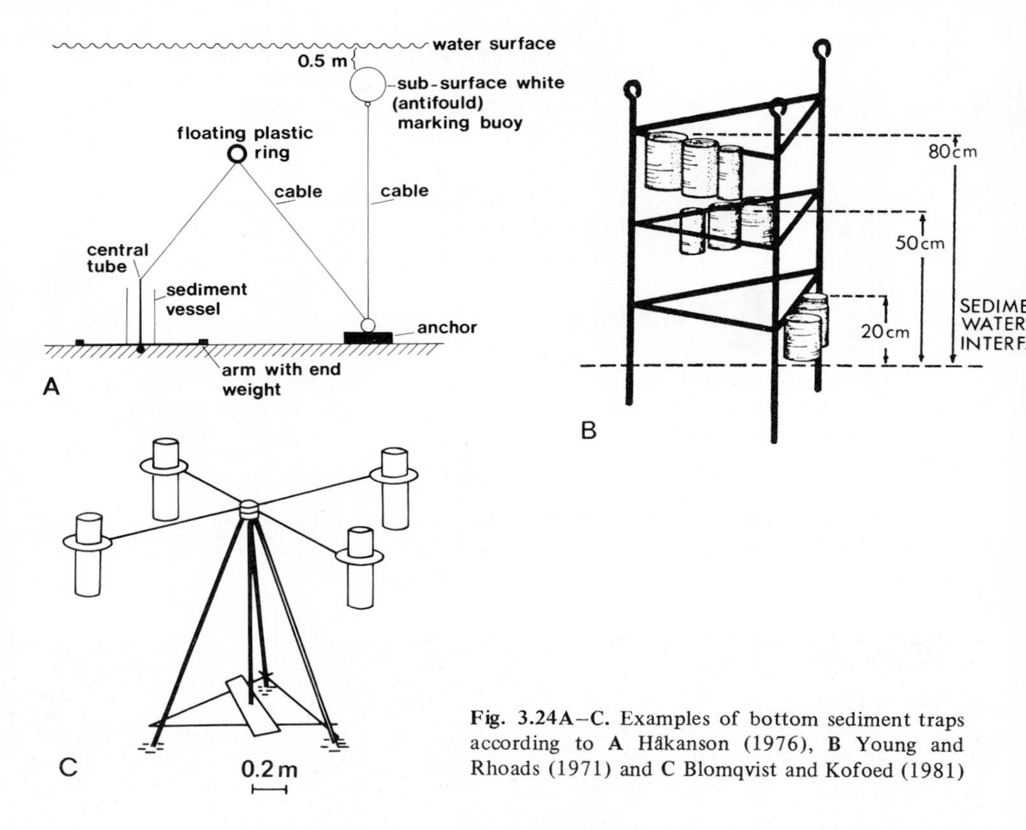

Fig. 3.24A–C. Examples of bottom sediment traps according to **A** Håkanson (1976), **B** Young and Rhoads (1971) and **C** Blomqvist and Kofoed (1981)

to an anchor (lead or concrete). A rope from the anchor to an anti-fauled marking buoy placed about 0.5 m below the water surface to prevent damage by ice or inquisitive persons. To empty the vessel, the anchor is first taken up, allowing the floating ring to come up to the surface. The cable is then run through the central hole in the lid, which is then lowered with a separate line onto the trap. When the lid is in place, the vessel can be hoisted up by the cable running through the central tube of the vessel.

3. Problems also exist, and will presumably never be solved, concerning marking and inquisitive people. This is, however, not only a serious drawback with the sediment trap technique, but a general problem encountered with all types of long-term registration in the field.

3.4 The Cone Apparatus for in Situ Determination of Sediment Types

The fall cone technique yields a simple, inexpensive and rapid measure of physical sediment character and bottom dynamics (erosion, transportation, accumulation) in all types of aquatic environments in an objective, numerical and reproduceable manner. The basic idea is to determine the sediment type by means of an instrument consisting of two or more cones of different shape and weight, whose tips are zero adjusted at the sediment surface (in situ or in a sampler) (Håkanson 1982b,c). The cones are then allowed to penetrate the sediments for a certain period of time, whereafter the penetration depth of the cones is measured. Thus, the instrument yields rough but objective data on the sediments.

The theory on cone penetration, i.e., a mathematical description of the physical forces acting on a cone penetrating sedimentary deposits, has been discussed by Hansebo (1957) and Källström and Håkanson (1982) and will not be dealt with in this context.

Figures 3.25 and 3.26 illustrate the construction of the in situ cone apparatus. Figure 3.25 shows that the cone apparatus consists of a bottom plate with three holes for the three cones, a direction plate for the cone axis, a stop/release plate with the stop/release mechanism, three cones with cone axis (1 m long) and a suspension axis for the whole apparatus. The diameter of the apparatus is 20 cm and the total weight 7.5 kg. The use of the cone apparatus can be described by the following steps:

A. Lowering. The cable attached to the suspension axis is taut, which keeps the stop/release wedge in up-position. The wedge presses the stop/release disk with its rubber stoppers against the cone axis so that they are kept still.

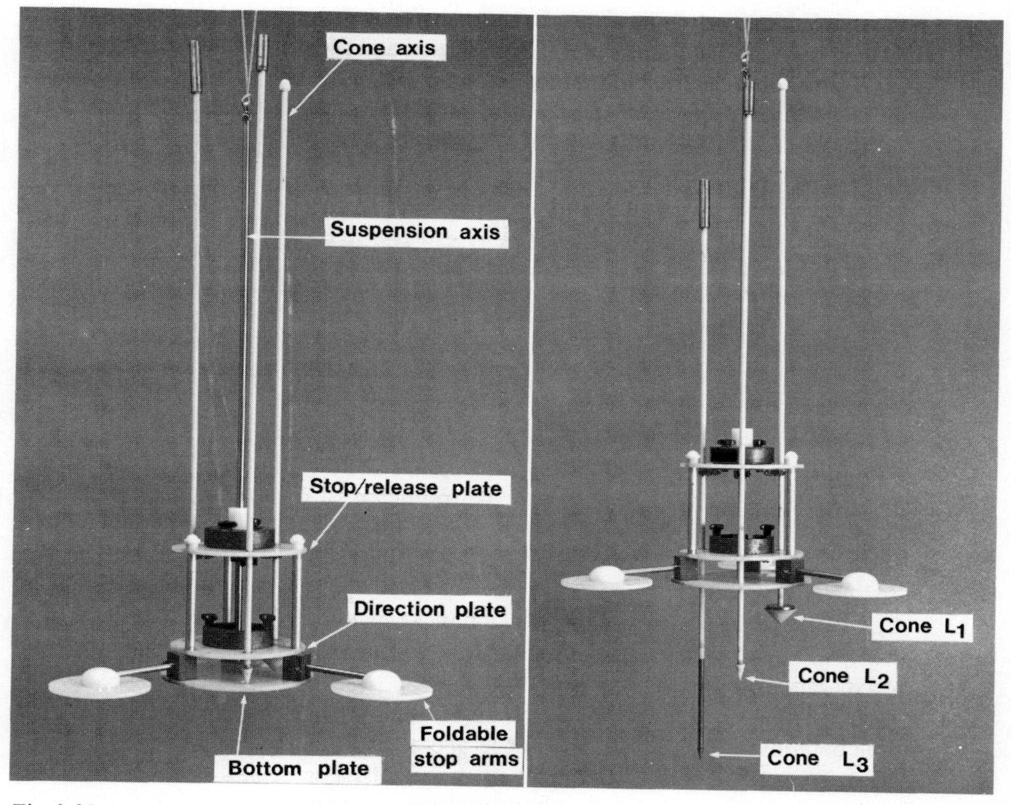

Fig. 3.25. The in situ cone apparatus (Håkanson 1982c)

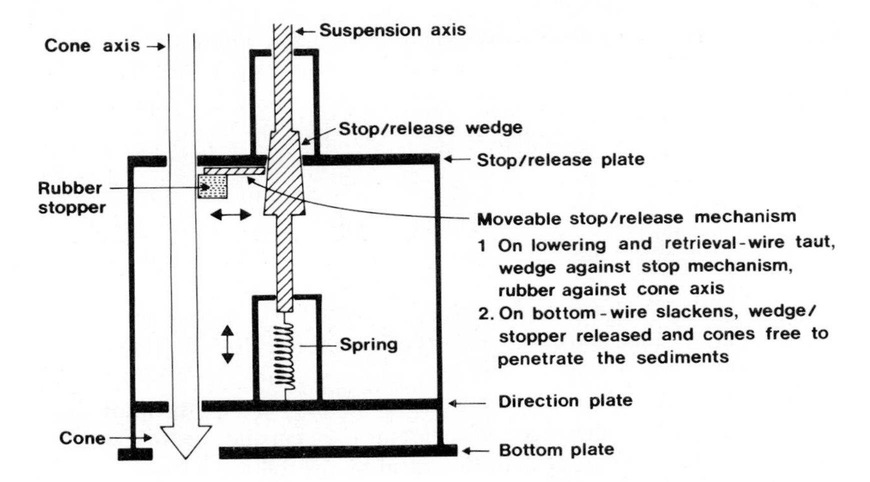

Fig. 3.26. Diagrammatic sketch of the mechanism of the cone apparatus (Håkanson 1982c)

B. At bottom contact. When the cable slackens the stop/release wedge attains down position (accentuated by the spring – see Fig. 3.26). Then the pressure of the rubber stoppers aginst the cone axis is released and the cones are free to penetrate the sediments. The widest and lightest cone (L_1) will fall the shortest distance.
C. Retrieval. When the cable is taut for retrieval, the stop/release wedge again attains up-position, whereby the rubber stoppers hold the cone axis in the position attained at the sediment penetration. When the apparatus is retrieved, the penetration depths of the three cones can be directly read. Repeated empirical tests and theoretical calculations have demonstrated that the depth of penetration equilibrium is reached after 1–3 s depending on cone type and sediment type; after this short time span the cones have obtained a practically fixed position.

To give stability when the bottom is reached and to minimize the penetration of the apparatus into the sediments, three foldable arms are attached to the bottom plate.

The widest and lightest cone, L_1 (top angle 90°, weight 250 g, height 3.0 cm), will penetrate a comparatively short distance in the sediments; a distance primarily depending on the physical character of the top deposits: it has a register from 0, on flat, hard bottoms, to about 5 cm, on very loose bottoms.

Cone two, L_2 (acute angle 30°, weight 300 g, height 3 cm) has a greater capacity for sediment penetration.

L_3, which is the heaviest and most pointed (angle is 30°, weight 500 g, height 1.5 cm), will have a register down to 80 cm in the depicted version (the length of the cone axis can, of course, be altered).

The potential bottom dynamic situation (erosion, transportation, accumulation) may be determined by means of the cone apparatus according to a method given in Chapter 7.3.2.2. Table 2.5 gave preliminary results concerning cone penetration depth, sediment type and prevalent bottom dynamic situation.

3.5 Methods of Defining Concentrations

A proper interpretation of elements and pollutants in sediments presupposes a relevant way of defining concentrations. Depending on differences in, e.g., sediment bulk density, practically opposite results may be obtained from the same material if the elemental concentration is given per dry substance (ds) or per volume unit (cm^3) wet substance (ws). This is illustrated in Table 3.6 for the zinc content in surficial sediment samples at sites at different distances (580 m, 1380 m, 2880 m and 4990 m) from the mouth of River Fyris in Lake Ekoln. Table 3.6 also gives data on Pb, Cu, Cr and Hg. Thus, from this example, it is clear that if the Zn-content is given in $\mu g\ g^{-1}$ ds (= microgram per gram = 10^{-6} g = ppm = parts per million), then the data indicate a decrease with distance from the mouth of the tributary. If, on the other hand, the Zn-content is given in $\mu g\ cm^{-3}$ ws, then these data would indicate an increase in sediment contamination with distance. Hence, the questions: How should concentrations be given? What alternatives exist?

Table 3.6. Zinc, lead, copper, chromium and mercury in surficial sediment samples at various distances from the mouth of River Fyris in Lake Ekoln. Concentrations are given in μg per gram dry sediments (ds) or in μg per cm^3 wet sediments (ws) (Håkanson 1973)

Distance m	Zn ds	cm^3 ws	Pb ds	cm^3 ws	Cu ds	cm^3 ws	Cr ds	cm^3 ws	Hg ds	cm^3 ws
580	421	141	144	48	117	39	71	24	2.01	0.80
1380	408	140	166	57	116	40	126	43	1.18	0.47
2880	344	177	124	64	54	23	147	76	0.78	0.19
4990	340	190	140	65	75	21	140	64	0.67	0.19

From a principle point of view, four alternative methods to present concentrations exist (see Förstner and Wittmann 1979 or Förstner and Salomons 1981):

1. Without corrections, per
   - dry substance (or sediments);
   - wet substance;
   - volume unit.
2. With grain size correction, e.g.
   - separation of grain size fractions: 204 μm-, 175 μm-, 63 μm-, 2 μm;
   - extrapolation from regression curves: % 16 μm, % 20 μm, % 63 μm-specific sur-surface area;
   - correction for "inert" minerals, e.g., metals in quartz-free sediments or carbonate/quartz-free sediments;
   - correction for, e.g., water content, bulk density, porosity, or any other physical sediment parameter showing natural physical linkage to the grain size.
3. With organic content correction, because most metals show high affinity for organic materials.
4. As quotients, relative to "natural" concentrations, preindustrial levels, etc.

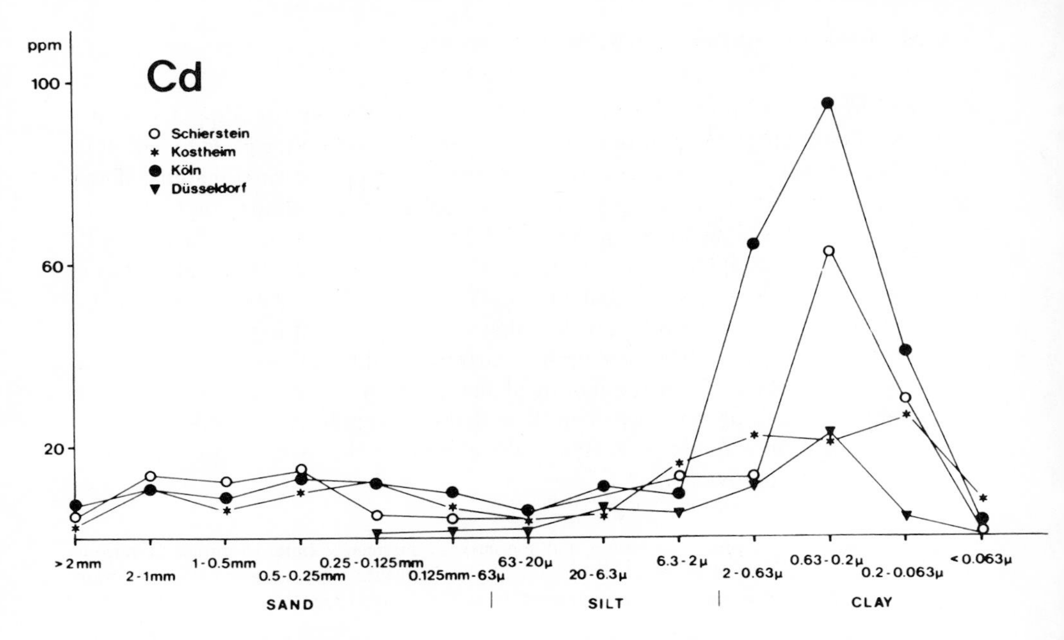

Fig. 3.27. Grain-size dependencies of cadmium concentrations in sediment samples from the Main and the Rhine rivers, West Germany (Förstner and Salomons 1981)

Figure 3.27 illustrates the relationship between grain size and concentration of cadmium in sediments of the highly polluted Rhine and Main rivers. Most Cd is found in the silt and clay fractions. Table 3.7 gives a short summary of some basic concepts concerning concentrations in sediments. All these alternative techniques to present concentrations have been used in different investigations and for different purposes, and they yield various degrees of applicability. Förstner and Salomons (1981) give the following well-founded arguments for the general use of the fraction less than 63 μm in contexts related to sediment contamination:

- "trace metals have been found to be present mainly in clay/silt particles;
- this fraction is most nearly equivalent to the material carried in suspension (by far the most important transport mode of sediments);
- sieving does not alter metal concentrations by remobilization (particularly when water of the same system is used);
- numerous metal studies have already been performed on the suggested < 63 μm fraction."

Table 3.7. Some basic concepts concerning concentrations of elements

kg = 10^3 g, g, mg = 10^{-3} g, μg = 10^{-6} g, ng = 10^{-9} g
ppm (= parts per million) = μg g^{-1}
ppb (= parts per billion) = ng g^{-1}
ds = dry sediments or dry substance
ws = wet sediments or wet substance

Most alternatives do not generally provide better data – just more work and/or more indirect information, which may be difficult to interpret.

The general approach to presenting sediment data is to give concentrations per gram of dry substance. Load calculations require concentrations per volume unit. Some contaminants are most clearly revealed when the concentration per gram organic material is used.

Subsequently, we will give the basic formulas used to transform these various expressions into each other.

The water content (W) is given as:

$$W = 100 \cdot \frac{\text{g ws} - \text{g ds}}{\text{g ws}} = 100 \cdot \left(1 - \frac{\text{g ds}}{\text{g ws}}\right). \tag{3.19}$$

The metal concentration in, e.g., $\mu\text{g g}^{-1}$ ds is given by:

$$\text{Me}_\text{d} = \frac{\mu\text{g Me}}{\text{g ds}} = \frac{\mu\text{g Me} \cdot 100}{\text{g ws}\,(100 - \text{W})}. \tag{3.20}$$

The metal concentration in, e.g., $\mu\text{g cm}^{-3}$ ws is:

$$\text{Me}_\text{w} = \frac{\mu\text{g Me}}{\text{cm}^3\ \text{ws}} = \frac{\mu\text{g Me} \cdot \rho}{\text{g ws}}, \tag{3.21}$$

where ρ = the bulk density (in g ws cm^{-3} ws).

Thus, the relationship between Me_w and Me_d, which is the key to the explanation of the "contradictory" results given in Table 3.6, may be given by:

$$\frac{\text{Me}_\text{w}}{\text{Me}_\text{d}} = \frac{\mu\text{g Me} \cdot \rho \cdot \text{g ws}\,(100 - \text{W})}{\text{g ws} \cdot \mu\text{g Me} \cdot 100} = \frac{\rho\,(100 - \text{W})}{100}. \tag{3.22}$$

This equation is illustrated graphically in Fig. 3.28. It enables direct comparisons between concentrations given per gram of dry substance and per cubic centimeter of wet substance.

The metal content expressed per gram of organic material (= loss on ignition) is given by:

$$\text{Me}_\text{IG} = \frac{\mu\text{g Me}}{(\rho\ \text{ds} - \text{gir})} = \frac{\mu\text{g Me} \cdot 100}{\text{IG}}, \tag{3.23}$$

where gir = gram ignition residual or inorganic residue; IG = loss on ignition.

A direct comparison of these expressions Me_d, Me_w and Me_IG, with empirical data from different sediment types in the central part of Lake Vättern, Sweden, is given in Fig. 3.29. From this example, it is evident that completely different interpretations could be made from the Hg_d and the Hg_IG-sequences. In coarse deposits (sand and coarser), with a low organic content, very high Me_IG-values can be obtained, since in Eq. (3.23) we divide by IG, and this may create a biased picture of the pollutional situation, especially for samples with low organic content. Thus, co-data on water content, organic content and also sediment depth/interval are often essential for a proper interpretation of sediment data.

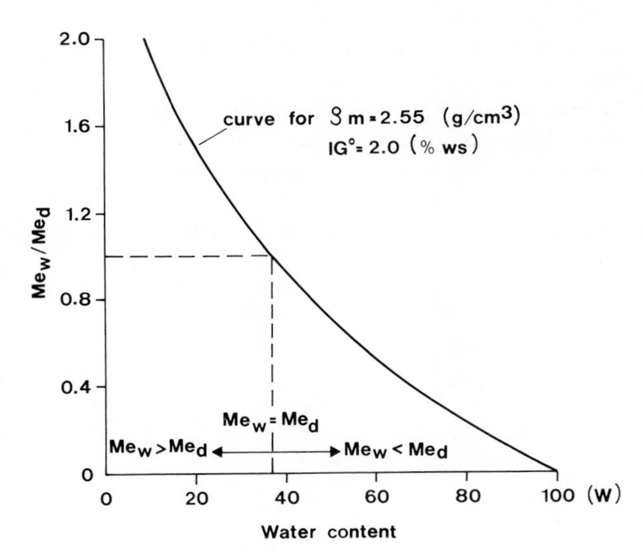

Fig. 3.28. Relationship between metal content per cm³ ws (i.e., cubic centimeter wet sediments, Me_w) and per g ds (i.e., gram dry sediments, Me_d) for various water content, W, of sediments. The curve is calculated for a density of inorganic material of 2.55 and an organic content of 2.0 (per gram wet sediments)

In this section, we will finally make an analogy with the results concerning the necessary number of samples (N) required to obtain a given statistical certainty in the determination of a mean value from various sedimentological environments, as given in Table 2.4. How would the *concentrations* depend on the defined steps (0 = direct analysis, 1 = wet-sieving through a 63 μm mesh and 2 = centrifugation) and corrections (W = water content and IG = organic content). The results are exemplified for copper in Table 3.8. From the previous discussions, we may conclude that the most representative data would be obtained after step 1, i.e., wet-sieving through a 63-μm mesh, without extra corrections.

Table 3.8. Concentrations of copper (in μg g⁻¹ ds) in three different sedimentological environments after various defined steps and corrections (see Fig. 3.11)

	Parameter: Cu Concentrations in μg g^{-1} ds		
	River	River mouth	Lake
0	54.9	16.2	35.0
0, W	75.6	31.0	35.0
0, IG	83.3	67.8	35.2
1	108.8	79.4	42.8
1, W	100.2	73.7	37.9
1, IG	120.8	68.2	36.8
2	115.7	66.8	36.4
2, W	128.0	75.2	36.9
2, IG	86.8	45.6	31.3

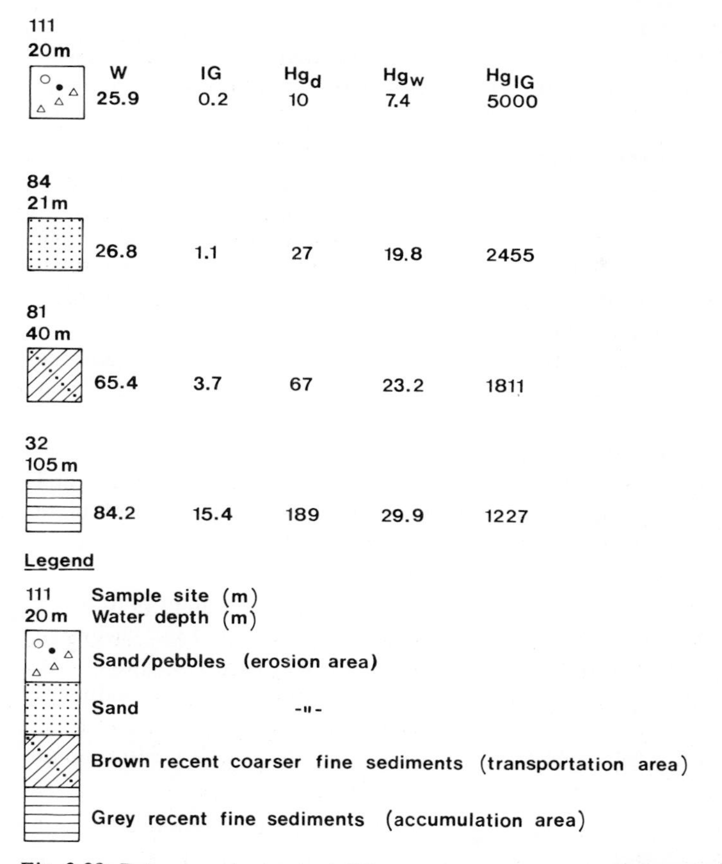

Fig. 3.29. Data on water content *(W)*, organic content *(IG)* and mercury concentrations in nanogram *(ng)* per gram dry sediments *(Hg_d)*, per cubic centimeter wet sediments *(Hg_w)* and per gram organic material *(Hg_{IG})*. Values from central Lake Vättern, Sweden (see Håkanson and Ahl 1976)

3.6 Sampling of Sediment Pore Water

The water content of surface sediments varies from about 30%–50% in minerogenic deposits from areas of erosion to approximately 95%–99% in highly organic sediments. Part of the sediment water is bound to crystals in chemical structures or forms films tightly adsorbed to sediment particles. The rest constitutes the mobile liquid medium, which surrounds the sediment particles and takes part in exchange processes between the particulate and dissolved phases, as well as in exchange processes between sediments and lake water. Sampling and analysis of this so-called *pore water* or *interstitial water* is therefore essential in investigations concerning transformation reactions in the sediments and for studies of the exchange of elements or compounds across the sediment-water interface.

The different procedures to separate pore water or substances dissolved in the pore water from sediment particles can be divided in three groups (see, e.g., Hesslein 1976, Mayer 1976, Robbins and Gustinis 1976, Brinkman et al. 1982).

1. Separation by centrifugation.
2. Separation by pressure or vacuum filtration.
3. Separation by dialysis.

It should be emphasized that, depending on the sampling technique, the yield would be in the order of 25%–50% of the total water content, and that the results, in terms of chemical composition, may be quite variable. Furthermore, the chemical stability of the sample is largely a function of proper handling before the analysis.

Thus, the pore water characteristics are defined by the sampling method and the sample handling, and interpretations of results or comparisons between different investigations should be made with due consideration to the prerequisites.

Centrifugation is the simplest way of obtaining a pore water sample. It is a convenient method for rapid processing of samples when the demands for accuracy and reproducibility are low. Centrifugation often entails disturbances from fine particulate matter (which, however, can be minimized by subsequent filtration) and chemical changes due to, e.g., oxidation of the samples or altered gas equilibria.

Filtration is probably the most commonly used technique to obtain pore water samples and several devices for use in the laboratory or for sampling in situ have been designed. Modern laboratory pressure filtration equipment is usually operated with N_2 gas, which makes it possible to avoid oxidation during processing of the sediment sample. The material extracted from the sediments is well defined by the porosity of the filter; membrane filters with 0.45 μm pores are a common choice for filtration of pore water.

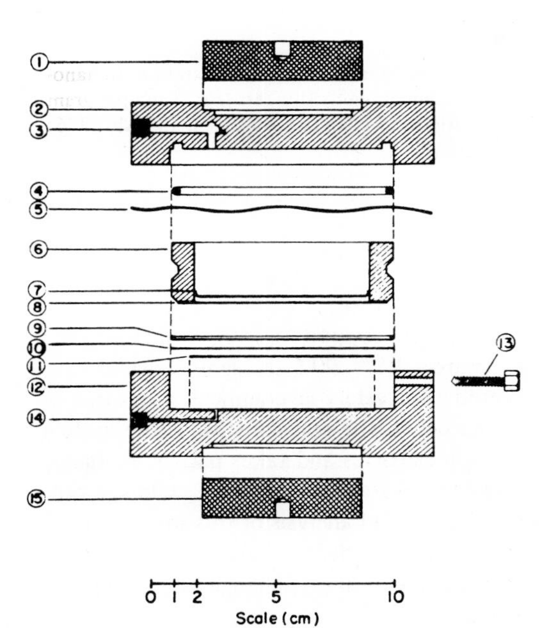

Fig. 3.30. A cross-sectional view of a sediment pore water squeezer. *1* top clamp plate; *2* cap; *3* gas inlet; *4* O-ring; *5* rubber diaphragm; *6* sediment cassette; *7* prefilter; *8* prefilter supporting screen cemented to the cassette with epoxy; *9* O-ring; *10* membrane filter; *11* supporting screen; *12* base; *13* one of the three positioning screws; *14* pore water outlet; *15* bottom clamp plate (Robbins and Gustinis 1976)

A sketch of a typical squeezer apparatus is given in Fig. 3.30. This type has a gas inlet at the top. The high pressure acts via a rubber membrane on the sediment sample, which is enclosed in a removable cassette. The pore water first passes through a coarse pre-filter and then a fine membrane filter.

The main disadvantage with the squeezer is the disturbances which may arise from changes in temperature and pressure during transport from the lake to the laboratory. To minimize artefacts, it may therefore be advantageous to carry out the pore water sampling in situ, which is usually maintained with probes pushed into the sediments. One such piece of equipment, where the pore water is collected in a tube after passing through a filter, is shown in Fig. 3.31. The pressure needed to force the pore water through the filter is simply gained from the hydrostatic pressure at the lake bottom or, alternatively, by a vacuum equipment connected to the sampler. The apparatus is flushed with N_2 gas before deployment to prevent oxidation. This particular apparatus was designed for use in shallow waters and is difficult to handle at water depths below 10 m.

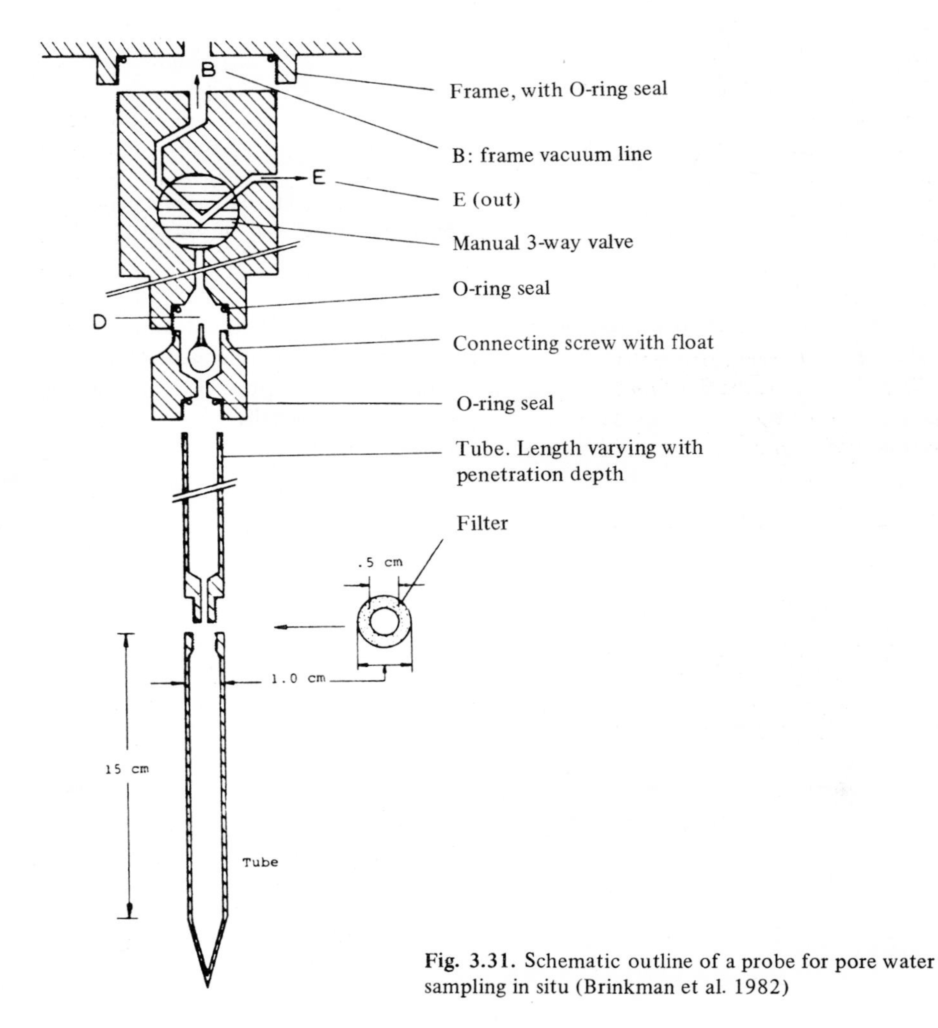

Fig. 3.31. Schematic outline of a probe for pore water sampling in situ (Brinkman et al. 1982)

Another principle for in situ sampling is to place dialysis chambers, or bags containing distilled water, in the sediments. The dialysis outfit is then left in the sediments as long as it would take to reach equilibrium, i.e., when the concentration of dissolved species in the dialysis bag is the same as in the surrounding pore water (usually 1 week is enough).

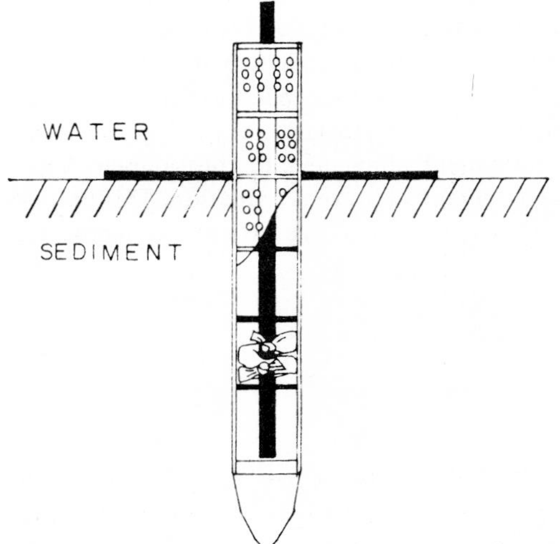

Fig. 3.32. Schematic outline of a pore water sampler with one dialysis bag (Mayer 1976)

With this technique, it is possible to obtain very pure samples free from, e.g., colloidal material. The probes can be constructed (Fig. 3.32) to collect dissolved substances from pore waters simultaneously at several sediment depths. If the fairly long interval can be tolerated, this technique is a good choice, especially when the concentration of dissolved gases in the pore water is to be analyzed.

4 Physical and Chemical Sediment Parameters

This chapter contains a discussion on definitions of basic physical and chemical sediment parameters. According to the Oxford Dictionary, *parameter* means a "quantity constant in case considered, but varying in different cases"; *variable* means different things in different sciences. Subsequently, we will not attempt to distinguish these concepts, but use parameter and variable synonymously.

4.1 Physical Parameters

Under this topic, the focus will be set on a limited number of basic physical sediment parameters, like water content, organic content, bulk density and grain size. We will not discuss the great variety of physical sediment parameters used, e.g., in geotechnical (se, e.g, Scott 1963), geomorphological (see, e.g., Goudie 1981), or agricultural contexts (see, e.g., Kohnke 1968).

Most other physical sediment parameters, like porosity, void ratio, permeability and compaction, may be defined and expressed in terms of these four basic parameters. A more thorough account of sedimentological procedures and statistical evaluations can be found in Müller's (1964) textbook entitled *Methoden der Sedimentuntersuchungen.* The use of these four physical parameters and the causal relationships regulating their distribution in lakes will be discussed more thoroughly in other sections of this book (see, e.g., Chapter 7.3).

4.1.1 Water Content

The water content, which is a key parameter in this context, can be defined in at least two ways:

1. As the ratio of the weight of water, W_w in g, to the dry weight of solids, W_s in g, i.e.:

$$W = \frac{W_w}{W_s} \cdot 100 = \frac{W_t - W_s}{W_s} \cdot 100, \quad (4.1)$$

where W = the water content in %;
W_t = the total wet weight (in g).

This definition of the water content yields W-values larger than 100% for loose deposits and will not be used here.

2. As the weight of water to the total wet weight, i.e.:

$$W = \frac{W_w}{W_t} \cdot 100 = \frac{W_t - W_s}{W_t} \cdot 100 = \frac{gws - gds}{gws} \cdot 100. \tag{4.2}$$

This definition provides W-values from close to 0 (rock) to 100% (pure water) and will, subsequently, be preferred.

The water content is generally determined by oven-drying of about 3–5 g wet sediments for 6 h (or to constant weight) at 105°C. This definition is based on an assumption that lake sediments are saturated with water and that the mass and volume of gases are neglected.

The water content, W, is a most important parameter in lake sedimentological contexts, which will be demonstrated in the sections on bottom and sediment dynamics. The water content is distributed in a typical way in lake sediments (Fig. 4.1). Low values are found in shallow waters, where coarser material often dominate and in

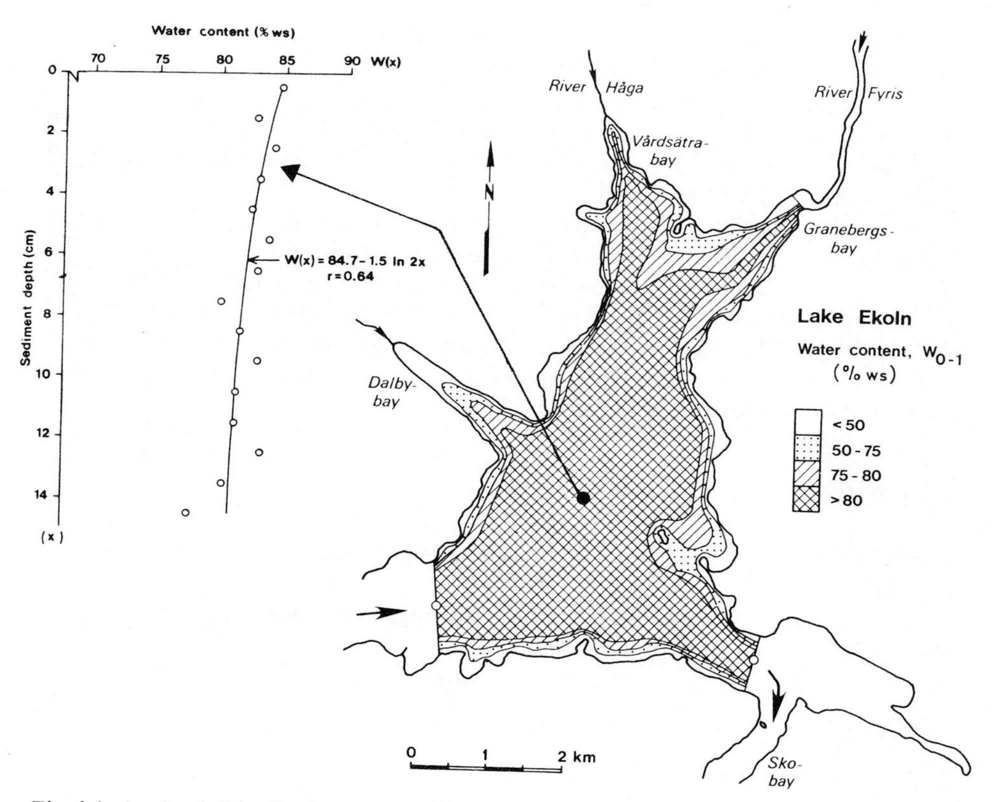

Fig. 4.1. Areal variation of water content in surficial sediments (0–1 cm) and vertical distribution in a sediment core from the central part (accumulation area) of Lake Ekoln (Håkanson 1981c)

river mouth areas, where larger grain sizes from tributaries are deposited. High water contents predominate in deeper parts of lakes. The water content decreases in sediment cores due to compaction (Fig. 4.1). The vertical variation of the water content depends on many factors, e.g.:

– The rate of sedimentation (present and past variations).
– The quality and character of the deposits.
– The degree of compaction.
– The degree of bioturbation and its variation in time.

No general dynamic models are available today which describe all possible interrelationships determining the vertical variation of the water content (and related physical sediment parameters) in a physically impeccable manner. Hence, one has to rely upon conceptual, empirical or semi-empirical models describing the vertical distribution of the water content. One such very simple relationship is:

$$W(x) = W_{0-1} + K_s \cdot \ln(2x), \quad (4.3)$$

where $W(x)$ = the water content of a 1-cm-thick layer at sediment depth x (i.e., $x \pm 0.5$ cm);
W_{0-1} = the water content of surficial sediments (0–1 cm);
K_s = an empirical sediment constant.

The sediment constant, K_s, is site as well as lake typical; the K_s-value depends on the prevailing bottom dynamics (erosion, transportation, accumulation); it varies generally comparatively little within open water areas (areas of accumulation) in lakes; it can vary widely between different lakes depending on the chemical, physical and biological characteristics; very loose sediments with a low degree of compaction have a low K_s-value and vice versa. Table 4.1 illustrates this with empirical data on the water content of surficial sediments (W_{0-1}), on the sediment constant (K_s) and the lake type, as given by the trophic status. Dystrophic lakes, here exemplified by

Table 4.1. Empirical data from sediment cores from some Swedish lakes illustrating the relationship between physical sediment character (water content and sediment constant) and lake type (as given by the trophic status)

Lake	Trophic level	Water content (W_{0-1})	Sediment constant (K_s)
Ingensjön	Dystrophic	95.2	– 0.41
Trösken	Dystrophic	95.4	– 0.83
Skälsjön	Dystrophic	97.4	– 0.64
Hjälmaren (average)	Eutrophic	90.4	– 2.99
Freden	Eutrophic	87.7	– 3.80
Väsman	Mesotrophic	95.9	– 3.95
St. Aspen	Mesotrophic	86.9	– 5.78
Vänern (average)	Oligotrophic	85.5	– 5.97
Vänern (Grums)	Oligotrophic	89.9	– 7.26
Vättern (Åmmeberg)	Oligotrophic	92.5	– 13.64

the Swedish lakes Ingensjön, Trösken and Skälsjön, have loose sediments with a high water content and a low K_s-value; eutrophic lakes, like lakes Hjälmaren and Freden, often (but certainly not always) have W_{0-1}-values in the range of 83–93 and K_s-values of about -2 to -5; mesotrophic and oligotrophic lakes cannot be distinguished from eutrophic lakes by the water content of the sediments, but often by the sediment constant, which generally attains higher numerical values with lower trophic level. This depends on many things, e.g., the supply of minerogenic material, which is larger in oligotrophic lakes, and on the bioturbation, which is generally higher (yielding lower K_s-values) in sediments rich in organic material and food for the bottom fauna.

4.1.2 Loss on Ignition (Organic Content)

The loss on ignition if frequently used as a measure of the organic content of lake sediments. If the dried sediment is heated at 550°C for 1 h, most of the organic substances, but also some chemically bound water and void compounds would evaporate. The loss on ignition is given by:

$$IG = \frac{W_s - W_r}{W_s} \cdot 100 = \frac{gds - gir}{gds} \cdot 100, \quad (4.4)$$

where IG = loss on ignition in percent of the weight of the solid particles (W_s);
W_r = weight of the inorganic residue;
gir = g inorganic residue;
gds = g dry substance (or sediments).

The loss on ignition may, under certain conditions (see, e.g., Mackereth 1966, Digerfeldt 1972, Cato 1977, Håkanson 1983c) be used to get an estimate of the content of organic carbon in the sediments. The correlation between the content of organic carbon (as determined by specific methods, see Skougstad et al. 1979) and the loss on ignition is generally very good. This is illustrated in Fig. 4.2 with surficial sediment data from three parts of Lake Mälaren, Sweden: Västerås Bay, Kungsåra Bay and Görväln (base data from Edgren 1978). The regression line for this particular set of data ($n = 58$) is:

$$C = 0.48 \cdot IG - 0.73. \quad (4.5)$$

In Fig. 4.2A this regression line is compared with the "standard" line, as given by the relationship $IG = 2 \cdot C$.

However, the differences between IG/2 and the C-content generally increase with decreasing IG-contents (see Table 4.2). If the regression line does not cross the origin, this would change the shape of the straight line rather drastically if we put the quotient IG/C on the y-axis instead of the C-content and keep the IG-content on the x-axis. This is shown in Fig. 4.2B. From this figure, and from the given literature, it may be concluded that:

- The carbon content can be roughly estimated as the loss on ignition divided by 2 only when the IG-content is larger than 10%.

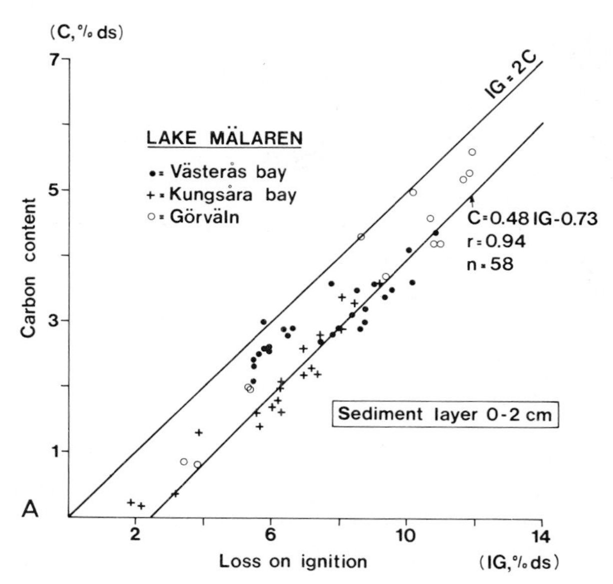

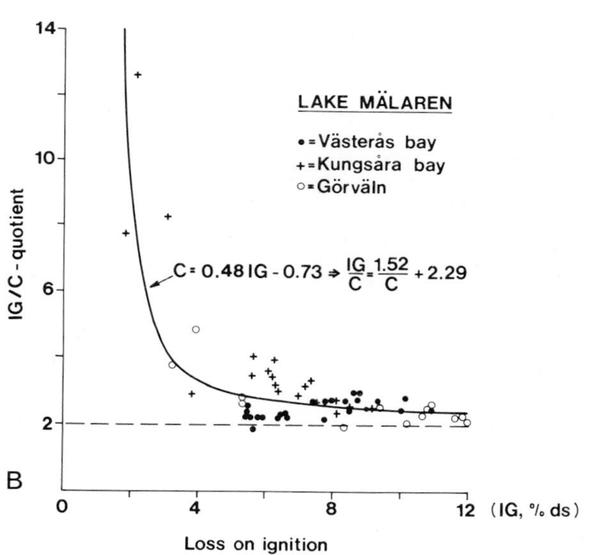

Fig. 4.2A, B. The relationship between carbon content and loss on ignition in surficial sediments from three bays in Lake Mälaren, Sweden

- The correspondence between the C-content and the IG/2-value is best for samples from areas of accumulation, i.e., when $IG > IG_{T-A}$, where IG_{T-A} stands for "critical" loss on ignition separating areas of transportation for areas of accumulation (see Håkanson 1981c and Chap. 7.3.1.2). It should be emphasized that IG_{T-A} can be significantly larger than 10% (e.g., in dystrophic lakes).
- The estimation IG/2 = C may be most erroneous for small IG-contents, i.e., the curve:

$$IG/C = a/C + b \tag{4.6}$$

can vary within vast limits with the empirical constants a and b.

Table 4.2. Examples from Lilla Ullevi Bay, Lake Mälaren, Sweden, on the relationship between loss on ignition (IG/2) and carbon content (C). (Base data from Ryding and Borg 1973)

IG/2	C	$\frac{IG/2 \quad C}{IG/2} \cdot 100$
2.0	0.2	90
2.0	0.2	90
2.5	0.3	88
3.4	0.9	74
3.6	1.2	67
7.8	4.8	38
11.2	9.2	18
11.8	10.2	14
12.5	11.0	12
13.2	11.8	11
13.6	12.3	9.6
14.5	13.5	6.9

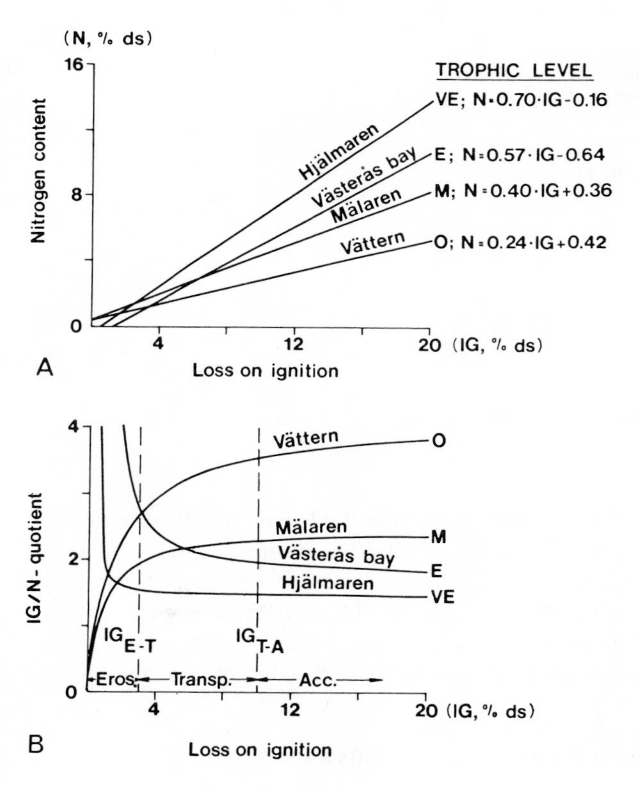

Fig. 4.3A, B. The relationship between nitrogen content and loss on ignition in surficial sediment samples from four lakes of different trophic level (Håkanson 1983c)

These characteristics of the regression line between the C-content and the IG-content would also hold if we put the N-content on the y-axis. The latter has bearings on the way the sediments reflect the trophic level of a lake (see Chap. 2.3). Figure 4.3 gives the regression line for four selected lakes of varying trophic level: Oligotrophic Lake Vättern, mesotrophic Lake Mälaren, eutrophic Västerås Bay and very eutrophic Lake Hjälmaren (data from the 1977 sampling programme, see Table 2.7). These lines show very nicely how the slope coefficient (BPN) increases with increasing trophic level, and they do not cross the origin. A regression line crossing the origin would be transformed to a straight line crossing the y-axis at 1/BPN in Fig. 4.3B.

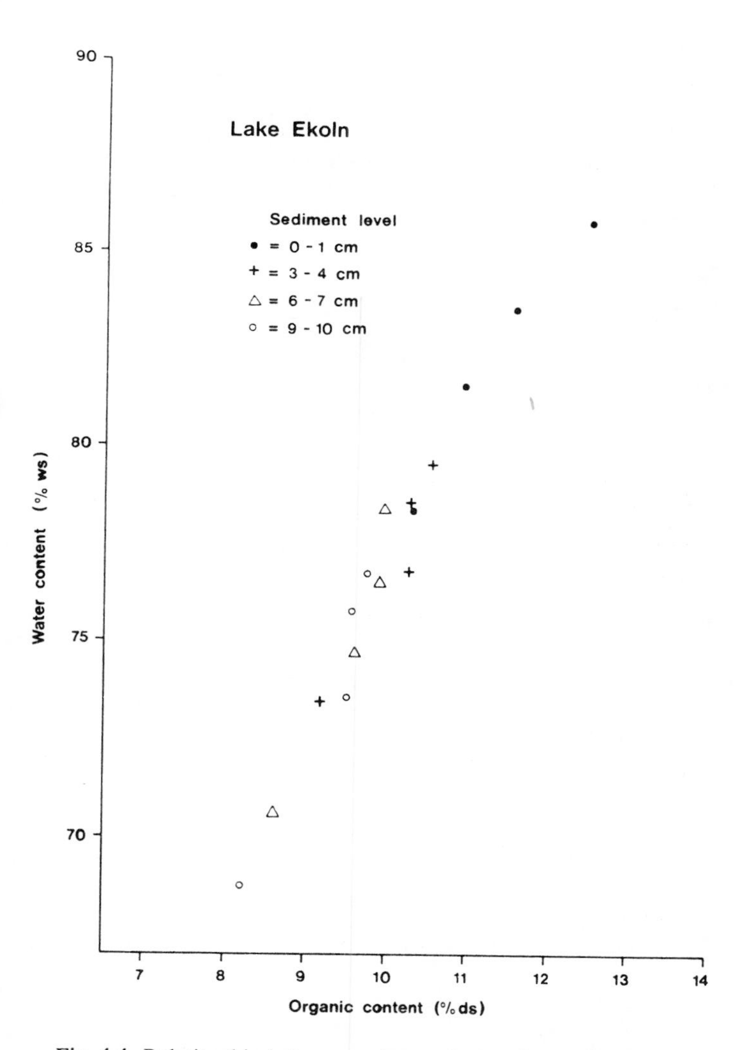

Fig. 4.4. Relationship between water content and organic content (loss on ignition) at different sediment depths in Lake Ekoln (Håkanson 1977a)

Carter (1973) states that high IG-values in sediments are favoured by the following three factors:

- anthropogeneous discharges of organic material (e.g., bark, wood chips);
- high lake productivity;
- reduced conditions in the sediments, which hampers degradation of organic materials.

The organic content is highly correlated to the water content (see Fig. 4.4) and hence also distributed in lake sediments in much the same way. Mineralization of organic materials in the sediments implies that the organic content generally decreases with sediment depth (Fig. 4.4).

4.1.3 Bulk Density

The bulk (or wet) density, ρ (in gws cm^{-3} ws), can be determined by, e.g., a Nutting pycnometer (see, e.g., Krumbein and Pettijohn 1938) or by the following formula (see Axelsson and Håkanson 1971):

$$\rho = \frac{100 \cdot \rho_m}{100 + (W + IG^0)(\rho_m - 1)} . \tag{4.7}$$

This formula is based on the assumption that the density of the water as well as the organic material in the saturated sediments can be set to 1.00 g cm^{-3}. Then the content of inorganic materials (W_s^0), in percent of the total wet weight (W_t), can be expressed in two ways:

$$W_s^0 = 100 - W - IG^0 \tag{4.8}$$

and

$$W_s^0 = \left(\frac{100}{\rho} - \frac{W}{1} - \frac{IG}{1}\right) \cdot \rho_m , \tag{4.9}$$

where IG^0 = the loss on ignition expressed in percent of the total wet weight (gws)
ρ_m = the density of the solid particles (g cm^{-3}).

Equation (4.7) is obtained by putting Eq. (4.8) equal to Eq. (4.9) and solving for the bulk density, ρ. Thus, Eq. (4.7) is a theoretically deduced formula which describes

Table 4.3. Densities of various sediment constituents (in g/cm^3) (Kohnke 1968)

Humus	1.3–1.5	Anorthite	2.7–2.8
Clay	2.2–2.6	Dolomite	2.8–2.9
Kaolinite	2.2–2.6	Muscovite	2.7–3.0
Orthoclase	2.5–2.6	Biotite	2.8–3.1
Microcline	2.5–2.6	Apatite	3.2–3.3
Quartz	2.5–2.8	Limonite	3.5–4.0
Albite	2.6–2.7	Magnetite	4.9–5.2
Flint	2.6–2.7	Pyrite	4.9–5.2
Calcite	2.6–2.8	Hematite	4.9–5.3

the relationship between the bulk density (ρ) and the water content (W), organic content (IG^o) and the density of the inorganic particles (ρ_m). Figure 4.5 depicts this equations in the form of a nomogram with different curves for different values of ρ_m (2.3, 2.4, . . . 2.8). This equation holds for limnetic (see Axelsson and Håkanson 1972) as well as marine (see Ericsson 1973) deposits. It is very simple to use and saves work as compared with the pycnometer, without major losses in terms of accuracy. But it requires that the ρ_m-value is known. From Eq. (4.5) it is clear that the ρ_m-value has rather little effect on the ρ-value in loose sediments, when the water content is larger than about 75%, i.e., for most deposits from areas of accumulation in lakes. So, for many purposes, it may be sufficient to estimate the ρ_m-value. Table 4.3 gives ρ_m-values for various minerals commonly encountered in soils and lake sediments and for humus.

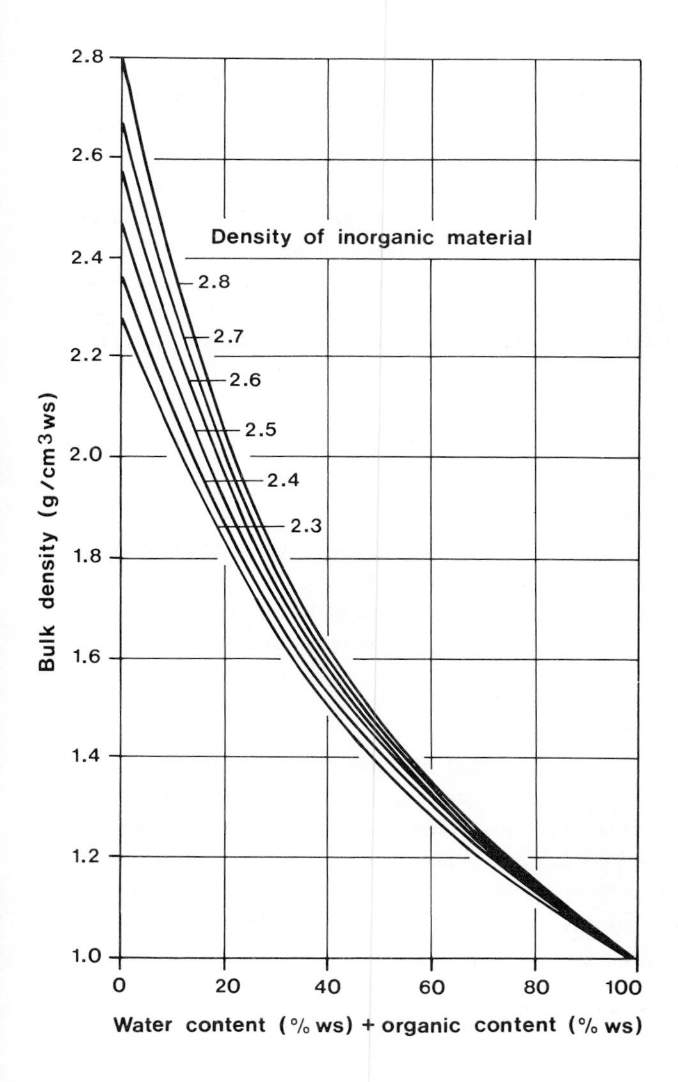

Fig. 4.5. Nomogram illustrating the relationship between water content, organic content, bulk density and density of inorganic material (Axelsson and Håkanson 1971)

Since the ρ_m-value has comparatively little influence on the ρ-value in loose deposits, since sand (quartz), with an average density of 2.65, predominates in zones of erosion in lakes, where the water content is often 50% or less, and since empirical data on cohesive silts and clays generally fall within a narrow range of 2.60–2.85 (see, e.g., Ingelman and Hamilton 1963, Keller and Bennett 1970, Richards et al. 1974), the value ρ_m = 2.6 may be used as a general rule of thumb in determinations of the bulk density. This means that Eq. (4.7) can be written as:

$$\rho = \frac{260}{100 + 1.60\,(W + IG^0)} \,. \tag{4.10}$$

Because the bulk density, the water content and the organic content are physically related to each other in lake sediments, it is quite natural that they show related distribution patterns. This is illustrated in Fig. 4.6 using data from a section and a sediment core in Lake Ekoln, Sweden.

4.1.4 Grain Size

The determination and interpretation of particle grain size has a fundamental role in hydraulics, geomorphology and sedimentology (see, e.g., Friedman and Sanders 1978, Goudie 1981). Here we will focus on classifications of grain size, certain measures (mean, sorting, skewness and kurtosis) of special relevance in lake sedimentological contexts and some applications related to sediment types and bottom dynamics. We will not deal with sedimentary properties of mainly geological and geomorphological interest, such as shapes (sphericity, roundness), fabric, packing, porosity, permeability or orientation.

4.1.4.1 Methods of Analysis

For the particles of interest here, ranging from $1-10^{-7}$ mm, a wide variety of analytical methods are available (see, e.g., Allen 1975):

1. Sieve analysis, which has a practical lower limit of 63 μm. Several standard mesh sizes exist, e.g., the German Standard DIN 1171 11934, and the British Standard BSS 410 11962. When *dry* sieving is used, it is important to minimize the moisture since as little as 1% of water would increase the adhesion forces of especially the finer fractions (Müller 1967). *Wet* sieving is preferable for lake sediments, dominated by fine materials (silt and finer).
2. Sedimentation methods:
   - Pipette method; based on the principle that the changes in concentration in a settling suspension can be measured by successive removal of definite small volumes using a pipette. This method is widely used and requires that carbonates are removed by dilute hydrochloric acid, organic matter by hydrogen peroxide and that some dispersing agent, e.g., sodium hydroxide, is added to replace the exchangeable cations and thus minimize coagulation. The calculation is based upon Stokes' law (see Chap. 6.1).
   - Hydrometer method; based on records of the variation in density of settling suspensions using a hydrometer (see Fig. 4.7).

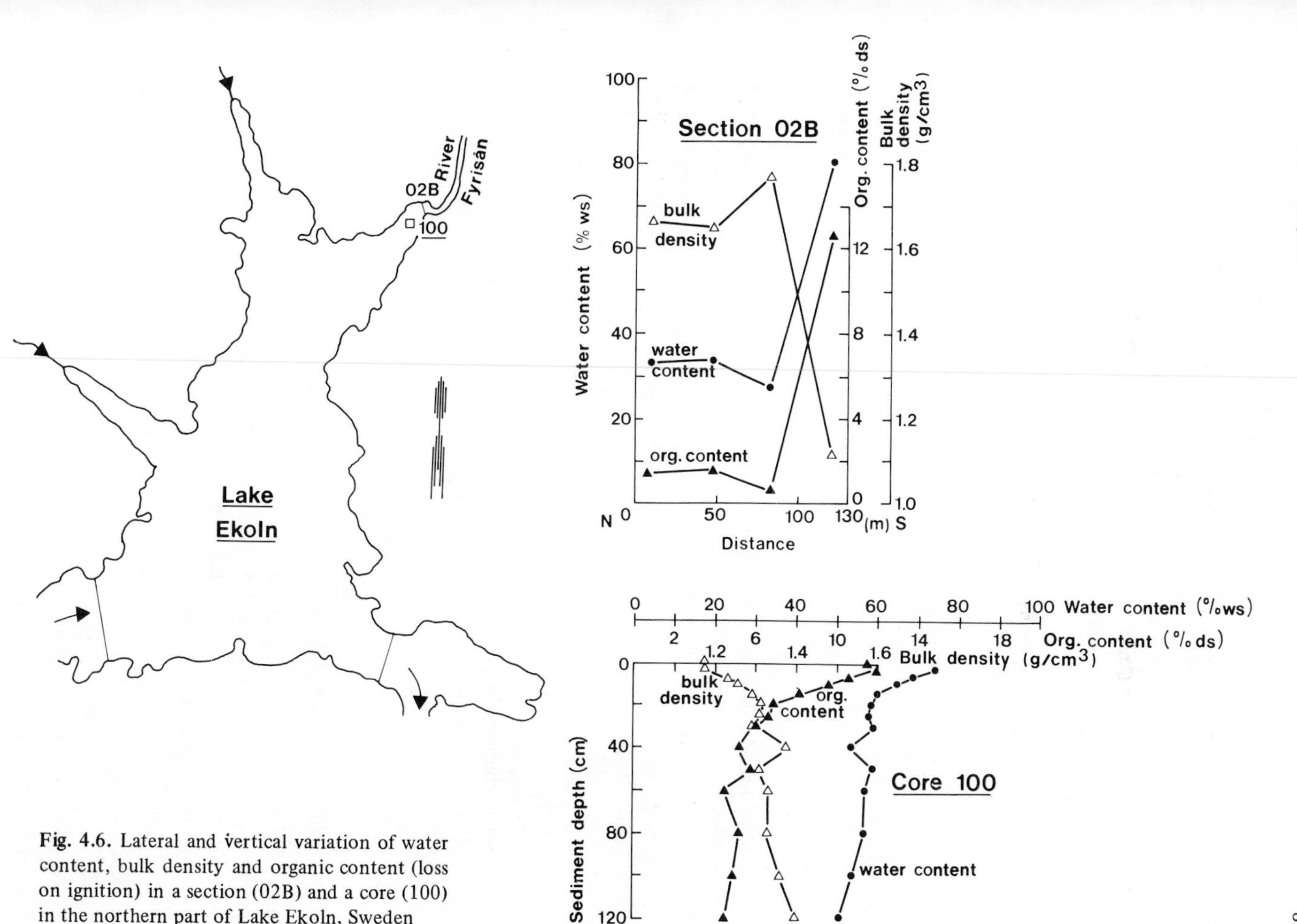

Fig. 4.6. Lateral and vertical variation of water content, bulk density and organic content (loss on ignition) in a section (02B) and a core (100) in the northern part of Lake Ekoln, Sweden

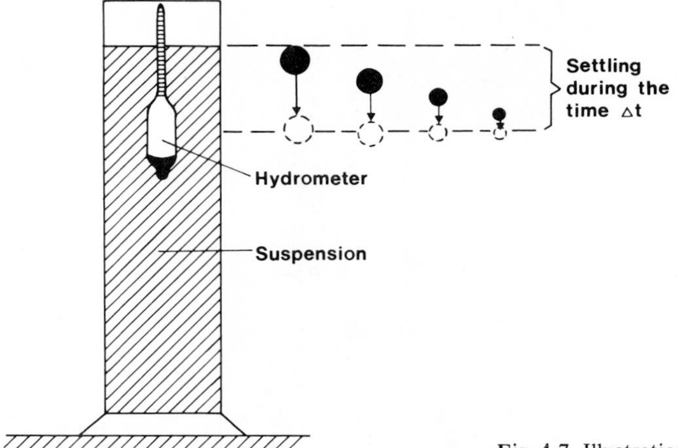

Fig. 4.7. Illustration of a hydrometer method

- Sedimentation columns; based on records of the rate at which a sample is settling out of suspension.

These sedimentation methods require inexpensive apparatus and cover a wide range of grain sizes. The hydrometer method is not applicable if less than 10% of the sample passes the 63-μm mesh.

3. Coulter counter method; based on a principle where particles suspended in an electrolyte pass through a small aperture with electrodes on both sides. The passing particles displace their own volume of electrolyte, whereby the resistance in the current is changed in proportion to the volumetric size of the particles. The number of changes per unit of time reflect the number of particles per unit of volume in suspension. Several types of instruments are commercially available covering a size range from 1–1000 μm. Table 4.4 summarizes the types of method to apply for various size fractions.

4.1.4.2 Grain Size Classifications

Many different systems to grade and classify grain sizes exist. Some of the most frequently used are given in Table 4.5., i.e., the Phi-scale (ϕ), the Millimeter scale (mm), the Micrometer Scale (μm) and the Wentworth grades. In lake sedimentology, the following three grain size classes (Müller and Förstner 1968a, b) are the most important:

sand $> 63\ \mu$m
silt 2–63 μm
clay $< 2\ \mu$m.

The mathematical definition of the ϕ-scale is:

$$\phi = -\log_2 (d) \tag{4.11}$$

or

$$\phi = -\log_2 \left(\frac{d}{d_0}\right) \tag{4.12}$$

where d = the particle diameter in mm;
d_0 = the standard particle diameter of 1.00 mm;
ϕ = (phi) = a dimensionless number of the grain size.

For example, a particle with a diameter of 8 mm = 2^3 mm, is written as -3ϕ; and d = 4 mm = 2^2 mm is -2ϕ; 1 mm = 2^0 mm is 0ϕ; 1/2 mm = 2^{-1} mm is $+1\phi$, etc. The main advantages of the ϕ-scale are (Friedman and Sanders 1978):

Table 4.4. Methods of grain size investigations. (After Kohnke 1968)

Size fraction	Diameter			Methods of investigation	Methods of mechanical analysis
	mm	microns	Å		
Gravel					
	2	2000			
Sand					Sieving
		200			
				Optical	
	0.06	50			Coulter Counter
				Microscope	
		20			
Silt					Sedimentation by gravity
	0.002	2	20,000	X-ray diffraction Electron microscope	
Coarse clay		Red*	7000		
	0.0005		5000	Limit of optical microscope	
		Violet*	4000		
	0.0002	0.2	2000	Infrared spectroscopy	
	0.00006	0.06	600		Sedimentation by centrifuge
Colloidal clay					
	0.00002	0.02	200	Limit of x-ray diffraction and of electron microscope	
	0.000005	0.005	50		
	0.000002	0.002	20	Limit of infrared spectroscopy	
True solution					

*Wavelength of visible light

- distribution of particle sizes can be plotted directly on arithmetic graph paper, thus obliviating the use of logarithmic paper;
- calculations of statistical parameters, like mean, standard deviation, skewness and kurtosis, are simplified;
- limiting particle diameters for size classes become whole numbers instead of fractions of millimeters;
- the negative log usage implies that on graphs larger sizes generally are plotted on the left and smaller sizes on the right.

It should be emphasized that an individual grain can be classified into a given size class, and that natural samples of particles generally include individuals of several size classes. Thus, it is essential to use a statistical approach in describing the grain size distribution of sediment samples.

Table 4.5. Size grades of sedimentary particles

Phi size (ϕ)	Millimetres (mm)	Micrometres (μm)	Wentworth grade
– 6.0	64	64,000	Cobbles
– 5.5	44.8	44,800	----- 60.0 mm -----
– 5.0	32	32,000	Coarse gravel
– 4.5	22.4	22,400	
– 4.0	16	16,000	----- 20.0 mm -----
– 3.5	11.2	11,200	Medium gravel
– 3.0	8	8000	
– 2.5	5.6	5600	----- 6.0 mm -----
– 2.0	4	4000	
– 1.5	2.8	2800	Fine gravel
– 1.0	2	2000	----- 2.0 mm -----
– 0.5	1.4	1400	
0.0	1	1000	
0.5	0.71	710	Coarse sand
1.0	0.5	500	----- 0.6 mm -----
1.5	0.355	355	
2.0	0.25	250	Medium sand
2.5	0.18	180	----- 0.2 mm -----
3.0	0.125	125	
3.5	0.090	90	Fine sand
4.0	0.063	63	
4.5	0.045	45	----- 0.06 mm -----
5.0	0.032	32	Coarse silt
5.5	0.023	23	----- 0.02 mm -----
6.0	0.016	16	
6.5	0.011	11.0	Medium silt
7.0	0.008	8.0	
7.5	0.0055	5.5	----- 0.006 mm -----
8.0	0.004	4.0	Fine silt
8.5	0.00275	2.75	
9.0	0.002	2.0	----- 0.002 mm -----
9.5	0.00138	1.38	Clay
10.0	0.001	1.0	

4.1.4.3 Statistical Definitions

In this section, we will discuss some standard measures to describe grain size distributions and also give examples how these measures can be used in lake sedimentological work to increase our knowledge of sedimentological principles. These measures are:

1. The *mode,* i.e., the most-frequent particle size. The mode is, simply, the peak of the frequency curve (see Fig. 4.8).

2. The *mean*, which can be defined in various ways, e.g., employing the 25th and 75th percentiles or the 16th and 48th percentiles, which represent sizes lying at one standard deviation (see Fig. 4.9) on either side of the mean in a normal distribution, i.e.:

$$\bar{x} = \frac{\phi 16 + \phi 50 + \phi 84}{3}. \tag{4.13}$$

The main grain size is an important tool for interpretations of sediment data in relation to bottom dynamics. Figure 4.10 illustrates the relationship between mean grain size and water depth in Lake Michigan; finer particles dominate in deep waters, where continuous accumulation prevails, here the spread is minimal and the material

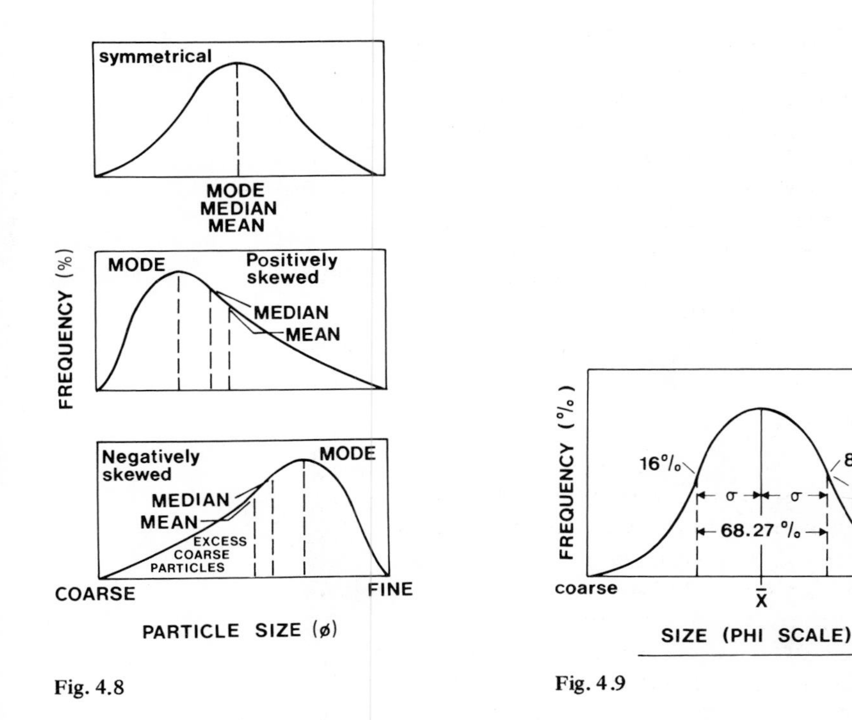

Fig. 4.8. Frequency-distribution curves having various skewness (Friedman and Sanders 1978)

Fig. 4.9. A normal frequency-distribution curve. $\bar{x}$ = mean; σ = standard deviation

well sorted; in shallow waters, where erosion and transportation dominate the topographically open areas and accumulation may prevail in sheltered zones, the material is generally coarser, the spread larger and the sorting poorer (see also Chap. 7.3.1.2).

3. The *median* value is the particle size in the middle (by weight) of the population.

$$M_d = \phi 50. \tag{4.14}$$

The median is smaller than the mean in positively skewed populations and larger than the mean in negatively skewed populations (see Fig. 4.8). For such skewed populations, the median provides more adequate a description of the most typical characteristic value than does the mean.

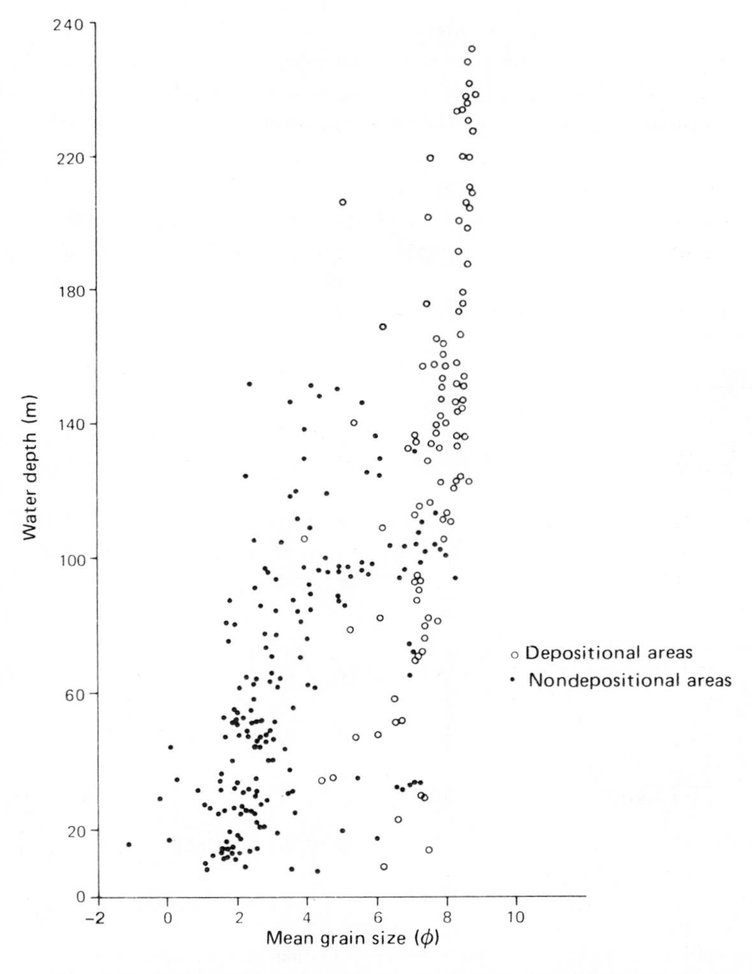

Fig. 4.10. Relationship between mean grain size of surficial sediment samples and water depth in Lake Michigan (Cahill 1981)

4. The *standard deviation* may be graphically determined as:

$$\sigma = \frac{\phi 84 - \phi 16}{4} + \frac{\phi 95 - \phi 5}{16}. \quad (4.15)$$

This concept is illustrated in Fig. 4.9. It gives a measure of the spread around the mean.

5. The *sorting* is defined as:

$$So = \frac{\phi 95 - \phi 5}{2}. \quad (4.16)$$

Descriptive terms for various So-values are given in Table 4.6; So $<$ 0.35 means very well-sorted grain size distribution, So $>$ 4.00 means extremely poorly sorted.

6. The *skewness* measure is defined as:

$$\alpha_s = \phi 95 + \phi 5 - 2\phi 50. \quad (4.17)$$

The skewness is graphically illustrated in Fig. 4.8. The nomenclature is given in Table 4.6. α_s in the range -1.0 to -0.3 means very negatively skewed; $\alpha_s = 0.3$ to 1.0 stands for very psositively skewed population. The skewness (and the kurtosis) provides a measure of the energy level in the bottom zone and the degree of particle mixing (see Thomas et al. 1972, Damiani and Thomas 1974, Sly 1977, 1978). Silt in sand gives a positive skew and silt in clay a negative skew. The relationship between mean grain size and skewness in surficial sediments (0–3 cm) of Lake Michigan is given in Fig. 4.11A.

7. The *kurtosis* or the peakedness of the population is defined by:

$$K = \frac{\phi 95 - \phi 5}{2.44\,(\phi 75 - \phi 25)}. \quad (4.18)$$

This concept is depicted in Fig. 4.12. Peaky distributions are called leptokurtic; flat populations named platykurtic; and normal distributions labelled mesokurtic, see Table 4.6.

The kurtosis gives numerical data of the sorting of the tails of the distribution curve relative to the central portion. Figure 4.11B illustrates the relationship between mean grain size and kurtosis in the surficial sediments of Lake Michigan.

Table 4.6. Descritpive terms for sorting, kurtosis and skewness as measured on the phi scale (Briggs 1977)

Sorting		Kurtosis		Skewness	
Very well sorted	< 0.35	Very platykurtic	< 0.67	Very negatively skewed	−1.0 – −0.3
Well sorted	0.35 – 0.70	Platykurtic	0.67 – 0.90	Negatively skewed	−0.3 – −0.1
Moderately well sorted	0.50 – 0.70	Mesokurtic	0.90 – 1.11	Symmetrical	−0.1 – 0.1
Moderately sorted	0.70 – 1.00	Leptokurtic	1.11 – 1.50	Positively skewed	0.1 – 0.3
Poorly sorted	1.00 – 2.00	Very leptokurtic	1.50 – 3.00	Very positively skewed	0.3 – 1.0
Very poorly sorted	2.00 – 4.00	Extremely leptokurtic	> 3.00		
Extremely poorly sorted	> 4.00				

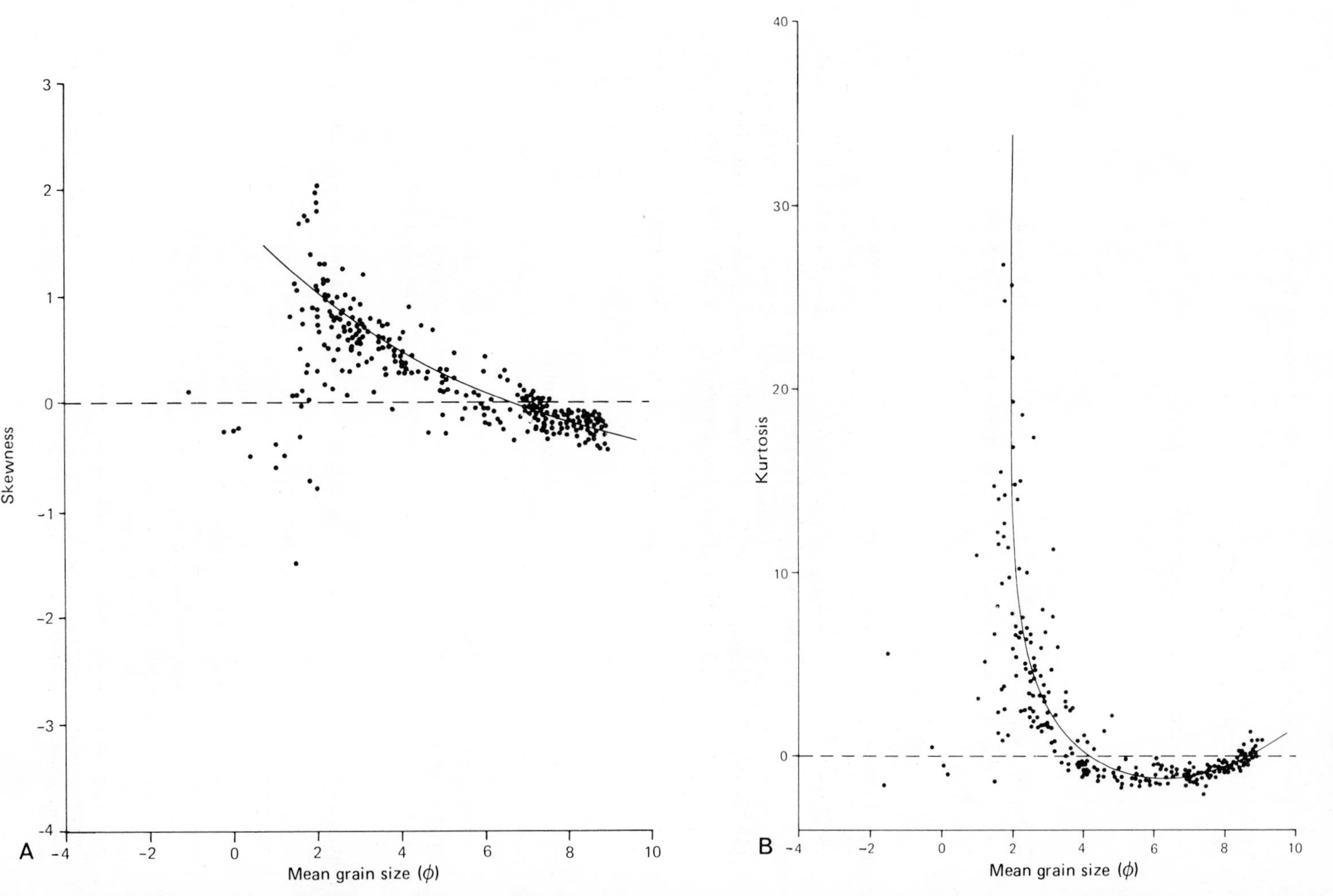

Fig. 4.11A, B. Relationship between A mean grain size to skewness and B mean grain size to kurtosis in surficial sediments in Lake Michigan (Cahill 1981)

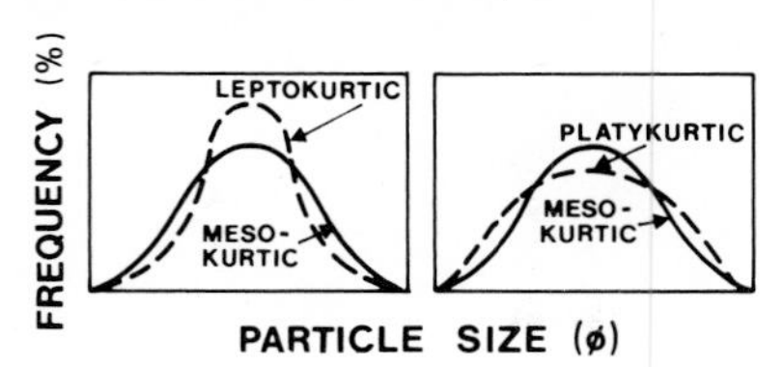

Fig. 4.12. Frequency-distribution curves having various kurtosis (Friedman and Sanders 1978)

4.1.4.4 Grain Size Interrelationships

The grain size of the material influences the sedimentation processes (see Chap. 6.1), the capacity for entrainment (see Chap. 7.2), as well as the capacity of the material to bind pollutants. The specific surface (S_s) of a spheric particle is given by:

$$S_s = \frac{4 \cdot \pi \cdot r^2}{4/3 \cdot \pi \cdot r^3} = \frac{3}{r}. \tag{4.19}$$

Figure 4.13 illustrates the relationship between particle size and specific surface. The potentially active binding area increases significantly with decreasing particle size. But, as has been pointed out, natural sediments consist of varying particle sizes, and the general way to describe the composition of grain sizes is to utilize the Shephard (1954) diagram, as illustrated in Fig. 4.14. With pure sand, silt and clay in the corners of the triangle, this textural classification of sediment grain size involves six pair-alternatives (silty sand, sandy silt, etc.) and one triple-alternative (sand – silt – clay) depending on the percentage by weight that each fraction makes up of the total sample weight.

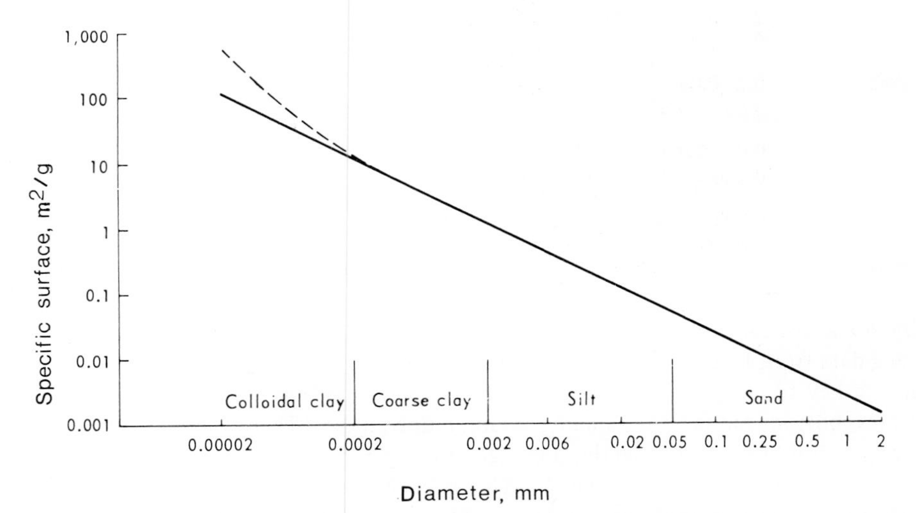

Fig. 4.13. Relationship between particle size and approximate specific surface. *Full line* is calculated for spheres of density 2.65 g/cm³. *Broken line* accounts for the fact that clay particles are more plate-shaped and that fine clays have a large internal surface (Kohnke 1968)

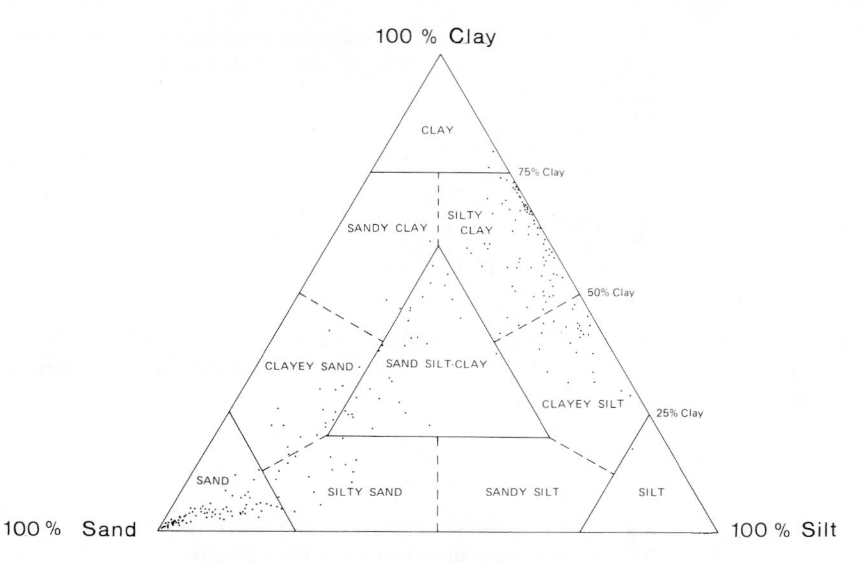

Fig. 4.14. Textural classification according to Shepard (1964) of surficial sediments in Lake Michigan (Cahill 1981)

Table 4.7. Mean grain size, number of samples (n), water content and bulk density of surficial sediments (0–1 cm) from Lake Ekoln, Sweden (Håkanson 1977b)

Mean grain size (mm)		n	Mean water content (% ws)	Mean bulk density (g cm^{-3} ws)
Coarse sand	2–0.6	2	28.0	1.77
	0.6–0.2	2	33.6	1.66
Fine sand	0.2–0.06	5	49.0	1.45
	0.06–0.02	10	68.3	1.23
Silt	0.02–0.006	13	74.8	1.16
	0.006–0.002	34	81.6	1.11
Clay	0.002–0.0006	5	84.5	1.09
	0.0006 <	0	–	–

The areal distribution of such sediment types is illustrated in Fig. 4.15, once again utilizing data from Lake Michigan. This type of areal distribution map may be looked upon as a key to most lake sedimentological interpretations concerning, e.g., bottom dynamics, bottom fauna or sediment contamination studies. There exists an evident, natural physical linkage between the grain size of inorganic particles in lake sediments, the water content and the bulk density. This is illustrated in Table 4.7 by means of data from 71 samples in Lake Ekoln (see Håkanson 1977a, b). The relationship between the mean grain size and the water content of lake sediments depends on many factors:

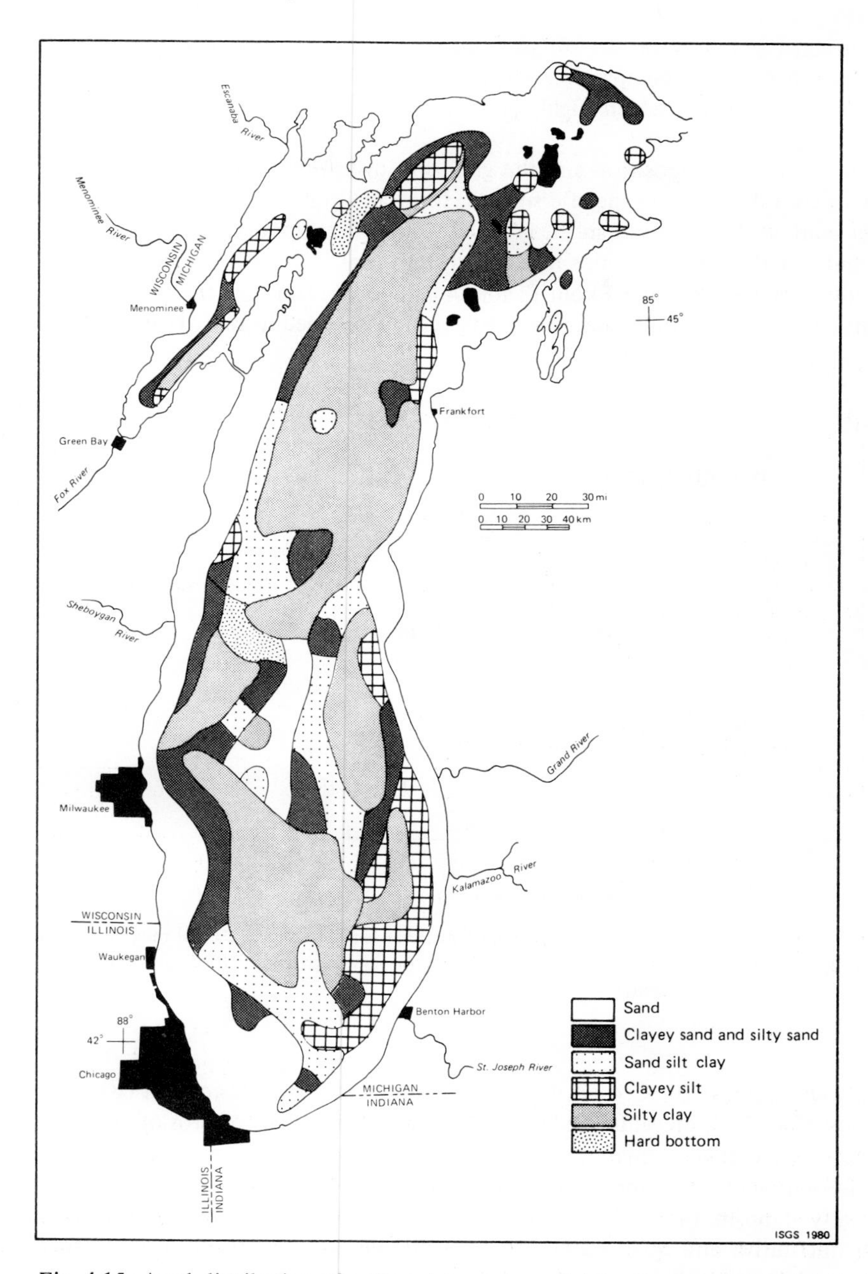

Fig. 4.15. Areal distribution of sediment types in surficial sediments of Lake Michigan (Cahill 1981)

- The utilized sediment interval/depth, since the water content depends on sediment depth/degree of compaction, but not necessarily the grain size.
- The organic content, which highly influences the water content, but not necessarily the grain size.
- The mineralogic composition and the grain-size distribution of the solid particles, which may influence the specific surface, form and cohesiveness, and, hence, the water content differently than the grain size. No general models exist today which account for these relationships in a strictly physical manner. However, based on empirical data from Lake Ekoln (Table 4.7) and focusing on the surficial sediments (0–1 cm), the following general rules can be assumed to apply (see Fig. 4.16).

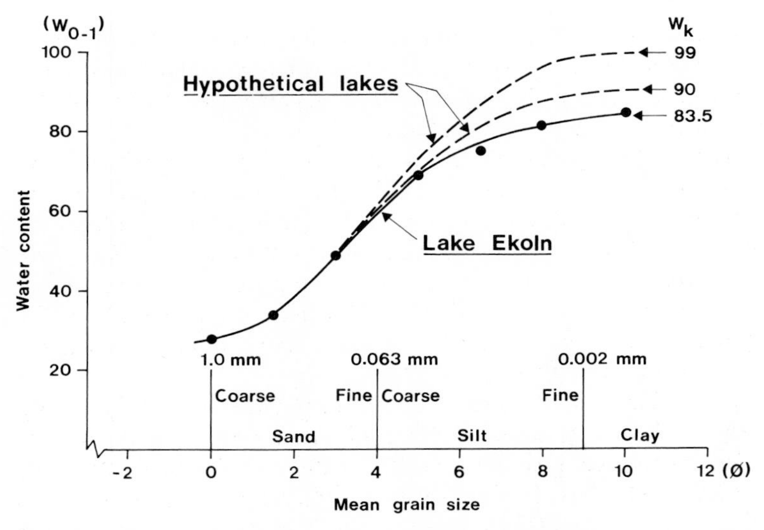

Fig. 4.16. Relationship between water content of surficial sediments (W_{0-1}) and mean grain size for Lake Ekoln data and for two hypothetical lakes with very high (W_K = 99%) and high (W_K = 90%) characteristic water contents

- For sandy deposits ($0 < \phi < 4$) with a naturally low organic content (IG generally less than 5%), prevailing in shallow waters within zones of erosion, one may expect comparatively small differences between different lakes concerning the relationship between water content and grain size.
- For silty deposits ($4 < \phi < 9$), whose organic content may vary widely, from both qualitative and quantitative aspects, we may assume increasing differences between lakes with increasing characteristic water contents (W_K) concerning the relationship between the water content and the mean grain size. This is illustrated in Fig. 4.16 with dotted lines for two hypothetical lakes with extremely loose (W_K = 99%) and quite loose (W_K = 90%) sediments.
- For clayey deposits ($\phi > 9$), one may assume that the correlation between the water content and the mean grain size becomes increasingly poorer with decreasing

grain size, since many factors other than the grain size regulate the water content, and that the curves asymptotically would reach the characteristic water content at about $\phi = 9-10$. *Mean* grain sizes smaller than about 12 ϕ can be expected to be quite exceptional in lake sedimentological contexts.

– The relationship between the water content and the grain size of very coarse sediment (gravel and cobbles) is of secondary interest from the perspective of this book.

4.2 Chemical Parameters

Since lake sediments are formed mainly from material supplied from the terrestrial surroundings of the lake and from material synthesized in the lake water, the chemical structure of a given lake sediment is a function of the characteristics of the catchment and the lake water (D.J. Roberts et al. 1982).

Consequently, chemical analyses of the sediments, if properly combined with correct sampling techniques, give a lot of useful information concerning the lake as well as its surroundings. Moreover, analyses of a sediment sample always mean integration over a certain time period, the length of which depends on the thickness of the sediment layer. Thus, sediment analyses represent a longer time period than a single situation analysis of a water sample, and instead of applying time- and money-consuming water-sampling programmes extended over several years, reliable results can sometimes be obtained by one sediment survey.

Chemical analyses of sediment cores can offer a good key to events in the history of the lake. In early sedimentological studies, the information stored in the sediments was also used mainly as a record of the lake life. In the recent decades, however, there has been an increased awareness of the fact that transport of substances across the sediment-water interface occurs both upwards and downwards. This means that the sediments not only *reflect* the quality of the lake water – they also *affect* the composition and processes of the superimposed water body. Thereby one more dimension is added when interpreting chemical data from lake sediments. This section gives data on the composition of sediments and discusses important features of elements and structural components. Exchange processes between sediment and lake water are treated in more detail in Chapter 9.

4.2.1 Elemental Composition

Any sediment sample consists of a highly complex mixture of discrete minerals and organic compounds to which a number of ions are more or less tightly associated by adsorption, absorption or complexing reactions. Complete analysis of elements in lake sediments does not exist. Mostly, the number of parameters studied are restricted, depending on the aim of the investigation. Table 4.8 exemplifies the elemental composition in lakes from different environments. Table 4.9 summarizes the main groups of elements in lake sediments. *Major* elements (Si, Al, K, Na and Mg) make up the

Table 4.8. Chemical characteristics of various lake sediments. (After B.F. Jones and Bowser 1978)

	Minnesota lakes (46)						African Lakes			
	Low organic group	High organic group	Intermediate group	High carbonate group	Low carbonate group	Linsley Pond	L. Kivu	L. Tanganyika	L. Edvard	L. Albert
Si	–	–	–	–	–	22.11	18.6	26.3	29.2	24.9
Al	–	–	–	–	–	2.91	4.97	9.95	4.29	12.1
Fe	5.01	3.80	2.86	2.21	2.46	8.46	5.32	5.04	2.38	6.22
Ca	0.90	1.95	6.35	15.2	6.76	0.86	9.51	1.21	3.00	1.36
Mg	0.73	0.43	1.02	0.85	1.67	0.60	1.15	1.21	1.21	1.39
Na	0.75	0.40	0.45	0.26	0.73	–	–	–	–	–
K	1.17	0.66	1.02	0.49	1.23	–	–	–	–	–
Mn	0.62	0.10	0.19	0.15	0.37	–	0.09	0.03	0.03	0.10
P	0.13	0.17	0.21	0.14	0.08	–	–	–	–	–
S	0.24	0.48	0.47	0.63	1.09	–	–	–	–	–
N	0.79	2.37	1.06	1.16	0.64	–	–	–	–	–
CO_2	1.43	2.20	7.66	16.96	9.16	–	–	–	–	–
Org C	7.59	21.00	9.20	9.85	6.10	–	–	–	–	–
L.O.1.	17.96	41.92	18.19	19.84	12.40	28.8	–	–	–	–
Si/Al	–	–	–	–	–	7.60	3.74	2.64	6.81	2.06
No. anal.	(8)	(13)	(10)	(10)	(5)	(1)	(381)	(31)	(51)	(92)

	L. Monona (calcareous)		L. Ontario	L. Erie	L. Michigan	L. Superior
Si	–	–	23.9	25.9	25.4	24.3
Al	0.06	3.25	5.05	4.79	2.75	2.38
Fe	0.68	1.68	3.74	2.76	1.48	2.45
Ca	10.1	20.7	0.40	0.35	10.8	1.22
Mg	1.06	1.32	1.28	1.28	4.08	–
Na	–	–	0.53	0.41	0.27	–
K	0.07	0.88	2.27	2.17	1.19	0.50
Mn	–	0.08	0.06	0.06	0.08	0.05
P	0.10	0.16	0.07	0.06	0.13	–
S	–	–	0.04	0.03	0.08	–
N	–	–	–	–	–	–
CO_2	–	–	–	–	–	–
Org C	–	9.71	–	–	–	2.30
L.O.1.	–	–	–	–	–	–
Si/Al	–	–	4.73	5.41	9.24	10.21
No. anal.	(1)	(1)	(1)	(1)	(1)	(1)

Table 4.9. Chemical classification of elements in lake sediments (Kemp et al. 1976, Håkanson 1977d)

1. Major elements (Si, Al, K, Na and Mg), make up the largest group of the sediment matrix.
2. Carbonate elements (Ca, Mg and CO_3-C), constitute the second largest group in sediments, about 15% of the materials by weight.
3. Nutrient elements (org.-C, N and P), account for approximately 10% in recent lake deposits.
4. Mobile elements (Mn, Fe and S), make up about 5% of the total sediment weight.
5. Trace elements (Hg, Cd, Pb, Cu, Zn, Ni, Cr, Ag, V, etc.), the smallest group, accounting for less than 0.1% of the sediments.

the largest group (about 80%) of the sediment matrix in most Nordic environments (compare major constituents (Ca, Mg, Na, K, HCO_3, SO_4, Cl and NO_3 for lake water), *carbonate* elements (Ca, Mg and CO_3-C), *nutrient* elements (org.-C, N and P) and *mobile* elements (Mn, Fe and S) make up approx. 20% of the sediments. The smallest group, the *trace* elements (Hg, Cd, Pb, Zn, Cu, Cr, Ni, Ag, V, etc.) constitute less than 0.1% of the sediments (see Kemp et al. 1976). Silicon, aluminium, calcium, carbonate-carbon plus organic carbon dominate most sediments, which is due to the fact that these elements (except org.-C) are the major structure components in many minerals of the earth's crust (Table 4.10). Calcium and carbonate dominate sediments from calcareous regions, while silicon and aluminium dominate in non-calcareous lakes. The content of organic carbon is variable and depends on the supply of organic material from the terrestrial surroundings and on the productivity of the lake itself.

Table 4.10. Concentrations (in μg/g) of various elements in igneous rocks, sedimentary rocks and dried soils (Bowen 1966, SNV 1976)

	Igneous rock	Sedimentary rock	Soils	
			Mean	Range
Ag	0.07	0.05	0.1	0.01–5
Al	82,000	4000–80,000	71,000	10,000–300,000
As	1.8	1–13	6	0.1–40
Ca	40,000	22,000–300,000	14,000	7000–500,000
Cd	0.2	0.035–0.3	0.06	0.01–0.7
Co	25	0.1–20	8	1–40
Cr	100	10–100	100	5–3000
Cu	55	4–50	20	2–100
Fe	60,000	4000–50,000	38,000	7000–550,000
Hg	0.08	0.03–0.4	0.03	0.01–0.3
K	20,000	3000–25,000	14,000	400–30,000
Mg	25,000	7000–50,000	5000	600–6000
Mn	1000	50–1000	850	100–4000
Mo	1.5	0.2–2.5	2	0.2–5
Na	25,000	400–10,000	6300	750–7500
Ni	75	2–70	40	10–1000
Pb	12	7–20	10	2–200
Si	281,500	73,000–368,000	330,000	–
Sn	2	0.5–6	10	2–200
V	140	20–130	100	–
Zn	70	15–100	50	10–300

Other common elements are iron and magnesium. The latter is mainly bound in inorganic minerals, while iron can be found both in mineral form and associated with organic compounds. Nitrogen in particulate form is always bound in organic structures. In highly organic sediments the nitrogen concentration can exceed 2%. Except for the major elements in Table 4.9, a number of minor constituents, such as phosphorus and heavy metals, are important components of many lake sediments. Although generally present in small amounts, they are essential nutrients or have toxic properties and can be used as tracers of pollution and anthropogenic influences.

4.2.2 Organic Carbon Compounds

Organic matter constitutes a large fraction of many lake sediments. Organic material which survives degradation in the water column is incorporated in the sediments, where it can be more or less preserved or subjected to further biological degradation. Reduced organic carbon provides the main energy source for heterotrophic organisms in the sediments.

Organic carbon compounds are either of *allochthonous* (transported to the lake from the surroundings) or *autochthonous* (produced within the lake) origin. The relative proportion of organic matter derived from these different sources is a function of the characteristics of the catchment area in relation to the productivity of the lake.

In principle, the share of allochthonous carbon increases with decreasing productivity of the lake. The extreme is represented by the dystrophic lake type, where nearly all organic material in the sediments is from terrestrial sources. The other extreme, i.e., mainly authochthonous carbon, is found in lakes with high input of nutrients resulting in high production of organic matter in the lake.

Major features and functions of important groups of organic material in lake sediments are presented below.

4.2.2.1 Humic Compounds

Humic matter is formed mainly as a result of the degradation of plants and can be of allochthonous as well as autochthonous origin. In humic-rich sediments, however, the main source of humic matter is the lake's terrestrial surroundings.

Independent of origin, humic compounds constitute a dominating part of the total organic carbon in many sediments and are most important for the metabolism and turnover of many inorganic elements, particularly metals and phosphorus.

The approximate elemental composition of humus is:

C 45–60%
O 35–40%
H 3– 5%
N 3– 5%
P 0.5%
S 0.5%

There seem to be no important structural differences between humic material in soils and aquatic environments, even though humus in lakes generally has a lower content of nitrogen than soil humus (Gjessing 1976).

Humic matter is formed by the microbial degradation of plant material. The most important compounds used for the synthesis of humus are lignin, carbohydrates, proteins and phenolic compounds. Initial humic-like substances of high molecular weight are synthesized from polyphenolic lignins. Further degradation gives humic acids, fulvic acids and a number of gaseous respiration products. It should be pointed out that apart from the "external degradation processes" maintained by bacteria, the autolysis (self-generated enzymatic breakdown processes following the death of an organism) provide a number of humic-like substances.

Humic compounds are broken down very slowly. The impact of humus on the metabolism of sediments is therefore not primarily as an energy source for heterotrophs but derives from the chemical properties of the humus. These properties in turn depend on the content of functional groups such as:

carboxyl: $-COOH$
methoxyl: $-OCH_3$
phenol: (ring)–OH
ketone: $>C=O$
quinone: (ring)=O

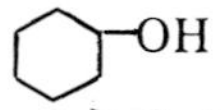

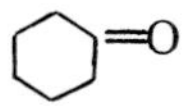

Particularly important are the carboxyl and phenol groups, which have acid-base properties and react with positive ions, giving humic compounds ion-exchange and metal-complexing capacity. The ability to bind metals is essential, since humic-rich sediments thereby can neutralize, by excessive complexation, the toxic effects of heavy metals, such as lead, cadmium, copper, etc.

Iron (III) is effeciently sorbed to humic compounds. Sediments rich in humic material therefore generally also express high concentrations of iron. The possible structure of the iron (III)-humus complex is (Schnitzer 1971):

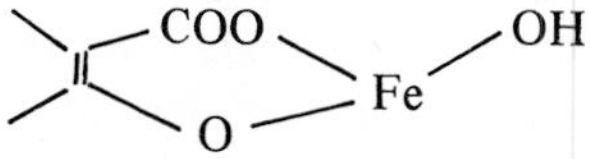

COO Fe OH
O

Iron associated with humus is not prevented from reacting with other negative ions. A typical example is the adsorption of negatively charged phosphates to humic-iron complexes. This reaction, which has been known for a long time, is important for the regulation of the nutrient state of sediments and lakes (Boström et al. 1982).

Humic matter reacts with other organic compounds to neutralize toxic substances, both man-made toxic agents and toxic compounds of natural origin. It has also been shown that humus from lake sediments can reduce the activity of enzymes, i.e., lysozyme (Povoledo 1972). Therefore, it is possible that humic compounds, besides being poor biological substrates themselves, also restrain some biological processes by reacting with extracellular enzymes.

Humic materials affect the pH of lake sediments. Humic-rich sediments are acid (pH 5–6), and since the humic material has acid-base properties, it may serve as a buffer. The buffering capacity of humus is low, however, compared to that of the carbonate system. But in acid environments, where carbonates are exhausted, humic compounds generally constitute the main buffer against pH changes. Therefore, humic-rich sediments should be less sensitive to acidification than comparable lime-poor, non-organic sediments.

The major influences of humus in sediments can be summarized as follows:
1. Humus can decrease toxic effects of metals and organic compounds.
2. Humus affects the nutritional state by binding phosphorus in humic-iron-phosphate complexes.
3. Humus serves as a pH-buffer.

Thus, the effects of humus on the biological activity in sediments may be positive – neutralization of toxic compounds – or negative – slowing down biological processes or reducing the nutrient supply. The full impact of humus on the chemical and biological reactions in sediments is more or less unknown and further research in this fields offers some of the most exciting topics in lake sedimentology.

4.2.2.2 Other Organic Substances

The occurrence of other natural organic compounds than humus depends on the prerequisites for biological life in lakes since most of these compounds have their origin in plankton remains, macrophytes or micro-organisms within the sediments. Consequently, analysis of the organic structure of lake sediments may, to some degree, reveal the relative share of organic input from different sources. Reviews on organic matter in lake sediments have been presented by Cranwell (1976) and Barnes and Barnes (1978). In this context, we will only briefly mention major groups of organic compounds, their possible origin and distribution in lake sediments.

Short chain *fatty acids* are generated from algae and bacteria and therefore common in sediments of productive lakes; long chain acids are typical remains of plant material of allochthonous origin (Cranwell 1974).

Hydrocarbons can be of either allochthonous or autochthonous origin. In oligotrophic lake sediments, the hydrocarbons are generally long chain compounds from terrestrial sources, while short chain hydrocarbons are characteristically derived from photosynthetic organisms (Barnes and Barnes 1978).

The content of *carbohydrates* in surface sediments is proportional to the productivity of the lake (Fleischer 1972). Simple sugars like glucose are rapidly metabolized by bacteria and therefore never accumulate to any large extent, except in highly eutrophic lake sediments.

Amino acids are most frequent in eutrophic lake sediments (Swain 1970). The results of Nissenbaum et al. (1972) from the Dead Sea indicate that decomposition of amino acids requires aerobic bacteria. Amino acids are generally the dominant nitrogen compounds in sediments, as illustrated by the investigations of surface sediments in Lake Ontario, where amino acid N accounted for approximately half the amount of total nitrogen (Kemp and Mudrochova 1973).

Photosynthetic pigments, like chlorophyll and its degradation product pheophytin, are best conserved in anaerobic sediments. These compounds are of essentially allochthonous origin and have been considered as indicators of lake productivity (see, e.g, Gorham 1960, Ohle 1964).

4.2.3 Minerals in Lake Sediments

Depending on the source, the minerals in sediments can be separated in three classes.

1. **Allogenic Minerals** are derived from outside the lake. These minerals are supplied from streams and surface flow, from shore erosion, atmospheric fallout, as well as from cultural sources. They are deposited in the sediments in the form in which they reached the lake water.

The allogenic mineral fraction gives useful information on the characteristics of the surrounding land, including the relative importance of man-made changes. Important allogenic minerals are silicates, clays and carbonates.

2. **Endogenic Minerals** are the result of chemical processes in the lake water leading to precipitates or flocculates which settle to the sediment surface. The endogenic minerals are important guides to chemical and biological conditions in the lake water. Settling of endogenic particulates often occurs only for short periods of the year. Examples are: (1) The sedimentation of silica together with diatom shells, which in temperate lakes often occurs after the spring bloom of diatoms. (b) Precipitation of $CaCO_3$ in calcareous lakes, which is often the result of a pH increase caused by algal photosynthesis. (c) Formation of iron (III) precipitates, which is often a function of temporal changes of pH and redox conditions.

3. **Authigenic Minerals** are formed within the sediments as a result of specific chemical and physical conditions. The process whereby allogenic or endogenic minerals undergo structural changes or new minerals are formed from solute species is commonly termed *diagenesis.* Such reactions are highly important since they affect the equilibrium between solid and liquid phases in the sediments, thereby changing the concentrations of elements which can be transported back to the lake water.

A number of minerals found in lake sediments in North America, Europe and Africa are listed in Table 4.11. Most of the minerals in sediments are allogenic or endogenic. Silicates, both clays and non-clay silicates, are derived mainly from allogenic processes while iron-phosphate and sulfur minerals are examples of minerals formed within the sediments. A few minerals (like calcite) can be of allogenic, endogenic, as well as authigenic, origin.

4.2.3.1 Carbonates

Carbonates are quantitatively important components of sediments. In lakes with calcareous bedrock, they often constitute the dominant single fraction. A major part of the carbonate species of many sediments originates from the erosion of rocks in the lake surroundings. This has been demonstrated for a large number of lakes in calcareous regions in central Europe, Scandinavia, Canada and the United States. By far the most common allogenic carbonates are calcite ($CaCO_3$) and dolomite [$CaMg(CO_3)_2$].

Table 4.11. Minerals from lake sediments (B.F. Jones and Bowser 1978)

Mineral	Type of source		
	Allogenic	Endogenic	Authigenic
Non-clay silicates			
Quartz – SiO_2	X		
Potash feldspar – $KAlSi_3O_8$	X		
Plagioclase – (Na, Ca) (Al, Si) Si_2O_8	X		
Mica – K $(Mg, Fe, Al)_3AlSi_3O_{10}(OH)_2$	X		
Amphibole – $(Ca, Mg, Fe, Al)_{3.5}Si_4O_{11}(OH)$	X		
Pyroxene – $(Ca, Mg, Fe)_2Si_2O_6$	X		
(Other heavy minerals, $\rho > 3.0$)	X		
Opaline silica (diatoms)		X	
Clays			
Illite – $K_{.8}Mg_{.35}Al_{2.26}Si_{3.43}O_{10}(OH)_2$	X		
Smectite – $X_{.3}Mg_{.2}Al_{1.9}Si_{3.9}O_{10}(OH)_2$	X		
Chlorite – $Mg_5Al_2Si_3O_{10}(OH)_8$	X		
Kaolinite – $Al_2Si_2O_5(OH)_4$	X		
Mixed clays, vermiculite – intermediate	X		?
Palygorskite – $(Ca, Mg, Al)_{2.5}Si_4O_{10}(OH) \cdot 4H_2O$	X		?
Nontronite – $X_{.5}Fe_2Al_{.5}Si_{3.5}O_{10}(OH)_2$	X		X
Carbonates			
Calcite – $CaCO_3$	X	X	x
Dolomite – $CaMg(CO_3)_2$	X		?
Aragonite – $CaCO_3$	x	X	
Mg-calcite – intermediate		x	X
Rhodochrosite – $MnCO_3$			X
Monohydrocalcite – $CaCO_3 \cdot H_2O$		X	?
Siderite – $FeCO_3$	?		?
Fe-Mn oxides			
Goethite, Lepidocrocite – FeOOH	X	x	X
Magnetite – Fe_3O_4	X		
Hematite, maghemite – Fe_2O_3	X		?
Birnessite – (Na, Ca) $Mn_7O_{14} \cdot 3H_2O$	?		X
Todorokite – $(Na, Ca, K, Ba, Mn)_2Mn_5O_{12} \cdot 3H_2O$	?		X
Psilomelane – $(Ba, K)(MnO_2)_{2.5} \cdot H_2O$			X
Ilmenite – $FeTiO_3$	X		
Phosphates			
Apatite – $Ca_5(PO_4)_3(OH, F)$	X		x
Vivianite – $Fe_3(PO_4)_2 \cdot 8H_2O$			X
Ludlamite – $(Fe, Mn, Mg)_3(PO_4)_2 \cdot 4H_2O$			X
(?) lipscombite – $Fe_3(PO_4)_2(OH)_2$			X
(?) phosphoferrite – $(Mn, Fe)_3(PO_4)_2 \cdot 3H_2o$			X
(?) anapaite – $Ca_3Fe(PO_4)_3 \cdot 4H_2O$			X
Sulfides			
Mackinawite – $FeS_{.9}$		x	X
Pyrite – FeS_2	X		x
Griegite – Fe_3S_4			X
Sphalerite – ZnS		x	
Fluoride			
Fluorite – CaF_2			X

Print size is varied to illustrate the freqᴜency of occurence. X in formulas refers to monovalent cation exchange

Carbonates are also supplied from endogenic processes. In regions with calcareous bedrock, the surface runoff, soil water and ground water are enriched with Ca^{2+} and HCO_3^- from weathering reactions between water, carbon dioxide and calcium minerals:

$$CaCO_3 + H_2O + CO_2 \rightleftharpoons Ca^{2+} + 2HCO_3^- . \tag{4.20}$$

$$CaMg(CO_3)_2 + 2H_2O + 2CO_2 \rightleftharpoons Ca^{2+} + Mg^{2+} + 4HCO_3^- . \tag{4.21}$$

Consequently, the lake water in such regions has high concentrations of Ca^{2+} and HCO_3^- or CO_3^{2-} and changes of the chemical or physical conditions may cause precipitation of calcite or other carbonate minerals.

Of the physical parameters, CO_2 and temperature are the most crucial for the solubility of calcite. As can be seen from Eq. (4.21), if CO_2 is exhausted, the carbonate system would buffer this loss by precipitation of calcite until a new equilibrium is reached.

Endogenic formation of calcite can occur in the photosynthetic zone of hardwater lakes due to the carbon dioxide consumption of algae or macrophytes. High water temperature promotes this reaction, since the solubility of calcite decreases with increasing temperature.

Carbonates formed by these processes do not necessarily reach the sediments, or become permanently incorporated in lake bottoms. In deeper lakes, calcite which settles through the thermocline comes into contact with colder water with higher CO_2 concentrations and may be dissolved again. High production of CO_2 due to breakdown of organic matter in sediments of productive lakes may dissolve calcite crystals which reach the lake bottom.

Another reaction, connected with photosynthesis, favouring the formation of calcite, is the pH-dependent calcite precipitation. The solubility product of calcite decreases with increasing pH and in productive lakes, where intense photosynthesis raises the pH value by as much as 2–3 units during the day, the solubility product of calcite is often exceeded. For similar reasons, as discussed above, calcite of this origin can be partially dissolved when settling to the hypolimnion or the sediment surface, since pH in these environments is always lower due to the decomposition of organic matter.

The optimal conditions for incorporation of endogenic carbonate minerals in sediments exist in shallow lakes, particularly where the allogenic contribution of carbonates is high enough to buffer the respiratory CO_2 production. The more productive the lake, the more carbonate is precipitated. This is balanced by the increased ability of the sediments to dissolve carbonates in situations of high supply of organic matter, such as exist in productive lakes. Organic compounds can also prevent the dissolution of carbonates. A number of dissolved organic species, like amino acids and humic matter, have a strong tendency to become absorbed on the surface of calcite crystals. According to Wetzel (1970, 1972), these organic coatings reduce the solubility of the carbonates and contribute to the preservation of these minerals in the sediments. If endogenic carbonate formation to a large extent depends on biological processes, authigenic carbonates, like Mg-calcite, are the result of specific abiotic conditions. Large amounts of Mg-calcite seem to be a common feature in sediments of lakes with high salinity (Callender 1968, Müller 1971, Degens et al. 1973).

4.2.3.2 Silicates

A major part of the earth's crust consists of silicate minerals, such as quartz (SiO_2) or more complex structures. In many silicates, aluminium substitutes part of the silicon atoms. Such minerals are termed alumino-silicates. Erosion and weathering of bedrock lead to significant transport of silicates with running water and the subsequent incorporation of these mineral particles in lake sediments. Hence the frequent abundance of silicon and aluminium among the elements in sediments.

Silicate minerals can be considered as exclusively allogenic. They do not change significantly in the aquatic environment and are deposited as particles of various size. Most lake sediments have a dominant portion of silicate material of silt ($> 2\ \mu m$) and sand ($> 63\ \mu m$) size (Müller and Förstner 1968a, b). The minerals in these size fractions are mainly quartz and feldspars.

Because of their allogenic origin, these minerals reflect the bedrock composition of the catchment area. Their stability towards structural changes makes them useful for evaluation of the physical processes involved in the transport and deposition of particles in lakes.

A number of silicates are found in the fine size fraction less than 2 μm defined as clay (Table 4.11). Clay minerals are formed from other silicates by hydrothermal processes (reaction with water at high temperature and pressure) or by weathering reactions. Like non-clay silicates, they should be considered as allogenic minerals reflecting the mineralogy of the lake surroundings.

Contrary to other silicates, which are more or less inert components of lake sediments, the clays have specific characteristics by which they affect the turnover of dissolved components in sediments and thereby also the exchange of substances between sediments and lake water. These properties are connected with the special structure of clays. Clay minerals (Fig. 4.17) are stacked layer structures where each layer is either a tetrahedral sheet (silicon atoms surrounded by four oxygen atoms) or a octahedral sheet (two layers of oxygen atoms in a hexagonal arrangement with aluminium or magnesium atoms in the octahedral sites). These separate sheets are stacked in units of two or three layers. Two-layer clays consist of one tetrahedral and one octahedral sheet and three-layer clays of two tetrahedral sheets surrounding one octahedral sheet (Fig. 4.18). The separate layers are often negatively charged, which means that cations and water can be absorbed between the layers to compensate for the negative charge. Thereby clays serve as cation exchangers and they also have a high water content. The fine grains, plus the porosity, give the clay particles a large surface area, which in turn increases the number of reactive sites. The property to absorb water "within" the mineral structure makes more reactive sites available for reaction with dissolved species.

The importance of a large surface area per weight unit is illustrated in Table 4.12. A clear connection exists between the area/weight ratio and the cation exchange capacity.

Clays not only sorb cations, but also efficiently bind negative ions like phosphate. This sorption probably involves two mechanisms: Chemical binding to positively charged Al^{3+} edges of clay plates and substitution of phosphate for silicates in the clay structures (Stumm and Morgan 1970). Clay particles are also often associated

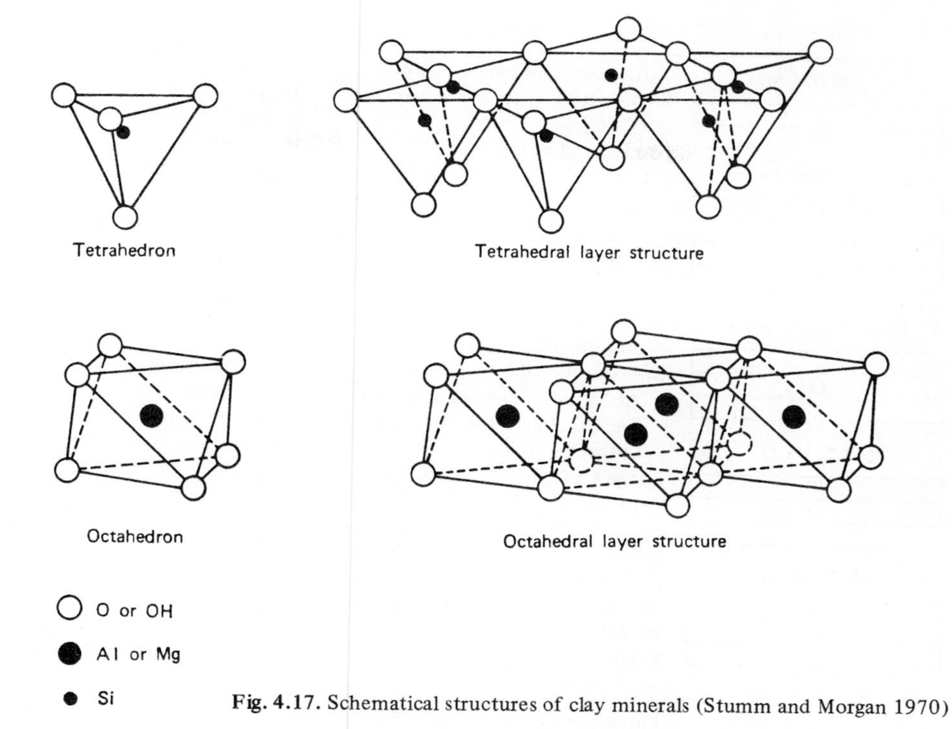

Fig. 4.17. Schematical structures of clay minerals (Stumm and Morgan 1970)

Table 4.12. Cation exchange capacity and surface area of various clay minerals (B.F. Jones and Bowser 1978)

	Kaolinite	Illite	Chlorite	Smectite	Vermiculite
Cation exchange capacity (meq/100 g)	3–15	10–40	10–40	80–150	100–150
Surface area (m^2/g)	16	56		38	
	12	52–82		203–250	
	114		92	234	75
	57		97	221	87
	19	127		778	
	20	139		775	237
				760	548

with organic compounds. According to Jenne (1977), the clay minerals can be regarded as mechanical substrates upon which organic material is precipitated.

Although the major fraction of silicon in lake sediments derives from the external contribution of silicate minerals, endogenic processes exist by which silicon is deposited in the sediments. By far the most important endogenic contribution is the sedimentation of SiO_2 incorporated in diatom frustules. In nutrient-rich lakes, diatoms are a common, sometimes dominant, component of the phytoplankton community.

Structure	Remarks	Names
A. Two-Layer Clays	Little isomorphous substitution	Kaolinite
T / O	Small cation exchange capacities (CEC)	Dickite
T / O	Nonexpanding	Nacrite
T / O		Halloysite (Interlayer water)
B. Three-Layer Clays	*1. Expanding (Smectites or Montmorillonites)*	
T / O / T	Substitution of a small amount of Al for Si in T-sheet and of Mg, Fe, Cr, Zn, Li for Al or Mg in O-sheet	Montmorillonite, Nontronite, Volkhonskyite, Hectorite
M^{+m}, nH_2O	Large CEC ($M^{m+} = Na^+, K^+, Li^+, Ca^{2+}, \ldots$)	Saponite, Sauconite
T / O / T	Swell in water or polar organic compounds	Vermiculite
M^{+m}, nH_2O	*2. Nonexpanding (Illites)*	
T / O / T	About $\frac{1}{4}$ of Si in T-sheet replaced by Al, similar O-sheet substitutions	Poorly crystallized Micas (muscovite, biotite, phlogopite)
	Small CEC	
	$M^{m+} = K^+$	
C. Chlorites	Three-layer alternating with brucite	
T / O / T	Brucite layer positively charged [some Al(III) replacing M(II)], partially balances negative charge on T-O-T (mica) layer	
O (Brucite)	Low CEC, nonswelling	
T / O / T		
D. Fibrous Clays	Different type of structural units consisting of double silica chains (tetrahedral) joined to one-dimensional O-layers and containing interstitial water	Attapulgite, Palygorskyite, Sepiolite

Fig. 4.18. Stacked layer structures of clay minerals (Stumm and Morgan 1970)

These algae assimilate silicon from lake water for incorporation in their frustules. After death, the algae lose their buoyancy and the frustules settle rapidly on the lake bottom. In some sediments diatom frustules are preserved for long periods, while in others dissolution and diagenetic processes make the existence of diatom silicon rare even though the deposition can be high. Typical examples of the first category with high content of diatom remains are the great African lakes, like Lake Tanganyika, and of the second type, Lake Michigan, where diagenetic remobilization of silica seems to be complete in the top 10-cm sediment layer (Jones and Bowser 1978) The reason for these different conditions is not clear. A possible explanation might be that high allogenic input of other silicate minerals plus a high and constant production of diatoms in the African lakes favour the chemical stability of silica in the sediment.

4.2.3.3 Iron

Next to oxygen, silicon and aluminium, iron is the most frequent element of the earth's crust, and consequently a common component of lake sediments (Table 4.10).

The most abundant iron minerals are oxides, hydroxides and sulfides. Iron is also bound in several silicates (Table 4.11). Particulate iron is deposited in the sediments as silicate grains, inorganic oxides or as oxide coatings on settling particles. Iron may also enter the sediments together with organic debris and humic colloidal matter. As seen in Table 4.11, a majority of the iron oxides and sulfides are authigenic, which is the result of the high chemical mobility of iron in lake sediments. The major controlling factors for the diagenetic transformation of iron are redox potential and pH (Fig. 4.19). In oxidized sediments, at slightly acid to alkaline pH, iron exists in the form of iron (III) hydroxides. When the redox potential decreases below about 200 mV at neutral pH, iron (III) is reduced to iron (II) and returned to solution in

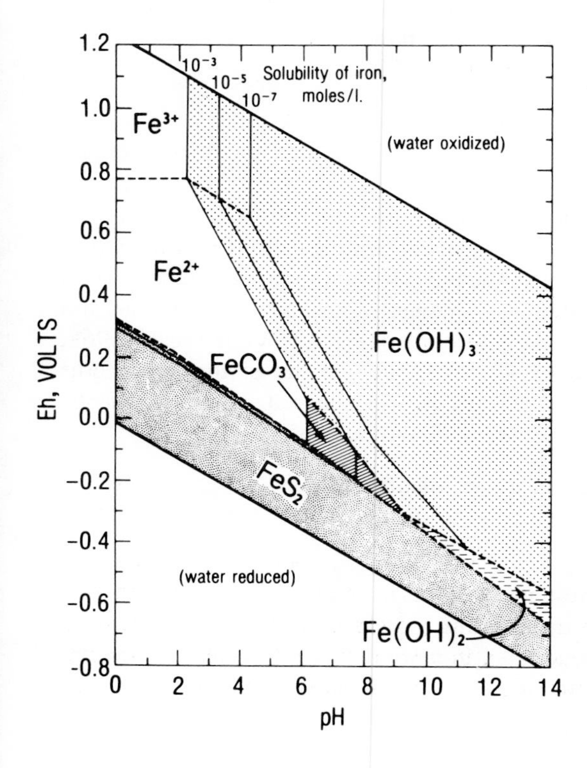

Fig. 4.19. Eh-pH diagram for iron in lake systems. The diagram is constructed for 25°C, 1 atm. and solute concentrations of 10^{-3} molal total carbon dioxide and 10^{-4} molal total sulfur (B.F. Jones and Bowser 1978, as modified from Hem 1977)

ionic form, and Fe^{2+} accumulated in the sediment pore water or transported back to the water column.

If the Eh decreases further, iron (II) can be precipitated as iron sulfide, which is a highly insoluble compound.

When a reduced sediment is oxidized, iron (II) ions are transformed to iron (III), and again precipitated. The stability of iron (III) hydroxides is favoured by high pH values (Fig. 4.19).

These redox-controlled transformations depend on several factors of which the productivity of the lake water above the sediments is by far the most important: the higher the productivity, the higher the loading of oxygen-consuming organic matter on the sediment surface. The breakdown of organic material can lead to reduced conditions in the surface sediments, which is particularly enhanced when the sediments are covered by stagnant bottom water. Reduced sediments with iron (II) species are therefore common below the thermocline in eutrophic and hypertrophic lakes in temperate regions. Oxygenation of the bottoms and related transformations of iron fractions can take place when the sediments are covered by circulating oxygenated water, e.g., when the water body is homothermal.

It is obvious that the existence of different discrete forms of iron minerals in lake sediments is the result of a highly complex pattern of interacting factors and processes. Dissolution, precipitation and transformation occur due to environmental conditions which change several times during the short period of the annual cycle.

The redox-dependent reactions are important not only for the mobility and distribution of iron in sediments. These processes greatly affect the phosphorus chemistry of lake sediments and are therefore of extreme importance for the exchange of this important nutrient across the sediment-lake water interface. These connections are discussed in detail in Chapter 9.

4.2.3.4 Phosphorus

Phosphorus is one of the most studied elements in lake sediments. This interest is primarily due to the fact that phosphorus is the most limiting factor for primary production in most fresh-water environments. The increased phosphorus loading to lakes during recent decades, from cultivated land, domestic and industrial sewage, has therefore been crucial for the eutrophication processes (see, e.g., Vollenweider 1968, Rohlich 1969, Golterman 1975, Wetzel 1975). Large amounts of the phosphorus entering a lake are metabolized in the lake water and then deposited in the sediments. The role of the phosphorus which reaches the sediments is twofold:

1. It can stimulate biological processes in the sediments itself.
2. Phosphorus may return back to the lake water and accelerate the eutrophication processes in the water body.

In this section, the sources of phosphorus and its main distribution forms in sediments will be discussed, while the mechanisms by which sediment phosphorus is returned to the water column are penetrated in more detail in Chapter 9. The main sources of phosphorus for lake ecosystems are:

1. Weathering of Bedrock, Mainly Apatite [$Ca_{10}(PO_4)_6(OH)_2$]. This source, together with the natural atmospheric deposition, constitutes the background phosphorus loading. The natural supply is very low; with few exceptions, as in volcanic regions, where bedrock is rich in phosphorus.

2. Arable Land. Drainage water from agricultural areas may be enriched with phosphorus because of the use of phosphoric fertilizers. The supply of phosphorus from farmland is, in terms of order of magnitude, about ten times higher than the background.

3. Point Sources (like municipal and industrial waste water). Municipal sewage has been an important cause of the rapid eutrophication of many lakes during the 20th century. Construction of phosphorus-reducing treatment plants during recent years has reduced the importance of this source. Considerable amounts of phosphorus are emitted from food industries, smelters and industries producing fertilizers and detergents.

Phosphorus is deposited in the sediments as:

A. Allogenic apatite minerals.
B. Organic associates; partly as structural elements of settling dead organisms, but also in humic complexes.
C. Precipitates together with inorganic complexes, like iron or aluminium hydroxides or as coprecipitate together with calcite.

The relative importance of these different fractions depends on the external supply of apatite minerals, organic matter, complex-forming agents (like iron) and to a large extent on the productivity of the system.

The distribution and forms of phosphorus in lake sediments have not been analyzed in great detail. So far the method used to separate and identify different phosphorus fractions is by chemical fractionation where one phosphorus component after the other is extracted in a series of steps.

One of the most relevant fractionation scheme has been developed by Williams et al. (1976) (Fig. 4.20). This fractionation distinguishes between organic phosphorus, apatite phosphorus and non-apatite inorganic phosphorus. Separations into additional fractions can be made, but the results of such analyses are often unreliable. Recent research, as discussed by Boström et al. (1982), indicates that even the simplified approach of Williams et al. (1976) cannot be used uncritically, since this type of extraction may underestimate the organic phosphorus fraction and the calcium-bound phosphorus.

Table 4.13 summarizes some investigations where fractionations (essentially of the Williams et al. 1976 mode) have been used to characterize the sediment phosphorus. The data include surficial sediments from very oligotrophic lake sediments (Barrow ponds, Alaska) to highly eutrophic systems like Lake Warner or Lake Norrviken.

Generally, the content of inorganic phosphorus is considerably higher than the content of organic phosphorus. The inorganic fraction consists mainly of non-apatite inorganic phosphorus, but in sediments in calcareous regions apatite phosphorus might be dominant.

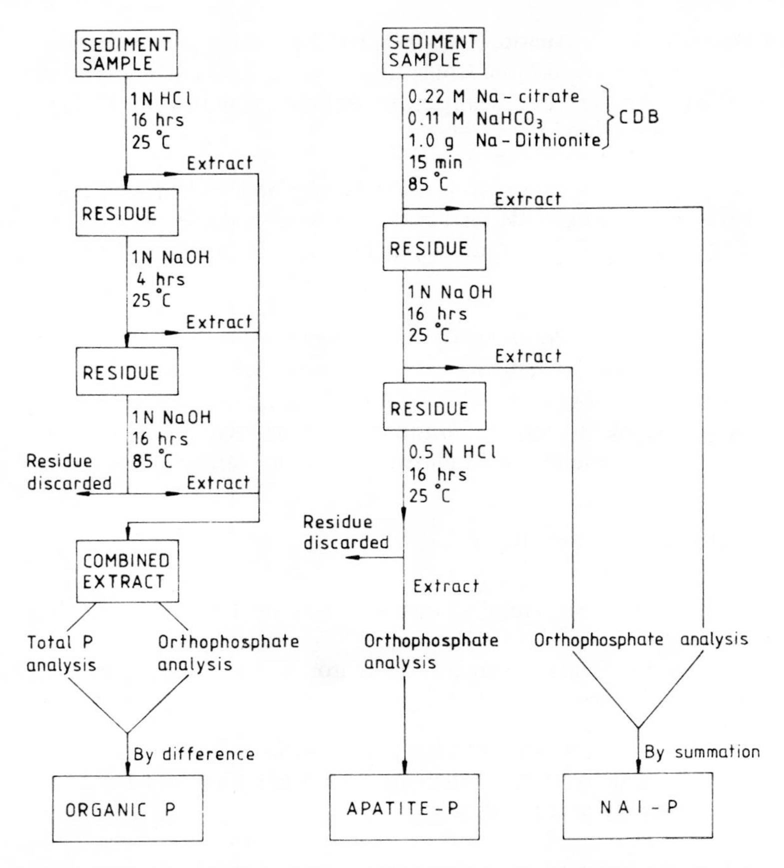

Fig. 4.20. Outline of analytical procedures to determine apatite-P, NAI-P and organic-P (Boström et al. 1982, as modified from Williams et al. 1976)

It should be pointed out that the total phosphorus content and the sediment phosphorus composition may be highly variable in different parts of a lake. The characteristics of the sediments depend on bottom dynamics, morphometry, hydraulic loading, water movements, trophic state and composition of allogenic material, etc. The total phosphorus concentrations in sediments generally increase with increasing water depth as a result of a more or less continuous transport of material from shallow zones of erosion and transport to the deeper accumulation areas. Consequently, different types of phosphorus compounds are found in different sediment types. In Lake Erie, for example, allogenic apatite phosphorus dominates the nearshore sediments, while non-apatite inorganic phosphorus and organic phosphorus accumulate in offshore areas of accumulation (Williams et al. 1976).

Diagenic processes seem to act towards a conversion of other phosphorus forms to apatite phosphorus in sediments. This means that apatite becomes increasingly dominant in older (deeper) sediments (Fig. 4.21).

Table 4.13. Concentrations of different fractions of phosphorus in some lake sediments. Values given in ppm ds and % of total P concentrations (Boström et al. 1982)

	Tot-P	Org-P		Pi		NH_4Cl-P[a]		NaOH-P		CDP-P		HCl-P		Res.-P		Sed. layer	References
	ppm	ppm	%	ppm	%	ppm	%	ppm	%	ppm	%	ppm	%	ppm	%	(cm)	
Stone Lake	3420	1600	47	1820	53					1460	43					0–1	Theis and McCabe (1979)
L. Charles East	2280	910	40	1370	60					1050	46					0–1	Theis and McCabe (1979)
L. Warner	2440	250	10	2190	90					1520[b]	62	190	8			0–3	Ku et al. (1978)
L. Wyola	1290	190	15	1100	85					410[b]	32	30	2			0–3	Ku et al. (1978)
L. Wingra	670	290	43	380	57	44	7	47	7	160	24	92	14	81	12	"surface"	Williams et al. (1971a)
L. Monona	1260	320	25	940	75	13	1	245	19	376	30	240	20	79	6	"surface"	Williams et al. (1971a)
L. Mendota	1370	423	31	947	69	2	< 1	380	27	342	25	133	10	92	7	"surface"	Williams et al. (1971a)
L. Delavan	1000	360	36	640	64	76	5	83	8	314	31	77	18	66	7	"surface"	Williams et al. (1971a)
L. Geneva	713	279	39	434	61	5	< 1	74	10	159	22	125	17	76	11	"surface"	Williams et al. (1971a)
L. Erie	880	110	13	770	87			13	1	317	28	445	51			0–3	Williams et al. (1976)
L. Esrom	4700	1100	23	3600	77					1050[c]	22	2550	54			0–1	Kamp-Nielsen (1974)
L. Esrom	1800	500	28	1300	72					200[c]	11	1100	61			4–5	Kamp-Nielsen (1974)
L. Norrviken	2000	750	38	1250	62			100	5	600	30	400	20	150	8	0–5	Ulén (1977)
Pond C	662	379	57	283	43			70	10	72	10	4	<1			0–1	Prentki et al. (1980)
Pond 6	904	374	41	530	59			143	16	210	23	48	5			0–16	Prentki et al. (1980)
Mean value:	1693	522	31	1170	69												

[a] Performed separately from sequential fractionation

[b] Oxalate extractions

[c] Fe + Al – P

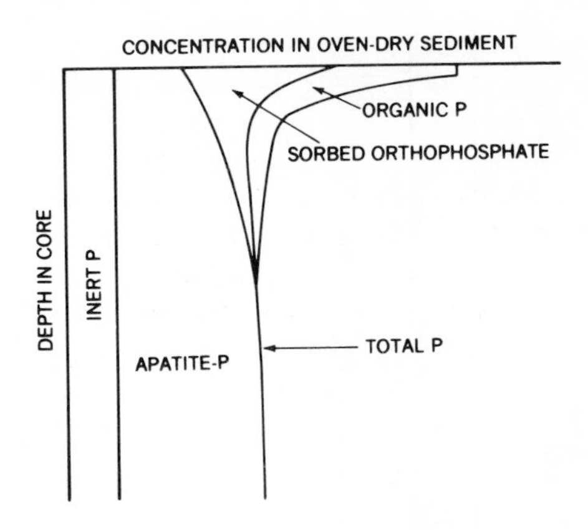

Fig. 4.21. Distribution of phosphorus in a sediment core from Lake Erie (B.F. Jones and Bowser 1978, as modified from Sweeney and Kaplan 1973)

4.2.3.5 Sulfides

As seen in Table 4.11, sulfides in lake sediments are primarily authigenic iron sulfides. The necessary prerequisites for the formation of sulfides is the reduction of SO_4^{2-} to S^{2-}. This is a microbiologically maintained reduction where the oxygen of sulfate is used for the oxidation of organic matter, or molecular hydrogen, and where one of the reaction products is H_2S. Presence of Fe^{2+} in the sediments leads to the precipitation of the highly insoluble FeS. The initial formation of sulfides needs completely reduced sediments.

In a series of sulfurization reactions, iron sulfide is transformed to pyrite (FeS_2) (Fig. 4.22). The sulfurization reactions, which have been described by Richards (1975), cannot proceed without elemental sulfur. The generally low amounts of solute sulfur in most sediment-water systems limit the transformation of FeS to FeS_2. It should be noted that sulfurization of FeS does not include the participation of microorganisms in contrast to most other processes in the sulfur cycle.

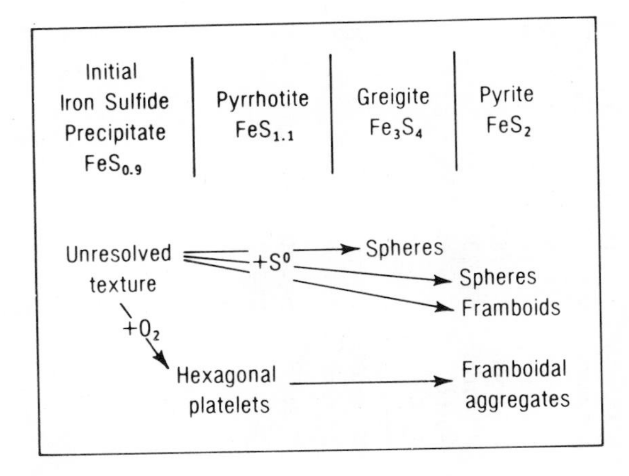

Fig. 4.22. Forms of iron sulfides in sediments. Pyrrholite has been reported only from marine environments (B.F. Jones and Bowser 1978, as modified from Sweeney and Kaplan 1973)

Since highly reduced conditions are necessary, sulfide minerals are formed mostly in the sediments of productive lakes where the oxidation of organic material is of such magnitude that the redox potential decreases below 100 mV, when completely anoxic conditions prevail. Since FeS is very hard to dissolve, iron bound as FeS is much more stable to changes of redox and pH conditions than, e.g, iron hydroxides. In fact, the formation of FeS is often an effective permanent sink for iron by which considerable amounts of the metal are withdrawn from the cycling between sediments and lake water and from reactions with other elements. FeS precipitation often leads to a lower capacity of the sediments to retain phosphorus, since the phosphorus fixation to sediments depends on the existence of reactive and mobile iron.

4.2.3.6 Heavy Metals

Heavy metals or trace metals are defined as metals with a density > 5 g cm^{-3}. These metals form oxides and sulfides which are very hard to dissolve, and they tend to be bound in stable complexes with organic and inorganic particles. Heavy metals are generally found in small amounts in sediments ($< 0.1\%$). They appear in the sediments with allogenic minerals or as endogenic or authigenic precipitates or complexes. The great interest in heavy metals, however, derives from the fact that these elements are supplied to aquatic systems in great excess by man. Furthermore, some heavy metals are hazardous to the aquatic life.

The sources of heavy metals can be divided into five categories (Förstner and Wittman 1979):

1. Geologic weathering.
2. Industrial processes of ores and metals.
3. The use of metals and metal components.
4. Leaching of metals from garbage and solid waste damps.
5. Animal and human excretions.

The main pathway to lake ecosystems are surface runoff, groundwater and waste outlets, and atmospheric deposition. Clays and shales have the highest metal content, while carbonate rocks are diluted with regard to heavy metals. The mean background values for lake sediments (see Chap. 10) are similar to those of shales, but it must be emphasized that the natural background values for heavy metals in sediments can be highly variable, depending on the bedrock structure of the catchment.

The increased supply of heavy metals to lakes and lake sediments is essentially connected with the industrial development (see Krenkel 1975). The onset of the "industrial revolution" during the 19th century is often reflected in increased concentrations of heavy metals in sediments (Fig. 4.23). Heavy metals emitted due to human activities include: Cr, Co, Cu, Zn, Cd, Hg and Pb. It is evident from Fig. 4.23 that the content of heavy metals in lake sediments offers an excellent key to the pollution history of the lake. In the same way, the horizontal distribution of metals in lake sediments can be used to evaluate the transport pattern and sedimentation rate of metals from point sources like inflows or sewage effluents (Fig. 4.24). For a more thorough account of metal distribution in lake sediments, see Chapter 10.

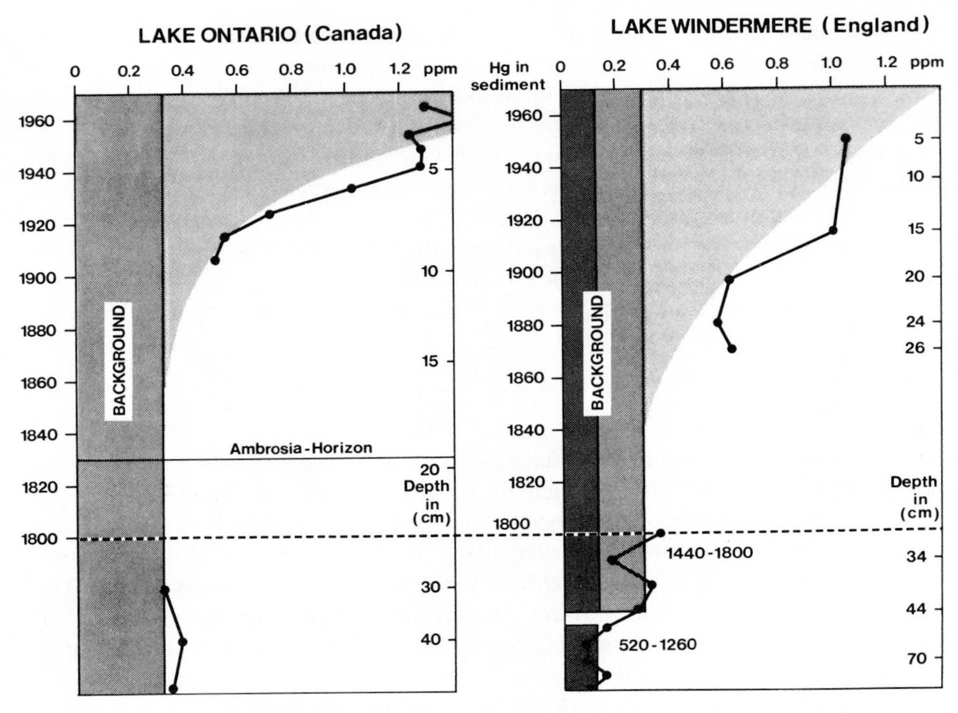

Fig. 4.23. Mercury in sediment cores from Lake Ontario and Lake Windermere (Förstner and Wittmann 1979, based on data from Thomas 1972 and Aston et al. 1973)

One of the main reasons why many heavy metals deserve so much attention is their toxicity. The main mechanisms behind the toxicity of heavy metals have been summarized by, e.g., Bowen (1966) and Förstner and Wittmann (1979):

A. Disturbance of enzymes or enzymatic systems because of the high electro-negative affinity for reactive groups on the enzymes, like amino or sulfhydryl groups.
B. Metals can form stable complexes with essential metabolites.
C. Metals can catalyse the breakdown of metabolites.
D. Permeability of cell membranes are disturbed because metals can be bound to the cell surface.
E. Metals can substitute other elements with important functions in the cell metabolism, whereby metabolic processes cease functioning.

The traditional way of determining the toxicity of metals is to establish the LC_{50} or LD_{50} value, where LC stands for lethal concentration and LD for lethal dose. The value is obtained for the concentration that exterminates 50% of a test sample relative to a control group of test organisms. More than 120 monographs on various toxicological test systems have been published (see, e.g., Cairns 1981). A fundamental point of critique against most of these test systems is that they very rarely account for the environmental factors that play an important role for the fate of metals on

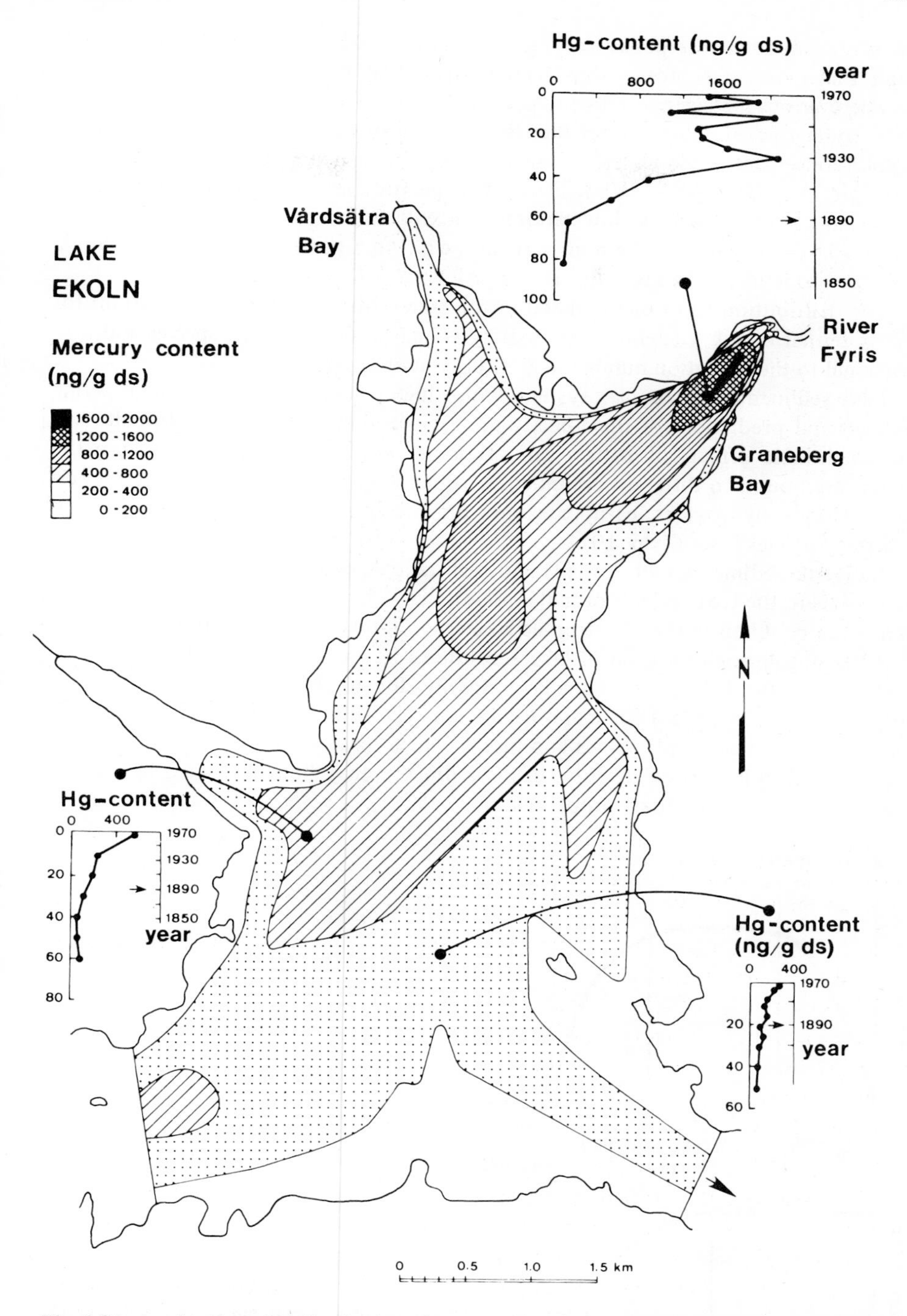

Fig. 4.24. Areal and vertical distribution of mercury in the sediments of Lake Ekoln (Lundin and Håkanson 1982, based on data from Axelsson and Håkanson 1972)

the way from dose to response (see, e.g., Holdgate 1979). In this context, we will not enter that maze of ecotoxicology, merely stress that there exist major differences between various metals in terms of toxicity in limnic environments.

A crude role of thumb states that the least hazardous elements appear with the highest concentrations in water, sediments and biota, and vice versa (the "abundance principle", see Håkanson 1980a). Elements appearing on the ppb (parts per billion) scale, i.e., with extremely low natural concentrations, are Hg, Ag and Cd (see Table 10.6). Elements on the ppm scale are, e.g., As, Co, Cr, Cu, Mo, Ni, Pb, Sn, V and Zn. Elements on the mg scale are, e.g., Al, Ca, Fe, K, Mg, Mn, and Na.

The distribution form of the metal in the lake environment is very important for the potential effect. Generally the toxicity is highest for the ionic species and proportional to the oxidation number, e.g., CrO_4^- is more toxic than Cr^{3+}.

Lake sediments are complex systems with a highly variable composition, implying that general predictions of the toxicity in any given context are very difficult. The potential toxic effects of metals can often be significantly reduced because the metals are bound to different compounds, which may camouflage the toxic properties. Sulfides, hydroxides, carbonates and organic complexes are examples of such "carrier particles" (see Chap. 10).

Inorganic sediments with low pH and high redox potential are therefore environments where the toxic effects of metal pollution can be most pronounced. Such sediments are common in the oligotrophic lake type. In this connection, it may be noted that the ongoing acidification of lake ecosystems in various parts of the world, e.g., Scandinavia, the U.S.A. and Canada, primarily due to the burning of fossil fuels, creates an indirect hazard for metal toxication. The solubility of most heavy metals increases at decreasing pH values and metals which have previously been bound and rather harmless in particulate form in the sediments may be recirculated to the lake water and express their toxic properties.

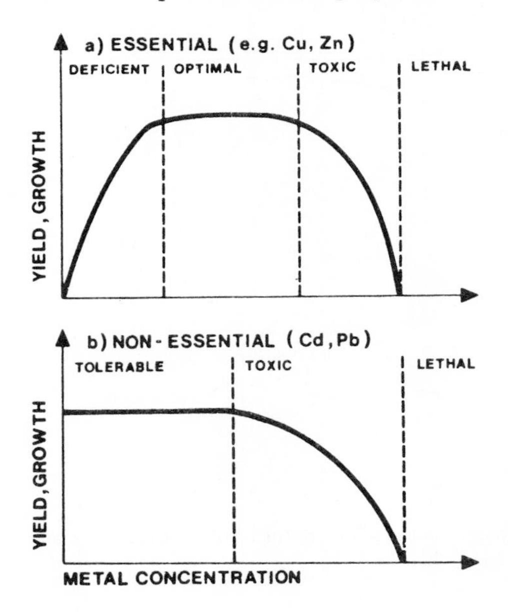

Fig. 4.25a, b. Schematic illustration of the growth of organisms as a function of the supply of essential and non-essential metals (Förstner and Wittmann 1979, as modified from Baccini and Roberts 1976)

The roles of some heavy metals are complicated by the fact that they are essential in small amounts for the growth of organisms. This means that a small increase of the metal concentration may stimulate growth while an increase beyond a certain limit may have inhibitory effects. A schematic illustration of the role of essential and non-essential metals is given in Fig. 4.25.

To conclude, heavy metals in sediments are often of anthropogenic origin, their turnover, residence time, and possible ecological effects are highly variable and depend on the distribution form and a number of environmental parameters.

5 Biological Parameters

Lake sediments often constitute a rich substrate for different types of organisms. All trophic levels are represented: photosynthetic plants, herbivorous animals, carnivorous animals and decomposers.

In some aspects the sediments are a much more suitable environment for biological life than the lake water. The supply of inorganic nutrients and organic matter is, for example, much higher in the sediments. On the other hand, the light intensity is generally low and restricted to the sediment surface of shallow areas. This implies that the main biological activity is maintained by heterotrophic organisms (animal and bacteria) which utilize the energy fixed in reduced carbon compounds supplied from settling organic particles. However, when light transmission allows photosynthesis in large parts of the total sediment area, the primary productivity on lake bottoms can be relatively high, sometimes greater than in the water column.

In this chapter, we will focus on descriptions of four main groups of benthic life: algae, macrophytes, benthic invertebrates and bacteria. It has been our intention to try to explain and exemplify the role of different organisms in the sediments in relation to the whole lake ecosystem. The section on benthic invertebrates is mainly descriptive, which reflects the fact that most of the research in this field has been taxonomical, descriptive and centred more on the animals themselves than on their role in the lake ecosystem. Most space is dedicated to bacterial processes, reflecting our opinion that this form of biological activity is of extreme importance for abiotic as well as biological features of lake ecosystems.

5.1 Sediment-Living Algae

Benthic algae utilize the lake bottom as their substrate. They are subdivided into *epipelic* algae, meaning forms which grow on or in the sediment surface, and *epilithic* algae, which grow on rocks and stones.

Several other concepts exist. Periphyton is, for example, a frequently used term which can comprise epipelic, epilithic or epiphytic (i.e., attached to other plants) algae as well as entire benthic communities. In its original sense, however, periphyton meant algae attached to artificial submerged substrates like boats, poles, etc. (Hutchinson 1975).

In this section, we will discuss only the algae living in the sediments, which are referred to as epipelic.

Studies on the epipelic algae have been concentrated on a limited number of dominating groups, particularly diatoms, and have mostly focused on taxonomical and autecological aspects. Investigations of entire algal communities, including taxonomy, productivity as well as ecosystem aspects, are very rare. Much of this earlier work is summarized by Hutchinson (1975). Round (1957, 1960, 1961) conducted a series of studies on epipelic algae in lakes of the English Lake District, where diatoms and blue-green algae dominate the total biomass. Green algae and flagellates were also found in substantial but much smaller amounts. No clear correlation between different nutrient parameters in the sediments and the algal biomass was found. It seemed, however, that calcareous sediments were more favourable to diatoms, while blue-green algae were more abundant on peat-like sediments with high organic content.

The fairly weak correlation between algal growth and nutrient state is probably explained by the fact that the growth of epipelic algae generally is limited more by low light conditions than by low nutrient supply. Evidence for this statement exists from subarctic lakes in the Kuokkel area, northern Sweden (Persson et al. 1977, Lundgren et al., unpublished material from the Kuokkel project). Blue-greens and diatoms dominated also in these lakes. The epipelic algae did not significantly change their productivity or biomass after whole-lake fertilization with nitrogen and phosphorus, but responded in other ways to the restricted light climate due to increased planktonic turbidity. The algae also increased their photosynthetic activity when exposed to higher light intensities than at their original sites of growth. This latter phenomenon was also demonstrated in Marion Lake, British Columbia (Table 5.1).

Table 5.1. Enhancement of photosynthesis of epipelic algal communities in Marion Lake, Canada, by moving intact sediments from various depths to 0.5 m (Wetzel 1975, with data from Gruendling 1971)

Sample depth (m)	Solar radiation (g cal cm^{-2} h^{-1})	Temperature (°C)	Incubated in situ at depth (ml O_2 m^{-2} h^{-1})	Incubated at 0.5 m depth (ml O_2 m^{-2} h^{-1})
0.5	32.6	23.0	19.9	19.9
1.0	23.3	22.0	22.7	26.4
2.0	14.4	17.0	19.4	28.8
3.0	10.6	14.0	14.5	19.4
4.0	5.5	13.0	2.4	5.7

To meet their demands for light, the epipelic algae are most probably capable of movements within the sediments. If not, the algae would be covered by settling matter or resuspended particles cutting off their energy supply. Migration ability is indirectly indicated by the fact that living algae are constantly found several centimeters down in the sediments, while light is completely extinguished only a few millimeters below the sediment surface. Regular diurnal migration patterns have also been shown for epipelic algae (Round and Happey 1965, Round and Eaton 1966). These studies showed that the cell numbers at the sediment surface were high in the midmorning and low in the evening.

The epipelic algae can probably utilize the nutrients stored in the sediments, making them relatively independent of the nutrient state of the lake water. Jansson (1978) showed, from turnover experiments with N-15 and mass balance calculations, that blue-green epipelic algae from the Kuokkel lakes took up ammonium nitrogen from the sediment pore water, and that the algae could easily meet their nitrogen demand from this source. It was also demonstrated that the benthic algae served as a transforming link in the transport of nitrogen from sediments to lake water. Thus, the blue-green algae assimilated ammonium from the pore water and excreted large parts of it to the overlying water as dissolved organic nitrogen, thereby considerably affecting the water chemistry of the lake. Consequently, epipelic algae should not be regarded as an isolated benthic community, but as a component of the lake ecosystem which in several ways may affect the lake metabolism. This is further emphasized by the results from the Kuokkel lakes, where the epipelic and epilithic algae were responsible for 50%–80% of the total primary productivity. Such high number are obtained only in oligotrophic clearwater lakes where the nutrient supply restricts the planctonic algal growth but not the algae in the sediments, and light penetration allows photosynthetic activity down to considerable depths.

In this connection, it is interesting to mention a few words about the development of epipelic algae in acidified lakes, a type of lake which is becoming increasingly common in parts of Scandinavia and North America due to acid atmospheric deposition (see, e.g., Overrein et al. 1980, Haines 1981). Correlated with the increasing light transmission, the sediments are invaded by *Sphagnum* and blue-green algal species. In typical cases, mosses and algae form a thick belt covering the bottoms down to 15–20 m depth; the primary productivity has been directed from the water column to the sediments.

Algal communities in the sediments generally have longer generations and grow in a more "stable" environment than planktonic species. Therefore the sudden, often dramatic changes in biomass and productivity, typical for planktonic algae, seldom occur within benthic communities. Biomass is generally low in winter. A slow and successive increase follows the increasing light intensities in spring and early summer. Sometimes only one biomass maximum occurs in the middle of the summer (Fig. 5.1), but the development of two maxima, one in spring and another in late summer, is

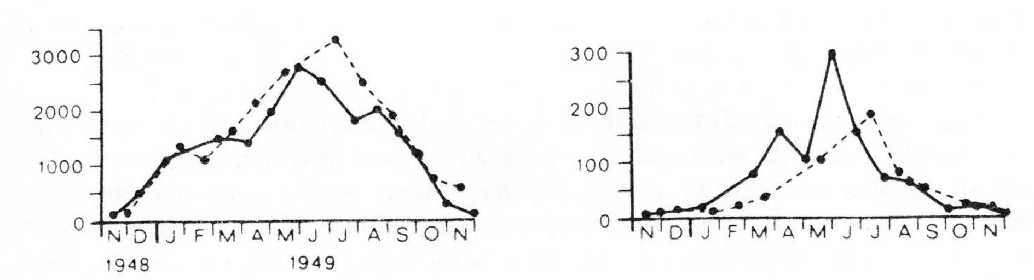

Fig. 5.1. Seasonal cell counts of epipelic diatoms *(left)* and blue-green algae *(right)*. All depth stations added for each sampling site in the photosynthetic zone (1–6 m). *Full line* data from Lake Windermere; *dashed line* value from Lake Blelham Tarn, England (Wetzel 1975, as modified from Round 1961)

not uncommon. It should be emphasized that this applies only to the total biomass and that the growth of individual species may exhibit highly variable patterns. It is also important to note that the algal flora found on the sediment surface does not always consist of truly epipelic forms, but may include living species or resting stages of planktonic algae. The mixture of benthic and planktonic forms may be particularly pronounced in shallow lakes where plankton algae are more or less continuously in contact with the sediment surface.

Some planktonic algae form resting stages in the sediments which can serve as inocula for growth in the lake water. One such algae is the large diatom *Melosira italica,* which settles out of the water column during winter and summer stagnation, and is resuspended at spring and autumn circulation (Lund 1954, 1955).

5.2 Macrophytes

The term aquatic macrophyte is used for macroscopic forms of aquatic plants, mostly referring to true angiosperms, but also including water mosses and sometimes even *Chara* and *Chladophora.* Macrophytic vegetation is in some respects essential for lake sedimentology. A detailed treatise of macrophytic communities lies, however, beyond the scope of this book. The interested reader is referred to the outstanding work of Hutchinson (1975), or to the shorter, but in many aspects complete, text by Wetzel (1975, pp. 355–388).

Macrophytes are either free-floating or form permanent stands with parts (roots and rhizomes) attached to the sediments. They are further separated into *emergent* plants, with green parts above the water surface, *floating-leaved* macrophytes, with leaves floating in or on the water surface, and *submerged* plants, which have all parts below the water surface.

The dense stands of macrophytes which are sometimes developed in shallow water create a special environment sheltered from wave turbulence and with a rich variety of environmental microzones. The biological activity in these sheltered organogenic sediments is high; often with dense populations of different benthic animals and with intense bacterial metabolism.

A more direct linkage between plants and sediments is via the root systems of many plants. Much of the nutrition occurs by uptake of essential nutrients in the roots and subsequent transport to the green parts above the mud. Where macrophytes occupy large parts of the total sediment area, as is often the case in shallow eutrophic lakes, the nutrient uptake by rooted plants causes considerable transport of these elements from the sediments to the plant biomass in the lake water. After death and senescence of the plants, large amounts of nutrients may be released to the lake water. Calculations of this type of translocation for phosphorus have shown that the release from dying plants can supply phosphorus to the lake water in amounts equal to 50% or more of the total external phosphorus loading (Barko and Smart 1979, Welch et al. 1979). Oxygen for respiration is transported in the opposite direction, from the water to the roots. This means that increased redox potentials can be maintained around macrophytic root systems. This, in turn, has pronounced effects on redox-dependent chemical and bacterial processes.

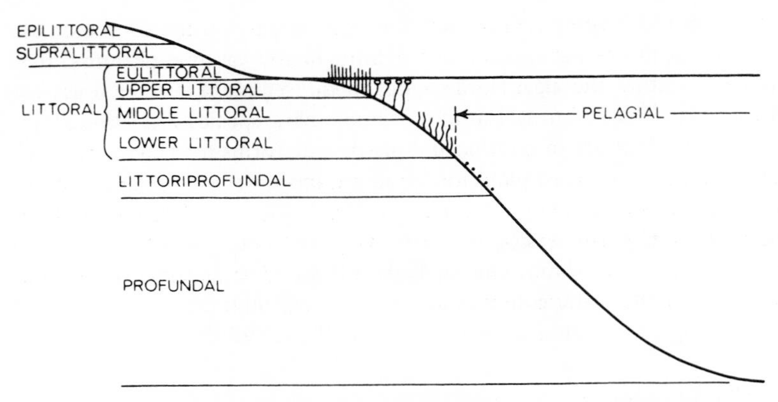

Fig. 5.2. Zonation of lake bottoms from the distribution of bottom vegetation (Wetzel 1975, after Hutchinson 1967)

Finally, it should be mentioned that macrophytes are used for lake classification purposes. Figure 5.2 shows the well-known and often-used system for zonation of lake bottoms based on the vegetation type. The deep sediments without vegetation are termed *profundal.* These are, in principle, the sediments below the thermocline. Above the profundal zone is a transition zone, scarcely inhabited by different forms of green plants called the *littoriprofundal.* This zone often concurs with the thermocline. The shallow bottoms (more or less densely populated with plants) constitute the *littoral* zone, which in turn is divided into several subzones. Thus, the lower littoral is generally colonized by submersed plants, the middle littoral by floating-leaved rooted vegetation and the upper littoral by emergent macrophytes. The so-called *eulittoral* is the area between the highest and lowest water level.

5.3 Benthic Invertebrates

The surficial layers of lake sediments are inhabited by a rich variety of invertebrates. Some animals spend their entire life in the mud. In many cases, however, the benthic fauna exists as such only for parts of their life cycle. Many species live partly in the mud and partly in the lake water; most insects spend larval and pupal stages in lakes while they emerge as adults outside the aquatic environment.

The benefits derived from the sediments are food and protection. The benthic fauna feeds on organic deposits, benthic plants, bacteria and animal species. In search of food and protection, they often live below the sediment surface, stationary or moving around. The movement of animals in the sediments causes turbulent flows and redistribution of particles. This physical phenomenon is called *bioturbation* and is often a highly important factor for the transport and turnover of particles and solutes within the sediments. The physical features of bioturbation are treated in detail in Chapter 8.2. Chemical aspects of bioturbulent mixing have been discussed by Berner

(1980). In many important aspects, however, the role of the benthic fauna in the sediment metabolism, and the interaction between bottom-living animals and other components of the lake ecosystem, is little known. A majority of the efforts spent on studies of benthic animals have been focused on taxonomy, and descriptive analyses of species distribution. Benthic animals have also been utilized for classification of lake types and as indicators of water quality. An excellent review of major topics in benthic animal research has been presented by Brinkhurst (1974). Here we will first give a brief presentation, partly extracted from Wetzel (1975, pp. 488–526), on major groups of benthic animals. Then the distribution of animals in lakes, and important contributions to lake type classifications, will be discussed.

5.3.1 Important Forms of Benthic Animals

Protozoans. Protozoans exist as both planktonic and benthic forms. Little is known about their ecology, food requirements or their role in lake sediments. They seem to have a wide tolerance of different environmental conditions and, although strictly aerobic organisms, are able to tolerate low levels of oxygen. They feed on algae, bacteria, detritus and other protozoans. Protozoans are particularly abundant in sediments rich in organic matter, suggesting that they might contribute to the overall biodegradation of organic matter in lakes. Because of their generally low biomass, the quantitative impact on biodegradation or bioturbation is considered to be low.

Nematodes. Nematodes or roundworms are mostly parasitic animals, but also represented by a large number of free-living benthic species. They are found in most lake types. Consequently, they have widely different food preferences and feed on detritus, living plant material, as well as other animals.

Oligochaetes. This is one of two major groups of worms living in the freshwater habitat (the other is the leeches), and also one of the major components of the benthic fauna.

Most freshwater oligochaetes are between 1 and 5 mm in length. They are segmented with bundles of setae on each segment (except for the first and terminal one). Various types of oligochaetes are found in most lake sediments.

Typical behaviour for many species, especially those of the tubificids, is that they burrow their head into the sediments while the tail remains in the water. The tail segments contain gills or other means of respiration and in this manner the animals can benefit from the nutrient-rich environment in the sediments, while simultaneously respiring in the oxygenated water above the sediment surface. Under favourable conditions, the tails of oligochaetes form dense stands on the lake bottom. Most oligochaete species burrow down to a depth of 2–4 cm in the sediments, but occasionally they can penetrate as deep as 15–25 cm. This behaviour makes oligochaetes, especially the tubificids, important "bioturbators" (see Chap. 8.2). Many oligochaetes can tolerate low oxygen levels; their respiration can proceed at oxygen saturation levels as low as 15%. Some species can even withstand short anoxic periods. These qualities make oligochaetes very suitable for a life in the sediments of nutrient-rich lakes and mass abundance is often connected with organic pollution.

Oligochaetes feed on organic particles containing bacteria as well as on other organisms. Studies on the feeding behaviour of oligochaete species (Brinkhurst et al. 1972) have indicated that oligochaetes are able to feed selectively on microorganisms in the sediments.

Hirundinea. Leeches are generally ectoparasites which consume blood or other body fluids from vertebrates like fish. Between feeding events (up to 200 days) leeches consume nothing, which entails a considerable loss of body weight. Some species of leeches are predators, continuously feeding on small invertebrates. Abundance of leeches is generally well correlated with the trophic state of the lake.

Crustaceans. Several groups of crustaceans spend their whole life cycle or parts of it in the lake sediments.

Ostracods are small primitive bivalved organisms often found in great numbers on most types of lake bottoms. They live on the sediment surface where they move by sweeping movements by their antenna. When undisturbed, they open the clam-like valves and feed by filtration on detritus, bacteria, algae or other small animals. Their role for the turnover of nutrients and energy in the sediments is unknown, but the sometimes high abundance ($>$ 50,000 ind m^{-2}) suggests that they can be important in this respect).

Mysids (opossum shrimp) have a maximum length of 3 cm. They can swim and express typical vertical migration patterns. Mysids spend the day on the sediment surface and during the night they swim to feed on planktonic algae and zooplankton.

The mysids generally found in freshwater are typical coldwater forms, not tolerating temperatures higher than 15°C. They also have high oxygen demands. Their distribution is restricted to oligotrophic lakes; particularly lakes large enough to have a hypolimnion where the temperature is favourably cold throughout the year.

Mysids are excellent food for fish, and have been introduced in reservoirs with positive effects on fish growth (Olsén 1980). Part of this effect is due to the fact that the ecology of mysids makes them less sensitive to artificial water level changes than other fish prey.

Isopods are up to 2 cm long and have a dorsoventrally flattened body, making them suitable for a life in exposed littoral areas. The most common freshwater form is *Asellus aquaticus,* which often occurs in large numbers on shallow bottoms, preferably in algal mats or other types of vegetation where they feed on plant material.

Amphipods are mainly marine crustaceans represented in freshwater by a few species of which *Hyalella azteca* and *Pontoporeia affinis* are the most important. These freshwater crustaceans are between 0.5 and 2 cm long with a compressed segmented body. *Hyalella* spend its entire life cycle in the sediments, feeding on organic deposits and benthic microflora.

In a detailed study of *Hyalella azteca* in Marion Lake, Canada, Hargrave (1970a–c) found a mutual benefit relation between *Hyalella* and benthic algae. The animal fed on the algae and these feeding activities had a stimulating effect on algal growth. The reasons for this stimulation, however, was not clear.

Pontoporeia affinis differs from other amphipods by its preference for upward migration in the water column during the night. In this and other respects, this animal resembles the mysids. *Pontoporeia* is cold stenothermal, and lives mainly in hypolimnion or metalimnion bottoms in eutrophic lakes. During the day *Pontoporeia* stays on or in the surface sediments feeding on bacteria and organic desposits. It feeds selectively on small particles ($< 0.05\ \mu m$) with high bacterial and organic content (Marzolf 1965). Contrary to mysids, *Pontoporeia* does not consume food during migration, and this therefore seems to have other functions than nutrition.

Mollusks. The mollusks of inland waters are divided in two groups: Snails, which are univalved, and the bivalved clams and mussels.

The shell of the snails is generally spirally formed, although a few species with conically shaped shells exist. Snails can be found on different types of bottoms, but are particularly abundant in the littoral zone. They occur in all types of lakes, except for acid waters where their calcareous shells are destroyed. Snails are sometimes seen at the water surface, which is due to their special means of respiration. Oxygen is taken up by gills; or by "lungs" in the so-called pulmonate snails; or by cutaneous respiration through body membranes. The "lung" of the pulmonate species is filled with air at the water surface when the oxygen content at the sediment surface is low. Other species fill their lungs with water and then a gas exchange takes place similar to that with gills. Snails feed on detritus, plant material and other benthic animals.

Clams and mussels have their body enclosed with two identical shells and, in contrast to snails, their body lacks head, tentacles, eyes and jaws. They respire by means of gills. Water passes through the body by a siphonic mechanism, which also supplies the animal with food particles. The primary food for clams and mussels is detrital particles and small zooplankton living at the sediment surface. Because of their respiratory mechanism, clams are less tolerant of low oxygen levels than snails.

Insects. Insects are represented in freshwater by a wide variety of classes and species with different characteristics and environmental needs. Insects are probably the best-studied group of the benthic invertebrates. Some insects spend their entire life in the aquatic environment, but most species have a life cycle in alternating environments. Metamorphosis is typical of insects, whereby the animals grow through several stages (larvae, nymphs, pupae and adults) with varying morphometric features. Some other insects (e.g., *Hemiptera, Odonata, Plecoptera* and *Ephemeroptera*) undergo gradual metamorphosis. The young stages (nymphs) gradually increase in size and wings develop in pads on their backs. The final nymph stage precedes the adult phase.

Complete metamorphosis is characteristic for the *Diptera, Trichoptera, Megaloptera* and *Coleoptera.* These species grow in larval and pupal stages before they hatch as adults. Wings develop internally in the larvae and emerge on the outside in the pupal stage.

A brief characterization of important insect groups is given below, followed by a summary of food types and the major mechanisms of feeding.

Odonata are large insects spending their aquatic stages mostly in the littoral zone. Eggs develop on submersed parts of aquatic plants or similar types of substrates. The nymphs live primarily in vegetation or burrowing in the sediment. They undergo 10–20 mouldings before hatching above the water surface. *Odonata* in the submersed stages have high oxygen demands.

Ephemeroptera or mayflies spent most of their life in water. After their emergence as adults they live to mate only for a couple of days. Eggs are laid in the water. After hatching they grow through as much as 20–40 mouldings before reaching their maximum nymph size, which is usually less than 2 cm. Their life cycle is 1 year (occasionally 2 years). Mayflies are most frequent in the littoral zone and occur in many different types of lakes. Their demand for oxygen is high, but they tolerate organic pollution to a certain degree.

Plecoptera (stoneflies) are most abundant in running waters, but are often found in oligotrophic lakes in exposed shallow areas. They live in water as nymphs for 1–3 years and resemble mayfly nymphs, but can easily be distinguished by having two saetae instead of the three which are characteristic for *Ephemeroptera* species.

Diptera include flies, midges and mosquitoes. They undergo complete metamorphosis and many groups constitute important parts of the benthic fauna in a variety of lakes. Chironomids, for example, are widespread and found in lake sediments all over the world. The larval stages last from a few weeks to more than 2 years in some species. Adults are terrestrial. Respiration is generally cutaneous, but a few species depend on atmospheric respiration and have different means for obtaining air at the water surface. Some chironomids have a haemoglobin-like component in their blood which make them tolerant to low oxygen conditions.

Other important and well-studied members of the *Diptera* group are the *Chaoborus* species. These carnivorous plankton midges often dominate the profundal benthic fauna in eutrophic lakes together with chironomids and oligochaetes. *Chaoborus* have terrestrial (adult) and aquatic life stages. The first instars after hatching are limnetic. The third instar lives in the sediments and sometimes overwinters in the mud. The fourth instar expresses a very typical migration pattern, spending the day in the sediments and the time between sunset and sunrise migrating up to and down from the water surface. This migration appears to be regulated by temperature, light and oxygen conditions at the sediment surface. *Chaoborus* feeds on limnetic zooplankton and small benthic animals.

Trichoptera (caddis flies) have a 1-year life cycle where the adults hatch during the entire summer in temperate regions. Many *Trichoptera* larvae are easily recognized by the characteristic tube-shaped shelters they build from plant debris, sand or small stones. Caddis flies are found in oligotrophic as well as eutrophic lakes.

Coleoptera (beetles) mostly have a 1-year life cycle including three larval stages. Eggs are often laid in the sediments and hatch within 3 weeks. The larvae and pupae develop rapidly and the beetles overwinter in the lake as adults. Beetles need atmospheric oxygen for their respiration and usually swim around with air bubbles on the ventral side of the body.

5.3.2 Feeding Mechanisms and Food Types Among Insects

As might be expected from the large number of groups and species that exist in aquatic environments, insects exhibit highly variable food preferences and means to obtain their food objects. Insects occupy the whole range from detrivores over herbivores and carnivores to less specialized omnivores. The scheme in Table 5.2 is instructive and gives fairly clear view of the feeding habits of insects.

It is obvious that feeding behaviour is not to any large degree related to taxonomic organization. Almost every insect order has representatives among the different classes of food ingestion habits.

The insect group with the greatest diversity concerning food intake is the large and highly complex group of chironomids.

5.3.3 Distribution of Benthic Fauna Within Lakes

The highly different environmental characteristics in sediments of nearshore areas, compared with the area of accumulation in the deeper parts of the lakes, are of course reflected in the diversity, abundance and biomass distribution of benthic animals.

Generally, the number of species decreases with increasing depth in all types of lakes (Fig. 5.3) for several reasons: The shallow littoral zone colonized by rooted

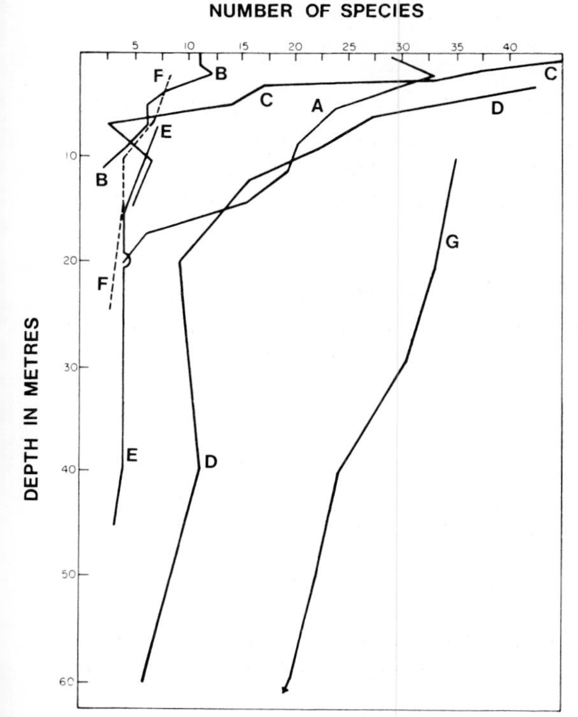

Fig. 5.3. Number of species in relation to water depth in various lakes. *A* Esrom Lake, Denmark; *B* Gribsø Lake, Denmark; *C* Lake Borrevann, Norway; *D* Lake Windermere, England; *E* Opeongo Lake, Canada; *G* Lake Neuchatel, Switzerland (Brinkhurst 1974)

Table 5.2. Trophic mechanisms and food types of aquatic insects (Wetzel 1975, after Cummings 1973)

General category based on feeding mechanism	General particle size range of food (μm)	Subdivision based on feeding mechanisms	Subdivision based on dominant food	Aquatic insect taxa containing predominant examples
Shredders	$> 10^3$	Chewers and miners	Herbivores: living vascular plant tissue	Trichoptera (Phryganeidae, Leptoceridae) Lepidoptera Coleoptera (Chrysomelidae) Diptera (Chironomidae, Ephydridae)
		Chewers and miners	Detritivores (large particle detritivores): decomposing vascular plant tissue	Plecoptera (Filipalpia) Trichoptera (Limnephilidae, Lepidostomatidae) Diptera (Tipulidae, Chironomidae)
Collectors	$< 10^3$	Filter or suspension feeders	Herbivore-detritivores: living algal cells, decomposing particulate organic matter	Ephemeroptera (Siphlonuridae) Trichoptera (Philopotamidae, Psychomyiidae, Hydropsychidae, Brachycentridae) Lepidoptera Diptera (Simuliidae, Chrionomidae, Culicidae)
		Sediment or deposit (surface) feeders	Detritivores (fine particle detritivores): decomposing organic particulate matter	Ephemeroptera (Caenidae, Ephemeridae, Leptophlebiidae, Baetidae, Ephemerellidae, Heptageniidae) Hemiptera (Gerridae) Coleoptera (Hydrophilidae) Diptera (Chironomidae, Ceratopogonidae)
Scrapers	$< 10^3$	Mineral scrapers	Herbivores: algae and associated microflora attached to living and non-living substrates	Ephemeroptera (Heptageniidae, Baetidae, Ephemerellidae) Triochoptera (Glossosomatidae, Helicopsychidae, Molannidae, Odontoceridae, Goreridae) Lepidoptera Coleoptera (Elmidae, Psephenidae) Diptera (Chironomidae, Tabanidae)

		Organic scrapers	Herbivores: algae and associated attached microflora	Ephemeroptera (Caenidae, Leptophlebiidae, Heptageniidae, Baetidae) Hemiptera (Corixidae) Trichoptera (Leptoceridae) Diptera (Chironomidae)
Predators	$> 10^3$	Swallowers	Carnivores: whole animals (or parts)	Odonata Plecoptera (Setipalpia) Megaloptera Trichoptera (Rhyacophilidae, Polycentropidae, Hydropsychidae) Coleoptera (Dytiscidae, Gyrinnidae) Diptera (Chironomidae)
		Piercers	Carnivores: cell and tissue fluids	Hemiptera (Belastomatidae, Nepidae, Notonectidae, Naucoridae) Diptera (Rhagionidae)

plants offers a much more heterogeneous environment than the deeper sediments, permitting the coexistence of a variety of species with different environmental needs. Many benthic animals are dependent on atmospheric oxygen for respiration, a demand which is more easily met from shallow bottoms. In the profundal zone, oxygen concentrations are often low at the sediment surface, which restricts the number of species to those that tolerate low oxygen conditions. Dominants among the fauna in the profundal are chironomids and oligochaetes, and in oligotrophic lakes also cold-water forms like Mysids and Pontoporeia. Thus, species diversity versus depth shows a similar pattern in most types of lakes. This, however, does not apply to the vertical distribution of the biomass/abundance, which is more variable. High numbers and high biomass of total benthic fauna are nearly always found in the littoral zone. Besides offering space for a number of different species, the littoral zone also supplies with plenty of food. In nutrient-poor lakes, the benthic animals biomass generally decreases with increasing depth, in direct relation to the food supply. In oligotrophic lakes, most of the energy bound in photosynthetic processes in the upper water layers is utilized in mineralization processes in the water column and the particles deposited in the deep sediments are not high-quality food.

Nutrient-rich lake sediments, on the other hand, often have a second biomass maximum in the profundal zone. This maximum is frequently dominated by oligochaetes, a feature becoming more pronounced with increased productivity of the lake. An idealized picture of the abundance of benthic animals at various depths in

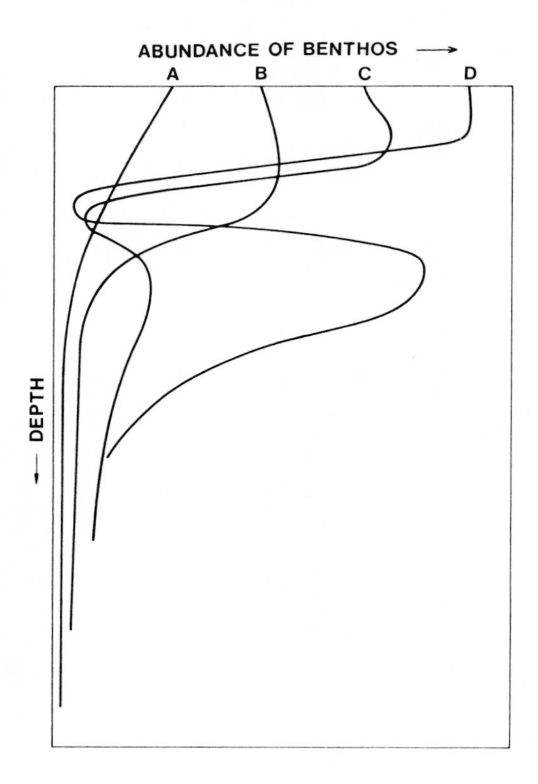

Fig. 5.4. Idealized depth distribution of benthos in relation to lake typology. *A* ultraoligotrophic; *B* oligotrophic; *C* mesotrophic; *D* eutrophic (Brinkhurst 1974, after Lundbeck 1936)

different types of lakes is shown in Fig. 5.4. The results of Lundbeck (1936) offer a typical example of the dominant theme in early benthic fauna research and numerous later investigations have contributed similar material.

The depth distribution curves in Fig. 5.4 are schematic and aberrations occur due to local conditions and season of the year. For example, insect hatching may reduce the biomass considerably in certain bottom zones for shorter periods.

5.3.4 Benthic Lake Typologies

During the 1920's and 1930's typology schemes based on the existence of different types of benthic fauna were established by a number of investigators, mainly in Germany. Distributions of different chironomid species were often used as the basis for such classification. The pioneering work was by Thienemann (Table 5.3) followed by Lenz (1925), Valle (1927), Lundbeck (1936), and Brundin (1949).

Table 5.3. Lake classification of Thienemann (Brinkhurst 1974, Berg 1938)

(1) *Tanytarsus* (oligotrophic) lakes	(2) *Chironomus* (mostly eutrophic) lakes			(3) *Chaoborus* lakes without *Chironomus*
Tanytarsus characteristic, *Chaoborus* absent, deep water oxygen levels always high (> 58% in summer, > 77% in winter). Relict crustacea present in Baltic regions	*Chironomus* larvae with tubuli characteristic, summer oxygen values < 40%, winter oxygen variable			*Chironomus* is absent from all regions, oxygen concentrations in deep water 6–12%, below ice 5–20%
	(a) No *Chaoborus*. Summer oxygen about 30–60%. Relict crustacea present in Baltic regions	(b) *Chaoborus* present. Summer oxygen values < 40%, winter oxygen variable		
		(b) (i) *anthracinus (Bathophilus)* lakes – with *Chironomus anthracinus (liebelibathophilus)* in profundal: 3 further subtypes dependent upon relative abundance of *C. anthracinus* and *C. plumosus* and oxygen values	(b) (ii) *plumosus* lakes – with *C. plumosus* in the profundal: 2 types depends upon presence of *C. anthracinus* and oxygen values	

The aim of the initial lake typology schemes was to relate the fauna to environmental parameters and to achieve a tool for characterization of lakes by indicator organisms. However, the choice of certain indicator organisms, such as chironomids was unfortunate; partly because they constitute only a limited portion of the total benthic fauna, and partly (and more importantly) because the taxonomy of chironomids is extremely difficult.

Among more recent attempts to create relationships between species composition of benthic fauna and lake water quality, the work of Saether (1979) and Wiederholm (1980) can be mentioned. Both authors found a good correlation between species composition of chironomids and trophic state of lake, as expressed by total phosphorus concentration divided by mean water depth (see Fig. 5.5).

It should, however, be stressed that patchiness, i.e., great variability in time and space, is a major problem in most contexts related to benthic communities and their use in classifying lakes, determining bioturbulent mixing or as indicators of pollution.

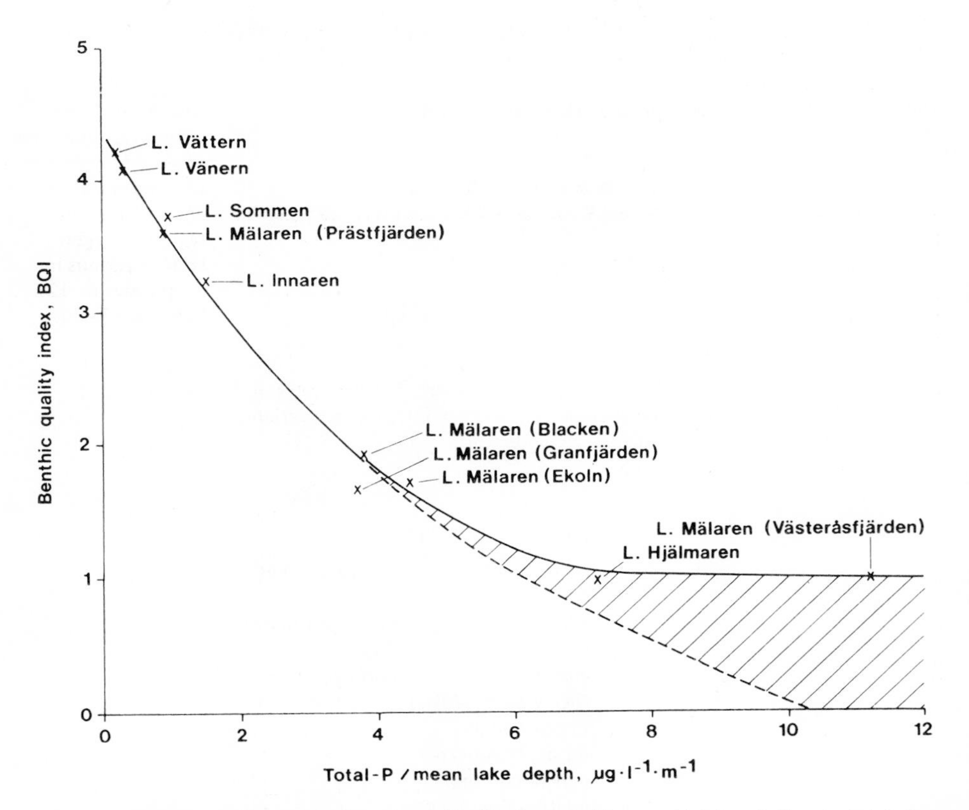

Fig. 5.5. Benthic quality index (based on distribution of chironomids) in relation to total phosphorus concentration in lake water divided by lake mean depth in some Swedish lakes. *Hatched area* eutrophic lakes with varying degrees of stability of the summer stratification (Wiederholm 1980)

5.4 Bacteria

The importance of bacteria and bacterial processes in the sediments is paramount. Sediment-living bacteria not only contribute to establishing the conditions in the sediment itself, the bacterial metabolism also largely affects the chemical milieu in the water. The impact of bacteria is enhanced by the fact that they utilize dissolved and particulate compounds not only for nutrition but also as sources of energy. Photosynthetic algae take energy from sunlight alone, while most bacteria can derive their energy from oxidation of reduced carbon compounds or reduced inorganic salts. Green plants depend on oxygen for their respiration; many bacteria are able to use a number of oxidized compounds as electron acceptors, e.g., NO_3^-, SO_4^{2+} and CO_2. Consequently, the bacterial activity in the sediment largely affects the turnover of essential nutrient elements like nitrogen and sulfur.

The magnitude of bacterial influence is, of course, dependent on the biomass and the bacterial activity. From Fig. 5.6, it is evident that the surface sediment is the prime site of bacterial metabolism in the lake ecosystem. The reason is that (although some bacteria utilize CO_2 as carbon source and derive energy from reduced inorganic salts), the main driving force for bacteria is the supply of organic matter which generally is much higher at the sediment surface than in the lake water.

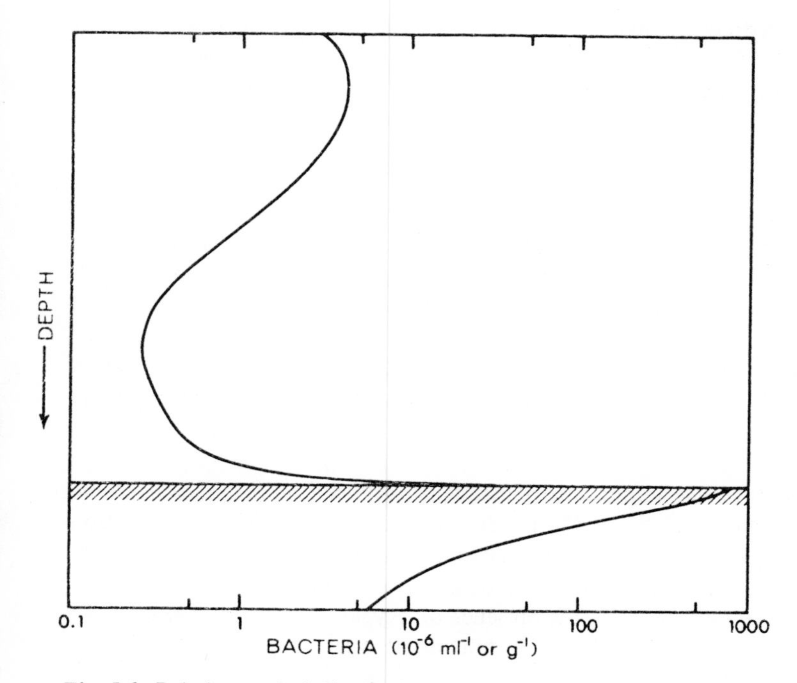

Fig. 5.6. Relative vertical distribution of bacterial numbers in water and sediments of a mesotrophic lake (Wetzel 1975)

It should be noted that bacteria, unlike algae or other green plants, do not generally create new biomass or fix energy. The bulk of bacteria utilize energy and structural compounds and their main role is as a catalyst for processes which permit the reutilization of essential compounds. Bacteria are often collectively referred to as "decomposers" and the result of their activity is called "decomposition" or "mineralization". These concepts tend to neglect the benefit derived from bacterial activity and it should be remembered that "recomposition" needs "decomposition", i.e., when a new structure is built with material taken from an old, this has to be disintegrated before the separate units can be put together again. The fact that molecules containing essential elements are utilized several times in biological production in lakes is largely the result of bacterial activity.

Subsequently, the text concentrates on the result of bacterial activity, mainly metabolism of general significance in sediment systems, which influence the turnover of essential elements in lakes. Important bacterial reactions are described and consequences discussed.

5.4.1 Functional Classification of Bacteria

Carbon Supply. According to classical nomenclature, bacteria (as well as other types or organisms) are divided in two groups depending on whether their nutritional needs are strictly inorganic or also include utilization of organic compounds. The first type of organisms, which use carbon dioxide (CO_2) as the only source of carbon, is called *autotrophs* ("self-feeding"); while the second class, which needs organic carbon, is called *heterotrophs* ("feeding on others"). The majority of sediment-living bacteria belong to the heterotrophs. It should be emphasized, however, that this division has lost much of its initial validity, because a great number of exceptions and overlappings exist.

Energy Supply. The demand for energy is met either by oxidation of reduced chemical compounds *(chemotrophy)* or by utilization of the energy of sun radiation *(phototrophy)*. Nearly all sediment bacteria of any significance are chemotrophic. Such bacteria are further divided into subclasses depending on the chemical substrate used to obtain energy.

Bacteria oxidizing inorganic compounds, like H_2S, CO_2 or Fe^{2+}, are called *chemolithotrophic* (Gr. lithos = mineral). Chemotrophic bacteria which oxidize organic substrates are termed *chemoorganotrophic*.

Electron Acceptor. It is also important to distinguish between *aerobic* and *anaerobic* bacteria. The first category utilizes O_2 as terminal electron acceptor in respiration reactions, while anaerobic bacteria use alternative acceptors, like NO_3^-, SO_4^{2-}, organic compounds or others. *Obligate* aerobes do not grow in the absence of oxygen, while *obligate* anaerobes can not survive in the presence of oxygen. *Facultative* aerobes and anaerobes are able to reproduce in both oxic and anoxic conditions.

5.4.2 Bacterial Turnover of Important Elements

5.4.2.1 Oxidation and Reduction of Nitrogen Compounds

Nitrification. Nitrogen is incorporated in organic material in reduced form, for example as amino groups in cell proteins. Thus, the primary nitrogen compounds released by heterotrophic bacteria in the decomposition of organic compounds is ammonia. In water solution at pH-values below 9 ammonia occurs mainly as NH_4^+. Therefore, in anoxic sediments the decomposition of organic matter leads to a considerable accumulation of NH_4^+ in the sediment pore water and often also in the bottom water of the lake.

However, in sediments superimposed by oxygenated water, ammonia is transformed to nitrate in a series of bacteria-mediated oxidation reactions:

$$NH_4^+ + 1.5\ O_2 \rightarrow 2H^+ + NO_2^- + H_2O \qquad (\Delta F = -66\ \text{kcal} \cdot \text{mol}^{-1}). \qquad (5.1)$$

$$NO_2^- + 1.5\ O_2 \rightarrow NO_3^- \qquad (\Delta F = -18\ \text{kcal} \cdot \text{mol}^{-1}). \qquad (5.2)$$

The oxidation to nitrate proceeds via intermediate products like NH_2OH and H_2N_2OH. All these intermediates are unstable and usually never found in substantial concentrations.

Bacteria of the genus *Nitrosomonas* are the primary executors of the oxidation of ammonia to nitrite. *Nitrobacter* is the best-known genus for the oxidation of nitrite to nitrate.

These bacteria are examples of chemolithotrophic autotrophs, where the energy derived from oxidation of reduced nitrogen compounds is utilized to reduce CO_2 for synthesis of organic matter. The total energy yield of the overall nitrification reaction is -84 kcal mol^{-1} of which the major part is generated from the oxidation of ammonia to nitrite.

Nitrification reactions need oxygen, but the processes can go on at O_2-concentrations as low as 0.3 mg l^{-1}. The demand for oxygen means that sediments, under the following circumstances, significantly contribute to nitrate formation: (1) In oligotrophic lakes where the surficial sediments are constantly covered with oxygenated water. (2) In eutrophic lakes during circulation periods or in shallow zones where oxygen is supplied continuously.

Nitrification reactions are pH-dependent and severely reduced at pH-values below 5. Therefore, in humic-rich sediments and in acidified lakes the bacterial nitrate production is very low.

Denitrification. Denitrification is the bacterial reduction of nitrate to molecular nitrogen which proceeds in the sequence:

$$NO_3^- \rightarrow NO_2^- \rightarrow N_2O \rightarrow N_2. \qquad (5.3)$$

This process is an example of dissimilative nitrate reduction, i.e., the reduced nitrate is not incorporated in the bacterial cell.

Denitrification is maintained by chemoorganotrophs which utilize oxidized nitrogen as electron acceptor in the oxidation of organic matter. Denitrifiers are found above all in the genus *Pseudomonas,* but also in the genera of *Actinobacter, Bacillus* and *Micrococcus.* These bacteria are facultative anaerobes, which means that they can switch from utilizing O_2 to NO_3^-. The energy derived from the oxidation of organic matter is about the same if O_2 or NO_3^- is used as electron acceptor. Denitrifiers, particularly *Pseudomonas* species, are very broad in their requirements for organic substrates. A strain of *Pseudomonas fluorescense*, for example, was able to oxidize more than a hundred widely different organic compounds (Golterman 1975).

Thus, denitrifying bacteria are very well adapted to meet the often rapidly changing conditions in lake sediments, which means that they are abundant and that their overall effect on nitrogen turnover in lakes may be very large. Since denitrification occurs in more or less anaerobic conditions (Eh $\approx$ 300 mV), the sediments are the prime site for denitrification in lake ecosystems. The end product of the process is N_2, which if not refixed (see below) is lost to the atmosphere. Denitrification is therefore largely synonymous with export of nitrogen from lake systems. The activity of denitrifying bacteria in lake sediments is sometimes so high that the production of N_2 corresponds to 50% or more of the total nitrogen input to the lake. Since nitrogen (together with phosphorus) is one of the key nutrients in lakes, the consequences of such high nitrogen losses are easily understood.

Denitrification is particularly important in productive lakes where the organic loading on the sediments is high and anoxic conditions frequent. In some cases dissimilative nitrate reduction ends with NH_4^+ instead of N_2, but this pathway is generally considered to be less important than denitrification. Besides the dissimilative reduction of nitrate, there is also assimilative reduction. This reduction is maintained in bacterial or algal cells, when nitrate is used as the source of cell nitrogen. This reaction is of less importance in lake sediments, where NH_4^+ is the prime nitrogen source.

Nitrogen Fixation. A special case of nitrogen reduction is the fixation of N_2 where gaseous nitrogen (N_2) is assimilated, reduced and incorporated in organic structures. N_2-fixation is maintained by species of *Clostridium* and *Azotobacter.* These bacteria are anaerobic heterotrophs and, consequently, most abundant in the sediments of productive lakes. By their action part of the N_2 produced in denitrification processes is refixed, and with an active population of N_2-fixing organisms in the sediments, a stratification in the lake water with low concentrations of N_2 at the sediment surface can occur.

Apart from heterotrophic bacteria, some blue-green algae, both planktic and benthic forms, are able to fix nitrogen. It is generally agreed that algal N_2-fixation is of greater magnitude than bacterial fixation.

5.4.2.2 Oxidation and Reduction of Sulfur Compounds

Biological structures contain sulfur in reduced form, e.g., as sulfhydryl (-SH) groups in protein molecules. Therefore, the heterotrophic decomposition of organic material leads to the release of dissolved reduced sulfur compounds which can be metabolized for energetic purposes by bacteria. The prime reduced degradation product is hydrogen sulfide (H_2S).

Chemolithotrophic bacteria, which derive energy from reduced sulfur, are of two types. The first, including the genera *Beggiatoa* and *Thiotrix,* oxidize H_2S to elemental sulfur, which is deposited inside the cell. When the H_2S supply is exhausted, the internal pool of elemental sulfur is further oxidized to sulfate, which is released to the ambient medium:

$$H_2S + 0.5\,O_2 \rightarrow S^o + H_2O \qquad (\Delta F = -41\text{ kcal} \cdot \text{mol}^{-1}). \qquad (5.4)$$

$$S^o + 1.5\,O_2 + H_2O \rightarrow H_2SO_4 \qquad (\Delta F = -118\text{ kcal} \cdot \text{mol}^{-1}). \qquad (5.5)$$

Since both reactions need reduced sulfur plus oxygen, these bacteria are restricted to boundary layers between aerobic and anaerobic environments such as the surface of a reduced sediment superimposed by oxygenated water.

A second type of sulfide oxidizing chemolithotrophs is represented by the *Thiobacillus* species which deposit elemental sulfur outside the cell and do not oxidize sulfur as far as to sulfate.

Oxidation of sulfur is also maintained by several photosynthetic bacteria. Their demand for light restricts their distribution in lake bottoms to shallow surficial sediments and their impact is greater in the water column than in the sediments.

Bacterial reduction of sulfur occurs in the degradation of protein compounds by the genus *Proteus.* These bacteria reduce sulfhydryl groups to H_2S and are abundantly distributed in lake sediments.

Hydrogen sulfide is also produced in other types of bacterial reactions where oxidized sulfur compounds, like sulfate, sulfite or thiosulfite, are used as electron acceptors in the anaerobic degradation of organic matter by heterotrophic bacteria. This dissimilatory sulfate reduction is analogous to the bacterial nitrate reduction discussed above. Sulfate reduction is performed by bacteria of the genera *Desulfovibrio, Desulfotomaculum* and *Desulfomonas:*

$$H_2SO_4 + 2(CH_2O) \rightarrow 2CO_2 + 2H_2O + H_2S. \qquad (5.6)$$

Apart from organic matter, the energy in sulfate reduction processes can also be derived from molecular hydrogen:

$$H_2SO_4 + 4H_2 \rightarrow 4H_2O + H_2S. \qquad (5.7)$$

Bacterial production of hydrogen sulfide from degradation of proteinaceous matter, or from reduction of sulfate, is a very important process in lake sediments with major consequences for the biological and chemical environment. High production and accumulation of H_2S occurs in anaerobic organic milieus, e.g., in the sediments of productive lakes. Hydrogen sulfide is a very toxic compound, which can cause extinction of bottom-living animals, plants and different types of bacteria. When diffusion of H_2S occurs to the water column faster than it is oxidized, the toxic effect is extended to the biota of the lake water.

Heavy metals are very reactive with sulfide and form precipitates of metal sulfides which often are hard to dissolve. These sulfides are blackish and give a typical dark colour to reduced sediments.

5.4.2.3 Oxidation and Reduction of Iron

The so-called iron bacteria utilize the energy derived from oxidation of iron (II) to iron (III). Some iron-oxidizing bacteria are chemolithotrophs *(Gallionella, Leptotrix, Thiobacillus* and *Ferrobacillus)* and derive energy from this particular process. The energy yield is, however, small – about 11 kcal · mol^{-1} Fe. Some of these bacteria, like *Thiobacillus thiooxidans* and *Ferrobacillus ferrooxidans,* are able to live in very acidic environments with pH values as low as 3. At such low pH non-biological oxidation of iron (II) by oxygen does not occur, and the bacterial oxidation is the only route for oxidation of iron.

At neutral or alkaline pH there is a "competition" between iron bacteria and oxygen for the reduced iron. This usually restricts the bacteria to transition zones rich in iron and with steep redox gradients.

Several iron bacteria deposit the oxidized iron as iron (III) hydroxides on their sheaths. Since the energy yield is low, fairly large amounts of iron are used in the process and thick precipitates, easily recognized by the typical red-brown colour, are formed. Typical sites for the formation of such precipitates are oxygenated sediment surfaces where reduced iron is supplied from ground water inflows.

Iron is also oxidized by heterotrophic bacteria like *Sphaerotilus* and *Siderocapsa.* These bacteria also deposit oxidized precipitates, but do not derive energy from oxidation of iron.

The reduction of iron (III) in reduced sediments has generally been considered an abiotic process where iron is reduced by organic matter or sulfide. However, bacterial reduction of iron (III) has been experimentally demonstrated (Ottow and Munch 1978, Sørensen 1982). These findings suggest that the reduction of iron is coupled to the ability of nitrate reduction and that nitrate-reducing bacteria can utilize iron (III) as an alternative electron acceptor when nitrate is exhausted. Sørensen's experiments (with marine sediments) indicate that biological iron reduction is more rapid and quantitatively more important than abiotic reduction processes. These findings are highly interesting and might offer new insights into the eutrophication processes of lakes since the phosphate-binding capacity of lake sediments is intimately related to the oxidation/reduction state of iron. If reduction of iron, which leads to release of phosphorus from lake sediments, is maintained by nitrate-reducing bacteria, this would imply that the iron-reducing potential in any given sediment is increased by high nitrate supply. Substantial increases in nitrate loading on aquatic ecosystems are today a common feature in agricultural areas because of the use of fertilizers and also in regions with high atmospheric deposition of nitrate due to burning of fossil fuels.

5.4.2.4 Fermentation

Fermentation refers to the anaerobic breakdown mainly of carbohydrates which results in the formation of acids and alcohols. A large variety of carbohydrate species are used as fermentable energy sources by a large number of microorganisms.

Characteristic for fermentation is that organic substances serve as both donors and acceptors of electrons, often with different parts of the same molecule. Unlike aerobic and anaerobic respiration, fermentation reactions proceed independently of inorganic electron acceptors and Eh during anaerobic conditions. Thus anaerobic degradation of organic matter can be divided into two principally different reactions: fermentation and anaerobic respiration. The end products of fermentation, alcohols and acids, either diffuse out of the sediments or are further processed in anaerobic respiration reactions.

5.4.2.5 Methane Formation

Methanogenesis is a bacterial process where CO_2 is used as electron acceptor in anaerobic respiration. Three genera: *Methanobacterium, Methanococcus* and *Methanosarcina* are known to mediate this process. The energy is derived either from H_2 :

$$4H_2 + CO_2 \rightarrow CH_4 + 2H_2O \qquad (5.8)$$

or from carbon compounds.

Some methane-producing bacteria oxidize one-carbon compounds, like methanol and formic acid, while others are able to utilize acetic acid. Thus, methane is produced from a few rather simple organic substrates. The processes are linked to anaerobic fermentation which provide the necessary substrates. The methane formation is therefore often described as a two-stage process:

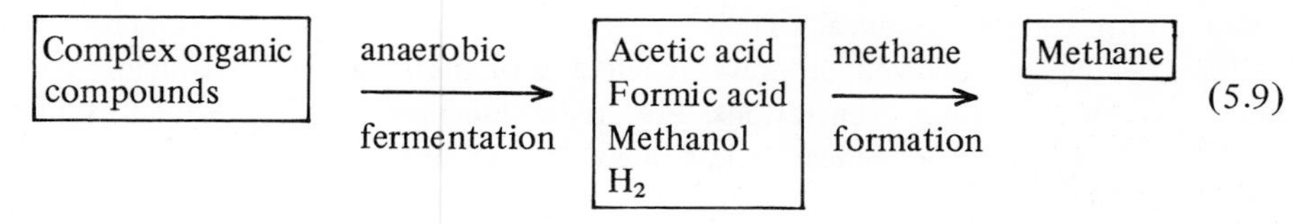

(5.9)

Methane-producing bacteria are strictly anaerobic. The reaction takes place at low redox potentials, –200 to –300 mV. In organic and highly reduced sediments, the methane production is a quantitatively important part of the anaerobic respiration (Table 5.4). During intense methanogenesis, methane bubbles are formed in the sediments and when the pressure becomes sufficiently high, the bubbles migrate to the lake water and the atmosphere. This process, called methane convection, may cause a significant turbulence in the sediments, to which has been attributed an essential role for the transport of solutes from sediments to lake water (Ohle 1958, 1978).

5.4.3 Decomposition of Organic Material – General Concepts

As seen above, the bacterial activity in the sediments is largely reflected in a number of reduction/oxidation reactions. Oxidation of reduced compounds takes place for energetic purposes and a number of different compounds are used as electron acceptors (are reduced) in these reactions. The major source of energy and carbon for sediment-living bacteria is organic matter supplied by settling material from the water column. The degradation of organic matter produces (apart from bacterial biomass) carbon dioxide, depletion of oxidized compounds and accumulation of reduced compounds. These changes are partly balanced by other bacterial reactions, where CO_2 is used as carbon source and reduced inorganic compounds are oxidized, but the net effect is always in favour of the chemoorganotrophic processes.

Since the highest bacterial activity takes place in the uppermost layer of the sediments (Fig. 5.6), elements utilized or produced in bacterial processes are partly taken from and released to the water column.

As long as oxygen is supplied, aerobic bacterial respiration leads to a rapid decomposition and a successive exhaustion of O_2. If the consumption of O_2 is faster than the supply, as it is in sediments superimposed by stagnant hypolimnion bottom water, aerobic bacterial metabolism leads to reduced conditions where anaerobic bacterial processes take over. In highly organic sediments, bacterial decomposition can cause deoxygenation in parts of, or in many cases throughout the entire hypolimnion.

After the exhaustion of dissolved oxygen, decomposition proceeds by utilization of alternative electron acceptors, like NO_3^-, Fe^{3+}, SO_4^{2-} and CO_2, and by anaerobic fermentation. The utilization of inorganic electron acceptors takes place in the order listed above and this type of respiration is redox-dependent; the approximate Eh ranges for each reduction together with the end products of the processes are shown in Fig. 5.7.

The sequential utilization of different electron acceptors means, in principle, that reduction of iron does not take place before nitrate is exhausted, sulfate reduction does not occur until iron (III) is reduced to iron (II), and methane production is delayed until sulfate is reduced. Therefore, one or several of the reduction processes in Fig. 5.7 may be delayed or prevented if any of the more easily reducible compounds occur in high concentrations. Fe^{3+} reduction may thus be delayed by high NO_3^- supply and methanogenesis is less likely to take place in sulfate-rich lakes, and so on. It should be emphasized that several of the bacterial reductions can proceed

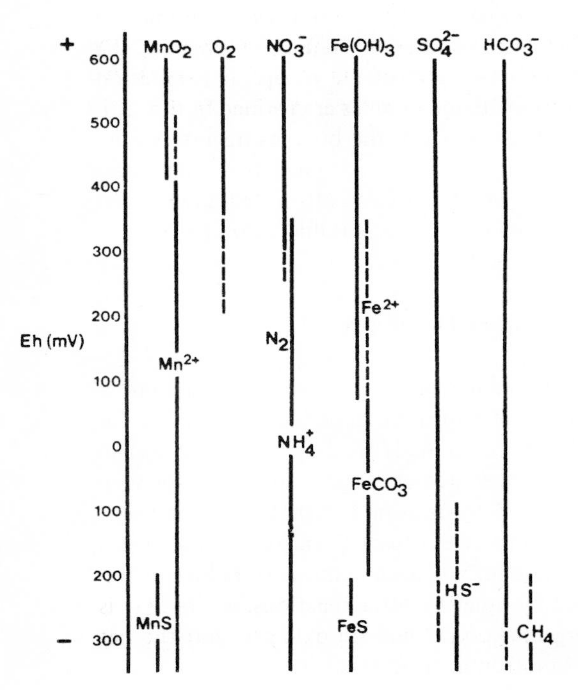

Fig. 5.7. Approximate Eh ranges of electron acceptors and their reduced products in freshwater (J.G. Jones 1982)

more or less simultaneously. This is possible partly because of the development of vertical redox gradients within the sediments and partly because of the heterogeneous structure of lake sediments which permits the coexistence of highly different micro-zones. Such zones can be expected particularly in sediments at, or above, the thermocline, where water movements create a changing physical and chemical environment.

The anaerobic bacterial respiration leads to a successive depletion of reducible agents and to the subsequent production of reduced reaction products. Some of these, like N_2, H_2S or CH_4, are gases transported from the sediments to the atmosphere. Therefore, the bacterial processes producing these gases are "elimination reactions", leading to substantial losses of essential nutrients from the lake ecosystem. Other products are dissolved salts or ions, e.g., NH_4^+, Fe^{2+} or S^{2-}, which escape to lake water by diffusion, or can be precipitated in inorganic chemical reactions in the sediments.

The result of bacterial degradation is often reflected in the water chemistry, particularly in the hypolimnion of stratified productive lakes. A schematic illustration of the sequential exhaustion and production of different compounds in the bottom water is given in Fig. 5.8.

The relative role of the different types of decomposition processes is, of course, highly variable between different lake sediments depending on the supply of oxygen and other electron acceptors, and few investigations have been made on this topic.

An attempt to give a rough evaluation of the importance of aerobic respiration versus several anaerobic respiration processes in the sediments of a eutrophic lake has been presented by J.G. Jones (1982) on the basis of net changes of bacterial metabolites in the hypolimnion during the summer stagnation (Table 5.4). It can be seen that O_2 consumption equals approximately 40% of the CO_2 accumulation, which indicates that, under these presuppositions, the aerobic degradation is the dominating "single process". It is also worth noting that the sum of the consumed products in Table 5.4 is not equal to the total CO_2 accumulation. According to

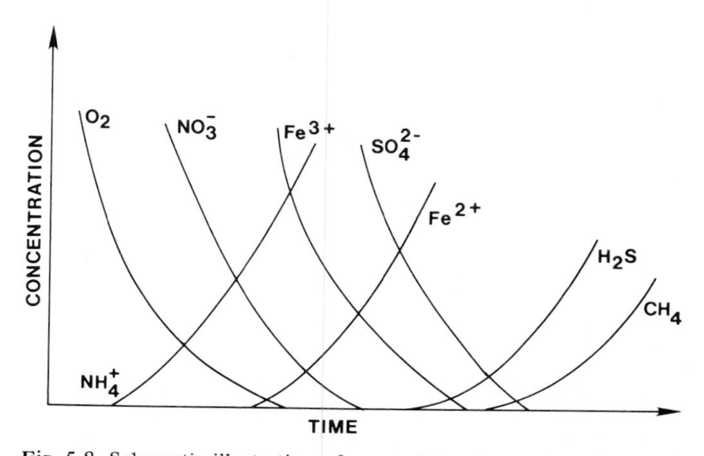

Fig. 5.8. Schematic illustration of sequential consumption of electron acceptors and production of reduced products as the result of bacterial respiration

Table 5.4. Summary of decomposition processes in the profundal zone of an eutrophic lake during summer stagnation. Results from field observations (J.G. Jones 1982, based on data from J.G. Jones and Simon 1980)

Standing crop of particulate material in the epilimnion (g atom m^{-2})	
Carbon	0.93 ± 0.18
Nitrogen	0.11 ± 0.03
Input to the hypolimnion as particulate material (g atom m^{-2})	
Carbon	6.3
Nitrogen	0.73
Output from the hypolimnion as gas bubbles (mol m^{-2})	
CH_4	0.28 (16%) [a]
N_2	0.124 (17%)
Net changes in the hypolimnetic water column (mol m^{-2})	
CO_2 accumulation	1.77
O_2 uptake	0.98 (42%)
CH_4 accumulation	0.16 (9%)
NO_3 consumption	0.174 (13%)
NH_4 accumulation	0.165 (12%)
S_2 accumulation	0.0139 (1.7%)

[a] These values represent the % contribution of each process to the production of CO_2, derived from the stoichiometric formulas of Richards (1965). Methane values are converted to % equivalents of the total CO_2 produced

J.G. Jones (1982), this emphasizes the possible important role of iron (III) reduction and fermentation, since these processes are not represented in the table.

J.G. Jones and Simon (1981) also tried to estimate how shallow littoral sediments and deeper profundal bottoms contributed to the overall decomposition in Glelham Tarn (Table 5.5). This calculation illustrates the interesting fact that the littoral

Table 5.5. Decomposition processes in littoral and profundal sediments during summer stagnation in an eutrophic lake (J.G. Jones 1982, based on data from J.G. Jones and Simon 1981)

		Littoral [a]	Profundal [b]	Relative contribution to the whole lake [c] Littoral:profundal
CO_2 evolution		1600	940–1800 [b]	1.2:1
O_2 uptake	d	1280	750–980 [b]	1.2:1
NO_3^- reduction	d	320	220 [b]–230	1.2:1
NH_4^+ accumulation	d	460	310 [b]–320	1.2:1
N_2 evolution	d	115	30 [b]–310	3.1:1
SO_4^{2-} reduction	d	15	30–55 [b]	0.3:1
CH_4 evolution		50	80 [b]–420	0.5:1

[a] With the exception of N_2 and CH_4 evolution, all rates for the littoral zone were derived from experimental systems

[b] The values for the profundal zone indicate the ranges over three seasons. The individual values marked are those which are directly comparable with the values in the littoral zone

[c] Based on the areas of anoxic and oxygenated sediments at the end of the period of stratification

[d] Converted to CO_2 equivalent according to the stoichiometric equations of Richards (1965)

sediments are the prime sites of decomposition. Only sulfate reduction and methanogenesis are quantitatively more pronounced in the profundal zone, the explanation probably being that high temperatures in combination with higher supply of electron acceptors, like O_2 and NO_3^-, accelerate bacterial metabolism above the thermocline.

In the sediments in deeper parts of the lake, the anaerobic decomposition becomes more important and the redox potential is driven low enough to favour sulfate reduction and methane formation. However, even in the littoral sediments the latter processes are significant, which stresses the fact already pointed out – that the conditions in this type of sediments are continuously changing, thus allowing different types of breakdown processes to occur in a seemingly simultaneous manner.

5.4.4 Strategies and Methods for Determination of Bacterial Activity

To measure, calculate or estimate the activity of sediment-living bacteria is extremely difficult. The problems are perhaps best understood by stating that an ideal analysis of the intensity of bacterial processes in any given sediment sample ought to include:

- Identification of importat species or genera.
- A measure of the biomass of different groups of bacteria.
- Accurate determination of the rate at which every important process proceeds in undisturbed sediments under natural conditions.

By adding that these requirements should be fulfilled while accounting for the highly variable conditions in the sediments at different depths and at different locations, it is obvious that the task is impossible. However, also imperfect methods can give valuable information and the combination of results from investigations where different strategies and analytical procedures have been used often allows a relevant picture to be obtained of the dominating causal relationships. Subsequently, we shall briefly present some possible means and parameters by which bacterial activity or parts of the bacterial metabolism can be studied in lake sediments.

5.4.4.1 The Whole-Lake Approach

"Whole-lake approach" is the concept used by J.G. Jones (1982) for the description of bacterial activity in sediments as judged from the chemical analysis of the lake water. In most cases this approach is applicable only for sediments superimposed by a "closed system", such as the hypolimnion of a stratified lake. One example of this approach has already been mentioned (Table 5.4).

Information from this type of study gives only the *net result* of all the different bacterial processes and will of necessity include the turnover of metabolites which occurs in the water body. It should be stressed that this strategy is the only "true" one in the sense that it reflects the activity of an undisturbed sediment system.

5.4.4.2 Bacterial Activity in Experimental Procedures

The most common way to study individual bacterial processes, or the result of overall metabolism, is by different kinds of experimental procedures where consumption or production of bacterial metabolites are examined for a limited period under controlled conditions. Experiments can be performed in situ or in the laboratory.

In situ incubation is done in order to imitate natural conditions as closely as possible. It can be achieved by using a sediment sample, usually from a core sampler, and making additions of bacterial substrates. Then the sediment core and a volume of overlying water is exposed in a cylinder at the lake bottom. After a chosen time of incubation the core is brought to the laboratory for analysis. A better but less convenient way is to place chambers in the sediments which enclose a sediment area together with a volume of lake water. Measurements are then generally performed on the water phase. This type of chamber can be handled by SCUBA divers, which restricts the use to relatively shallow clear waters.

A special type of chamber was developed by Edberg and von Hofsten (1973) for measurements of O_2 consumption in the sediments. This Plexiglas cylinder (see Fig. 5.9) was then successfully also used for experimental studies of denitrification rates in sediments (Tirén 1977), where N-15 labelled nitrate was supplied to the chamber through a hose from the lake surface. Samples were extracted in the same manner and analyzed for oxygen, nitrate, nitrite and nitrogen gas.

The great advantage with this type of in situ experiment is that natural conditions are simulated in a more realistic way than in the laboratory, and the use of a chamber, as in Fig. 5.9, minimizes the disturbances of the sediments. The disadvantage is that the fairly complicated facilities tend to reduce the number of studies. Thus, if the disturbances can be tolerated or kept to a ninimum, it is often more convenient to bring the sediment samples to the laboratory. This provides the opportunity to

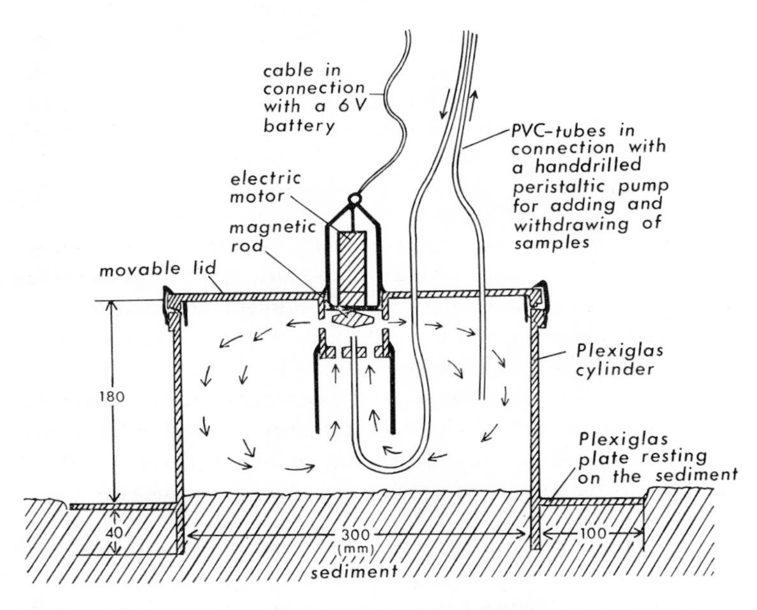

Fig. 5.9. Construction and function of a Plexiglas cylinder used for in situ incubation in lake water/sediment systems (Tirén et al. 1976)

control and manipulate the experimental conditions, and makes it possible to study a great number of sediment samples simultaneously.

Most laboratory studies are performed with intact cores in the Plexiglas cylinders containing a volume of overlying water from which samples are usually collected. These experimental set-ups are used with or without stirring. The rationale for studying intact cores is, of course, to approach natural conditions. However, it should be stressed that rates of bacterial activity obtained in experimental set-ups can seldom be extrapolated to nature. In many cases it may therefore be just as good or better to study a sediment suspension in the laboratory. Simple laboratory procedures have their greatest value in the possibility of comparing bacterial processes in different lake sediments or in sediments from different parts of a given lake. Laboratory studies also offer the simplest means of studying the influence of different environmental parameters on bacterial processes. We will conclude this section by presenting some parameters which can be related to different types of bacterial activity.

5.4.4.3 Parameters Reflecting Total Bacterial Activity

CO_2 Production. Decomposition of organic matter, whether aerobic or anaerobic, yields CO_2 as end product. The carbon dioxide production in sediments therefore reflects the total bacterial decomposition activity. Interpretation is complicated by the difficulties involved in accurate determining of CO_2 levels and by the fact that CO_2 can be re-utilized by sediment-living bacteria, either as a carbon source or as electron acceptor in the methane production at low redox potentials. Furthermore, CO_2 is produced from respiration of other organisms than bacteria. Significant CO_2 uptake by epipelic algae can also be a complicating factor. CO_2 production as a measure of decomposition has been utilized, for example, by Beyers (1963), Wetzel et al. (1972) and J.G. Jones (1982).

Redox Potential. The redox potential in the sediments is roughly a function of the availability of suitable electron acceptors and the respiration activity of bacteria. The higher the activity of bacteria, the faster is the exhaustion of electron acceptors. It has been demonstrated earlier in this chapter (Fig. 5.7) that the utilization of electron acceptors is connected with decreasing Eh-values. Thus, the redox potential in the sediments can be a measure, or at least an indication, of the extent and type of bacterial activity. A typical vertical redox gradient in lake sediments is shown in Fig. 5.10.

It should be emphasized that low redox potentials do not necessary imply higher bacterial activity than high values. Sediments superimposed by circulating, well-oxygenated water will always have higher Eh-values and higher bacterial activity than sediments superimposed by stagnant bottom water, provided that the supply of degradable organic matter is comparable. A detailed presentation of the connection between redox potential and microbial activity has been given by Fenchel (1969).

Electron Transport System Activity (ETSA). The metabolic activity of respiring bacteria is reflected in the activity of the electron transport system. This can be used for analytical purposes by introducing an artificial electron acceptor with such properties

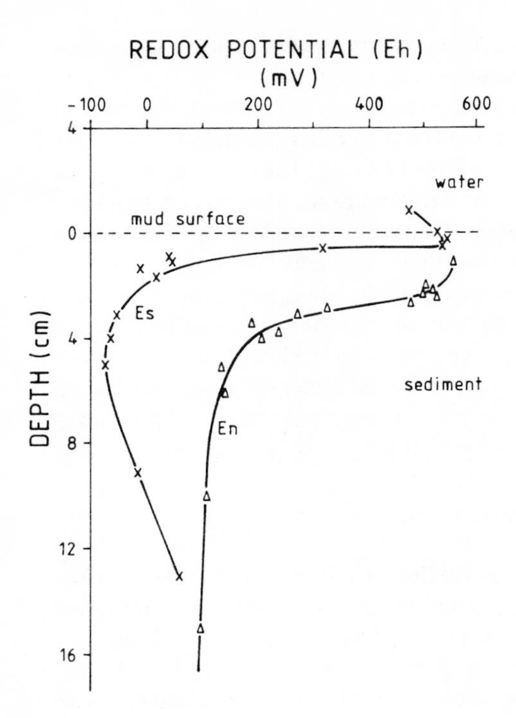

Fig. 5.10. Vertical distribution of redox potential in sediment cores from two different types of lakes. *En* Ennerdale water, oligotrophic, depth 40 m; *Es* Esthwaite water, eutrophic, depth 14 m (Boström et al. 1982, as modified from Mortimer 1942)

that the reduced product is easy to detect. In the ETSA method 2-(p-iodophenyl)-3-(p-nitrophenyl)-5-phenyltetrazoliumchloride, INT, is used as electron acceptor. INT is reduced by the coenzyme UQ-cytochrome b complex in the electron transport chain (Fig. 5.11) and the product, INT-formazan, has red colour and may be analyzed spectrophotometrically. The method yields the maximum potential respiratory activity since INT is introduced in excess and therefore does not limit the oxidation rate of organic substrate, as may occur in situations where there is a shortage of O_2

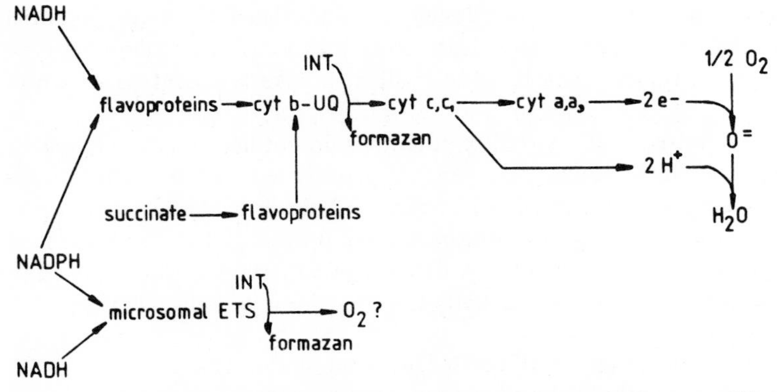

Fig. 5.11. Scheme of the electron transport chain (Broberg 1980, as modified from Kenner and Ahmed 1975)

or other electron acceptors in nature. ETSA can be used as a measure of the total anaerobic and aerobic respiration. Fermentation is not included since this process does not use cytochromes for electron transport.

It should also be stressed that respiratory activity of other sediment-living organisms than bacteria may confuse the results.

ETSA is a comparatively new method and has been used in sedimentological studies by, among others, Christiansen and Packard (1977), Broberg (1980) and J.G. Jones (1982).

5.4.4.4 Factors Reflecting Defined Parts of Bacterial Activity

O_2 Consumption. Measurement of O_2 consumption, whether done directly in the lake or indirectly in experimental procedures, is the classical way to measure bacterial activity. Sometimes O_2 consumption is used more or less synonymously with total bacterial activity in sediments. It should, however, be remembered that a decline in oxygen concentration reflects total respiration only when oxygen is utilized for aerobic respiration plus reoxidation of the products of anaerobic respiration. Generally, aerobic decomposition is the most important "single process", but in organic sediments the anaerobic degradation, though slower and yielding less energy can dominate.

Dehydrogenase Activity. In principle, dehydrogenase activity measurements are comparable with ETSA. Both measure the activity of the electron transport chain and the analytical procedures are quite similar. Dehydrogenase activity measurements, as utilized on lake sediments (e.g., Pamatmat and Bhagwat 1973), are based on the use of the artificial electron acceptor, TTC, which is reduced later in the electron transport chain than INT (cf. ETSA method). Therefore, in contrast to INT, the reduction of TTC is disturbed by the presence of oxygen which makes the method less suitable as a measure of aerobic respiration. Consequently, dehydrogenase activity analyzed by the reduction of TTC can be used to measure the approximate respiration capacity during anaerobic conditions.

Other Bacterial Metabolites. Naturally, it is also possible to measure the utilization of other electron acceptors than O_2 and the production of other metabolites than CO_2. Nitrate exhaustion in sediment-water systems generally provides a good estimate of denitrification rates. Similarly, sulfate exhaustion or H_2S production and CH_4 production reflect the activity of sulfate-reducing and methane-producing bacteria, respectively (Table 5.4).

In addition, the activity of enzymes other than those in the electron transport chain can be utilized as a measure of bacterial metabolism. Buikema et al. (1980) found the catalase activity in sediments to be a good indicator of the microbial sensitivity against metal pollution. J.G. Jones (1979) measured the activity of hydrolytical enzymes, like protease and amylase in sediments of Lake Windermere and Blelham Tarn, and found this well correlated with the supply of organic substrates.

6 Sedimentation in Lakes and Water Dynamics

The processes of sedimentation in lakes are closely related to the hydrological flow patterns, and to the topography of the basin, which influences the hydrodynamical regime. In this context, however, we will neither repeat fundamental concepts of lake hydromechanics, which have already been pedagogically summarized, for example, by Hutchinson (1957), Csanady (1978), Graf and Mortimer (1979), Lindell (1980) and Simons (1980), nor the basics of physical processes of sedimentation in general, since good texts have been published, for example, by Allen (1971), Smith (1975) and Friedman and Sanders (1978). Instead, the focus in this chapter will be on sedimentation in lakes and the causal relationships regulating areal and temporal variations in rate of deposition.

6.1 Physics of Sedimentation in Lakes

In calm waters particles denser than water will fall at a constant velocity, where the force of downward motion is equal to the drag force resisting motion. The force causing settling for a sphere is given by:

$$F_s = \frac{\pi \cdot d^3}{6} (\rho_s - \rho) \cdot g, \tag{6.1}$$

where F_s = the force causing motion (= mass · acceleration = volume · density difference · acceleration due to gravity) $[ML^{-2} T^{-1}]$;
d = particle diameter $[L]$;
ρ_s = particle density $[ML^{-3}]$;
ρ = density of water $[ML^{-3}]$;
g = acceleration due to gravity $[MT^{-2}]$.

The drag force is given by:

$$D_f = C_D \cdot \frac{\pi \cdot d^2}{4} \cdot \rho \cdot v^2 \cdot \frac{1}{2}, \tag{6.2}$$

where D_f = total drag force due to friction and particle form (= mass · acceleration) $[ML^{-2} T^{-2}]$;
C_D = drag coefficient, defined as the quotient between actual and hypothetical force [dim. less];
v = settling velocity $[LT^{-1}]$.

Smith (1975) explained the difference between *laminar* and *turbulent* flow in the following manner: "If a lit cigarette is left lying in a room free of draughts, the smoke rises initially in straight lines and the motion has a regular filamentous structure. Subsequently this coherent structure breaks down and the motion appears to be a confusion of swirling eddies. The initial phase is referred to as laminar or viscous flow and the later stage is termed turbulent flow." For particles in motion in water, the Reynolds Number, Re, expresses the relationship between the size (d = diameter), settling velocity (v) and the kinematic viscosity of the water (ν), i.e.:

$$\mathrm{Re} = \frac{v \cdot d \cdot \rho}{\mu} = \frac{v \cdot d}{\nu}. \tag{6.3}$$

If Re < 0.5, the flow around the falling particle is laminar; if Re is larger than 0.5 the flow is turbulent. In the latter case, the drag coefficient (C_D) is independent of Re, while C_D is a function of Re for spheres as well as for particles of other forms for laminar flow (Fig. 6.1). The well-known Stokes' law for spherical particles is derived by putting F_s equal to D_f if C_D is set equal to 24/Re, i.e.:

$$v = \frac{(\rho_s - \rho) \cdot g \cdot d^2}{18 \cdot \mu}, \tag{6.4}$$

where μ = coefficient of absolute viscosity.

The ratio between the absolute viscosity (μ) and the density of the fluid (ρ) defines the kinematic viscosity (ν), i.e.:

$$\nu = \mu/\rho. \tag{6.5}$$

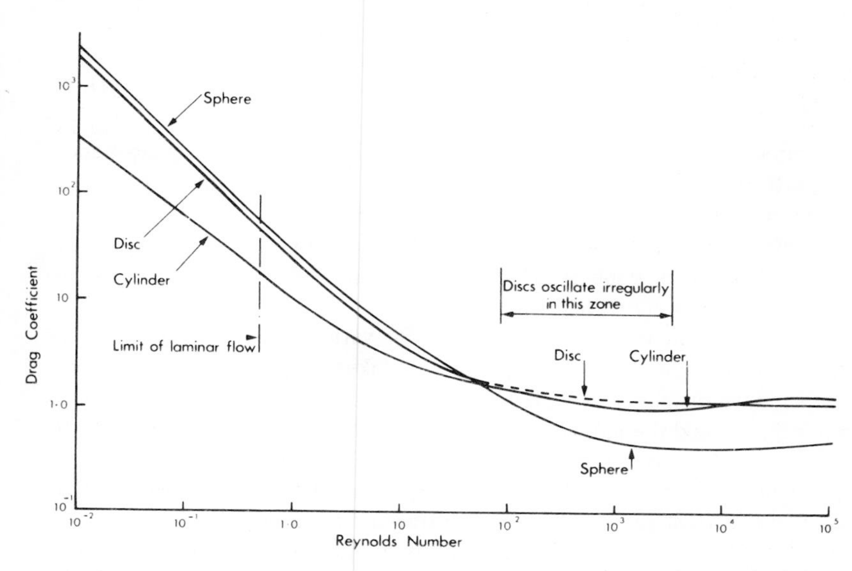

Fig. 6.1. Relationship between drag coefficient of falling particles and Reynolds number (Smith 1975)

Stokes' law (Stokes 1851) expresses (see Fig. 3.12) the settling of small, spherical particles during laminar flow conditions, i.e., the relationship between particle size (d), settling velocity (v) and submerged or excess density (i.e., $\rho_s - \rho$) (see Fig. 6.2).

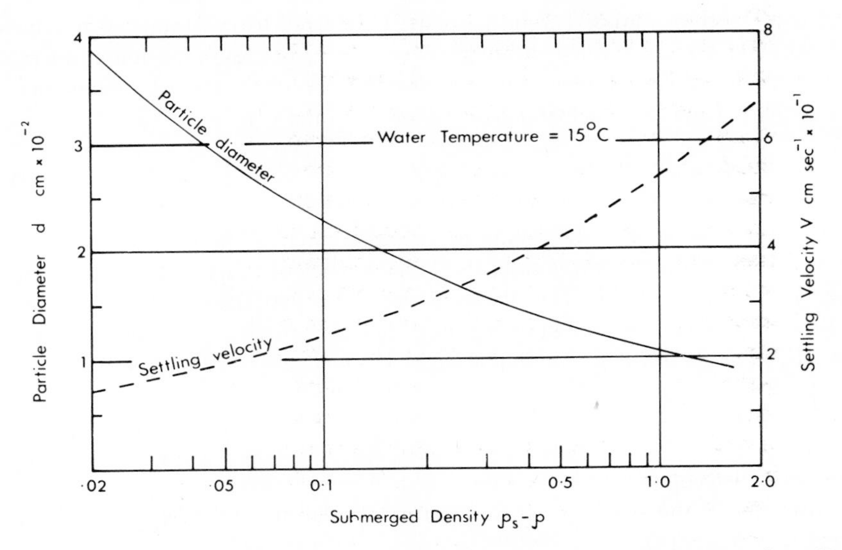

Fig. 6.2. Limiting conditions for spherical particles to settle under laminar flow (Re < 0.5) (Smith 1975)

For particles other than spheres, e.g., discs, cylinders or algae of various shapes, Stokes' equation becomes:

$$v = \frac{(\rho_s - \rho) \cdot g \cdot d_s^2}{18 \cdot \mu \cdot \phi}, \tag{6.6}$$

where d_s = diameter of a sphere having the same volume as the irregular-shaped particle;

ϕ = coefficient of form resistance.

Fig. 6.3 illustrates the relationship between ϕ and particles of various shapes. ϕ is equal to 1 in the case of a sphere. The sedimentation of planktonic materials has been discussed thouroughly by Hutchinson (1967).

Table 6.1 gives data on vertical sinking rates for various particles/aggregates as compared to the much higher horizontal velocities in lakes. Clays and organic-mineral aggregates, which have surface electric charges, form flocs, which generally settle more rapidly in water than would the individual particles as given by Stokes' law [Eq. (6.4)]; flocs may also drag down other types of suspended particles during the course of the downward movement. Flocculation is enhanced in river-mouth areas, due to a rapid decrease in turbulence, by high concentrations of suspended inorganic and organic materials, by increased salinity, ion strength and high algal concentrations. Smaller particles flocculate more easily than larger particles, due to their larger relative surface area and proportionally greater adhesive force. As progressively larger

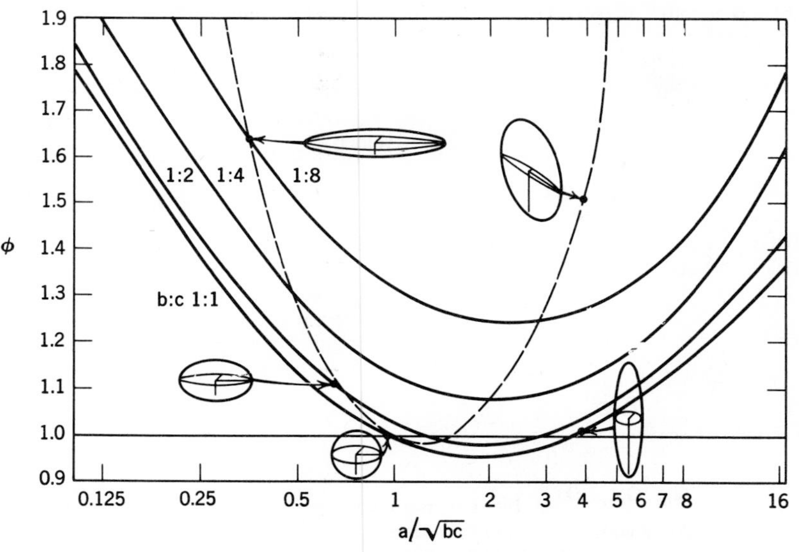

Fig. 6.3. Relationship between coefficient of form resistance (ϕ) and geometrical form of settling particles (Hutchinson 1967)

flocs are formed, the particle collisions become less frequent and further flocculation eventually ceases. (For further details on flocculation see, e.g., Mason 1966, Kranck 1973, Chase 1979, Gibbs 1982). Equations (6.4) and (6.6) describe the settling velocity up to Reynolds number of 0.5. At larger Re-values, the form resistance and inertia become increasingly important and the drag coefficient (C_D) decreases. For large spheric particles, with Re-values larger than about 1000, the settling velocity is approximately proportional to $\sqrt{d}$, i.e., the square root of the particle diameter.

To determine the sedimentation of materials in lakes, the following factors must be taken into consideration: the shape and volume (the morphometry) of the lake, the allochthonous sediment discharge, the autochthonous production, the sediment characteristics, delta development, turbidity currents, resuspension and hydrodynamical flow pattern (seiches, stratifications, Coriolis force, currents, etc.); all these factors (except the morphometry) vary in time and space. There is, to the best of our knowledge, no standard method available which describes all these extremely complicated processes in an adequate manner.

The simplest possible model is illustrated in Fig. 6.4, where:

$$V \cdot \frac{dC}{dt} = Q \cdot (C_{in} - C) - K_T \cdot V \cdot C, \tag{6.7}$$

where V = volume of the lake $[L^3]$;
dC/dt = rate of change $[ML^{-3}\,T^{-1}]$;
C_{in} = input concentration $[ML^{-3}]$;
C = concentration in the lake, equal to output concentration $[ML^{-3}]$;
Q = water discharge $[L^3\,T^{-1}]$;
K_T = turnover rate of the given substance (C) $[T^{-1}]$.

Table 6.1. Vertical sinking rates of particles in relation to horizontal movements in air and water (Bloesch and Burns 1980)

Particle	Diameter	Sinking velocity	Wind speed or horizontal current velocity	Proportional difference in velocity
Rain drops	0.05–5 mm	2.3–9.3 m s^{-1}	< 10 m s^{-1}	< 10
Snow flakes	–	0.5 m s^{-1}		
Sedimenting particles	1 μm	10^{-4} cm s^{-1}	< 200 cm s^{-1}	
Sedimenting particles	10 μm	10^{-1} cm s^{-1}	< 20 cm s^{-1}	10–10^6
Fecal pellets	–	0.04–1.0 cm s^{-1}	< 30 cm s^{-1}	
			< 5 cm s^{-1}	

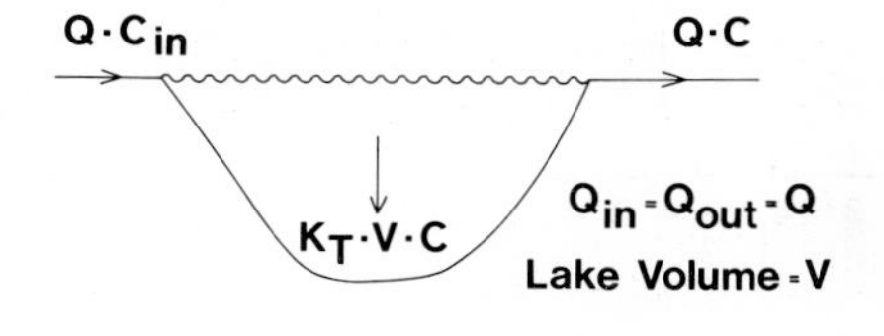

Fig. 6.4. Basic concepts for the mass balance equation for a lake

The equation is based on an assumption of steady state, i.e., that $Q_{in} = Q_{out}$. The residence (or retention) time of water in a lake (T_w in time units), i.e., the average time it would take to refill the lake if it were emptied, is defined as:

$$T_w = \frac{V}{Q}. \tag{6.8}$$

The residence time of a chemical or a given fraction of particles (T_r in time units) may be defined as:

$$T_r = \frac{V \cdot C}{Q \cdot C_{in}}. \tag{6.9}$$

The relationship between T_r and T_w is most important in lake sedimentological contexts. T_w is equal to T_r for water and conservative substances; $T_r < T_w$ for most allochthonous particles and pollutants, which are distributed in a typical manner with lobes of decreasing concentrations with distance from the source of pollution. This fundamental relationship can be derived from the steady-state solution of Eq. (6.7), which is:

$$C = \frac{Q \cdot C_{in}}{Q + K_T \cdot V} \tag{6.10}$$

or

$$C = \frac{C_{in}}{1 + K_T \cdot T_w}. \tag{6.11}$$

It follows, from Eq. (6.8), (6.9) and (6.11), that:

$$T_r = \frac{T_w}{1 + K_T \cdot T_w} \,. \tag{6.12}$$

Equation (6.12) is depicted in Fig. 6.5 for various values of the turnover rate (K_T). K_T = 0 for water and ideal conservative substances; K_T attains successively higher values with increasing grain size and settling velocity of particles. It should be emphasized that the knowledge of K_T-values for various particles, aggregates, algae or "carrier particles" (e.g., Fe/Mn-oxides, detritus, carbonates, humic substances, see Förstner and Wittmann 1979) is at present rather meagre. One example where K_T-values have been determined is Lake Ekoln, Sweden. Table 6.2 gives mean K_T-values based on monthly data from the years 1967–1977 for 20 traditional water chemistry parameters. From this table, we can see that NH_4-N, transparency (measured as optical density) and suspended matter are the three most reactive of the investigated parameters, while alkalinity, chloride and calcium are strictly conservative parameters in this particular sub-basin of Lake Mälaren. The residence time of the water (T_w) is 2.073 years for the given period. It should be stressed that these given K_T-values could not be generally applied; they are valid for this particular basin during this particular period. Equation (6.11) can be used to calculate an expected concentration (C) in a lake from input data (C_{in}) and data on T_w and K_T. This is illustrated in Fig. 6.6 with data from Lake Ekoln. A C_{in}-value of 6 (of given dimension) would, e.g., imply the following C-values for the given parameter: susp. = 3.3 < Tot.-P = 4.0 < Org. matter = 4.4 < Colour = 5.2 < Mg or SO_4 = 5.5 < Alkalinity, Cl or Ca = 6.0. Important theoretical aspects of the mass-balance Eq. (6.7), its applicability to, e.g., trace metals and its further mathematical refinement, have been given by, e.g., Imboden and Lerman (1978).

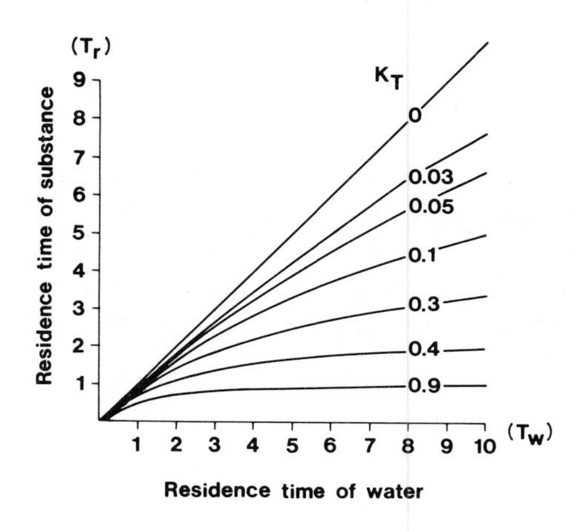

Fig. 6.5. Nomogram showing the relationship between residence time of water *(T_w)* and residence time of substances *(T_r)* with different turnover rates *(K_T)*

The deposition of suspended materials is, as has been pointed out, a most complex matter, and existing models are only applicable under given assumptions; they generally require both qualified input data and trained personnel. The following sedimentation

Table 6.2. Mean values for the ratio T_r/T_W and K_T for 20 traditional water chemical parameters in Lake Ekoln for the period 1967–1977. (Data from the National Swedish Environment Protection Board, Uppsala, Sweden)

	Parameter	Symbol	T_r/T_W	K_T
Conservative ↑	Alkalinity	$A'(HCO_3)$	1.1	0
	Chloride	Cl	1.0	0
	Calcium	Ca	1.0	0
	Sodium	Na	0.99	0.005
	Magnesium	Mg	0.90	0.048
	Sulfate	SO_4	0.90	0.048
	Potassium	K	0.88	0.058
	Colour	Opt-F	0.83	0.082
	Organic nitrogen	Org-N	0.69	0.15
	Phosphate-P	PO_4-P	0.66	0.16
	Organic matter	$KMnO_4$	0.64	0.17
	Total-N	Tot-N	0.55	0.22
	Nitrate-N	NO_3-N	0.55	0.22
	Total-P	Tot-P	0.50	0.24
	Nitrite-N	NO_2-N	0.48	0.25
	Silicon	Si	0.47	0.26
	Particulate-P	Part-P	0.34	0.32
	Suspended matter	Susp.	0.19	0.39
	Transparency	Opt-D	0.16	0.41
Reactive ↓	Ammonia-N	NH_4-N	0.15	0.41

Water retention T_W = 2.073 years

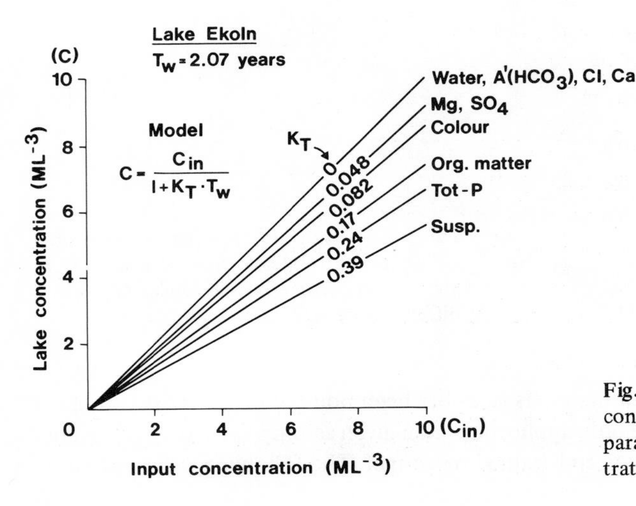

Fig. 6.6. Relationship between input concentrations of various water parameters (C_{in}) and lake concentrations (C)

model, which has been derived by Sundborg (1956), is included in this context to illustrate central concepts.

$$\overline{C}_t = \overline{C}_0 \cdot e^{-v \cdot C_b \cdot t/\overline{D}}, \tag{6.13}$$

where $\overline{C}_t$ = mean vertical concentration of suspended sediment of a particular grain size (settling velocity = v) at time t [ML^{-3}];
$\overline{C}_0$ = mean vertical concentration at time 0;
C_b = concentration of the material in question just above the bed; this value may be obtained graphically from curves for the vertical distribution of suspended matter;
$\overline{D}$ = mean depth of the lake [L]; equal to V/A.

The ratio $t/\overline{D}$, [TL^{-1}], may be replaced by $\Delta L \cdot \overline{B}/Q$, [$TL^{-1}$], where ΔL is the length of the stretch, $\overline{B}$ = the mean or characteristic width of the lake and Q the water discharge. The crucial point in applying this equation is to have access to adequate data on C_b and to be able to make a qualified guess about the flow conditions in the lake. Axelsson (1967) has demonstrated the use of this formula in Lake Laitaure, Sweden (Fig. 6.7). His calculations of sediment discharges at the outlet of Lake Laitaure show a good fit with empirical data, except for coarser particles which are present in larger amounts than predicted by the model. Figure 6.7 illustrates very nicely that sand and coarse silt are deposited in the lake and that very small amounts of these coarser fractions are transported through the lake; that most of the finer fractions (fine silt and clay) are transported through the lake; and that deposition is highly governed by the water discharge.

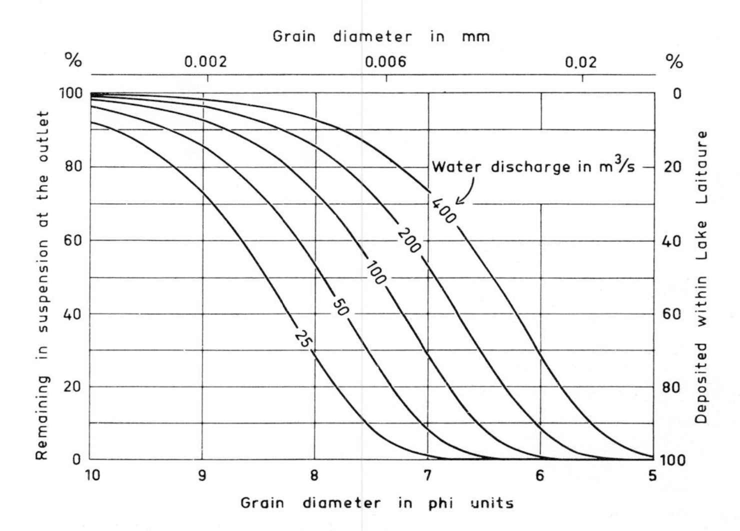

Fig. 6.7. Calculated amounts of material in suspension at the outlet of Lake Laitaure, or deposited in the lake, in percentage of the total input to the lake. The calculation assumes that, on average, 1/3 of the lake volume takes part in the through-flow (Axelsson 1967)

6.2 Geography of Sedimentation in Lakes

The aim of this section is to discuss the causal relationships regulating the areal and temporal variations, i.e., to give a geographical perspective, of the sedimentation of materials in lakes. Sly (1978) has given a broad description of these relationships and presented the figure depicted here as Fig. 6.8. From this figure, it is evident that the most important forms of driving forces or energy input, in terms of sedimentological response, are: winds, river inflow and atmospheric heating. The effects of these forms of input are controlled by the morphometry of the lake, the relief of the surroundings and the prevailing hydrological regime of the lake.

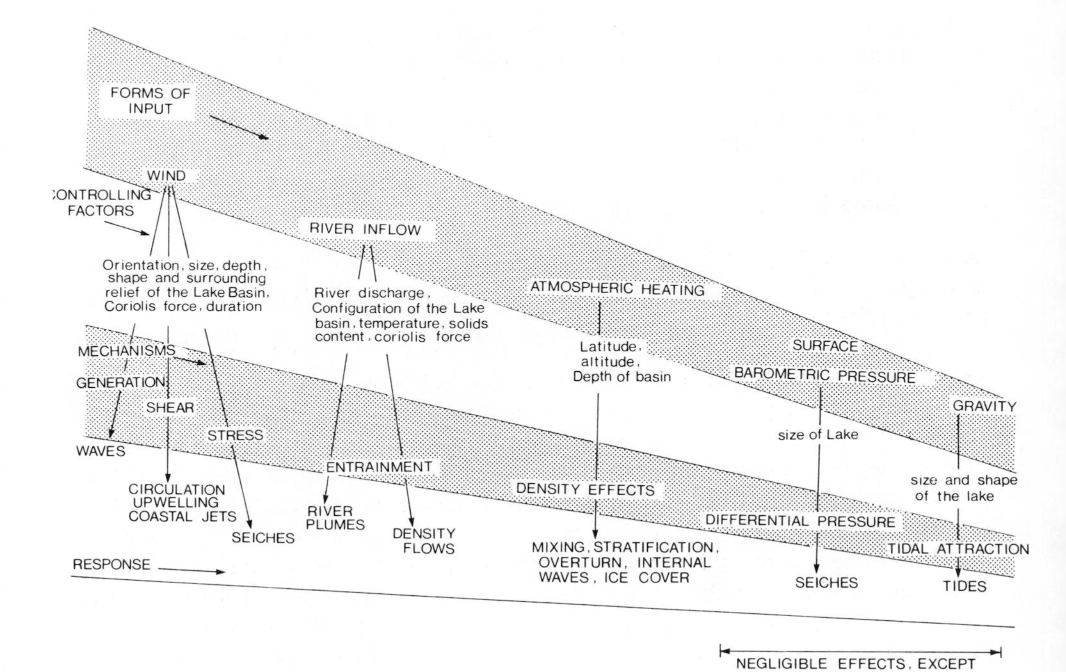

Fig. 6.8. Lake response to various forms of physical input (Sly 1978)

In lake sedimentological contexts, it is often useful to distinguish between areas dominated by river action and areas dominated by wind/wave action. Delta sedimentation and river plume sedimentation are exclusive for areas dominated by river action, while "pelagic" sedimentation and turbidific sedimentation (i.e., sedimentation connected to various mass movements of already deposited material) are not restricted to river-mouth areas. In this section, we will not discuss turbidific sedimentation (resuspension, turbidity currents, etc.), which will be in focus in the Chapter 7, dealing with lake bottom dynamics.

The following four factors regulate the rate of deposition in lakes:

1. A *deposition* factor describing the capacity of a given lake to act as a sediment trap – the larger the lake volume the higher the entrapment capacity, if all other factors are held constant.
2. A *production* factor describing the autochthonous production, i.e., the total internal bioproduction.
3. A *pre-trapping* factor describing the fact that lakes "downstream" receive less allochthonous materials than lakes "upstream", which act as sediment traps for suspended particles/aggregates.
4. A *load* factor describing the natural load of allochthonous materials (bed load, saltation and suspended load) and the anthropogenic load (from industries, urban areas and agricultural activities) on a given lake.

Subsequently, we will discuss the geographical aspects of sedimentation in lakes, i.e., what particles are deposited where, and why.

6.2.1 River-Mouth Areas

6.2.1.1 Delta Sedimentation

Many papers and books have been published on sedimentation in river-mouth areas and deltas (see Axelsson 1967, Coleman 1981); here we will attempt to sketch only the basic principles.

The hydraulic conditions change rapidly at the mouth of rivers entering into lakes. The flow pattern in delta areas is very complex but bears similarities to the expansion of submerged jets with complementary zones of reverse flow (Jopling 1960). Figure 6.9 illustrates some fundamental concepts on delta sedimentation. The form of the delta front depends both on fluvial and shore processes, on bed load and suspended load, and on the original topography of the basin. The fluvial processes give rise to fan-shaped, lobate deposition patterns. The shore processes generally tend to smooth out irregularities, and the delta shore line is, hence, often straight or cuspate. Deltas are generally divided into a *topset* area, a *foreset* area constituting the delta front and a *bottomset* area, beyond the foreset slopes (Fig. 6.9). The deposits within the foreset area are generally characterized by cross-bedding, whereas the bottomset beds have horizontal bedding. Bed-load transportation of particles larger than 0.18 mm is a necessity for the formation of foreset slopes and such materials, often highly sorted, constitute the bulk of the foreset deposits. The size of the foreset area depends on the grain size of the material (the coarser the material the larger the foreset area) and on the depth conditions of the river-mouth area. When bed-load transportation ceases, the deposition of the suspended load (particles finer than 0.18 mm) will systematically reduce the height of the foreset slopes. This transitional link is often referred to as a *toeset* deposit. The inclination of foreset slopes are usually in the order of 30°–35°, depending, for example, on grain size, degree of sorting, particle form and particle density. The presence of finer particles (silt, clay) reduces the dip angle of foreset slopes and increases the dip angle of bottomset slopes (Fig. 6.9B). The bottomset slopes are, generally, much more stable than the foreset slopes, where

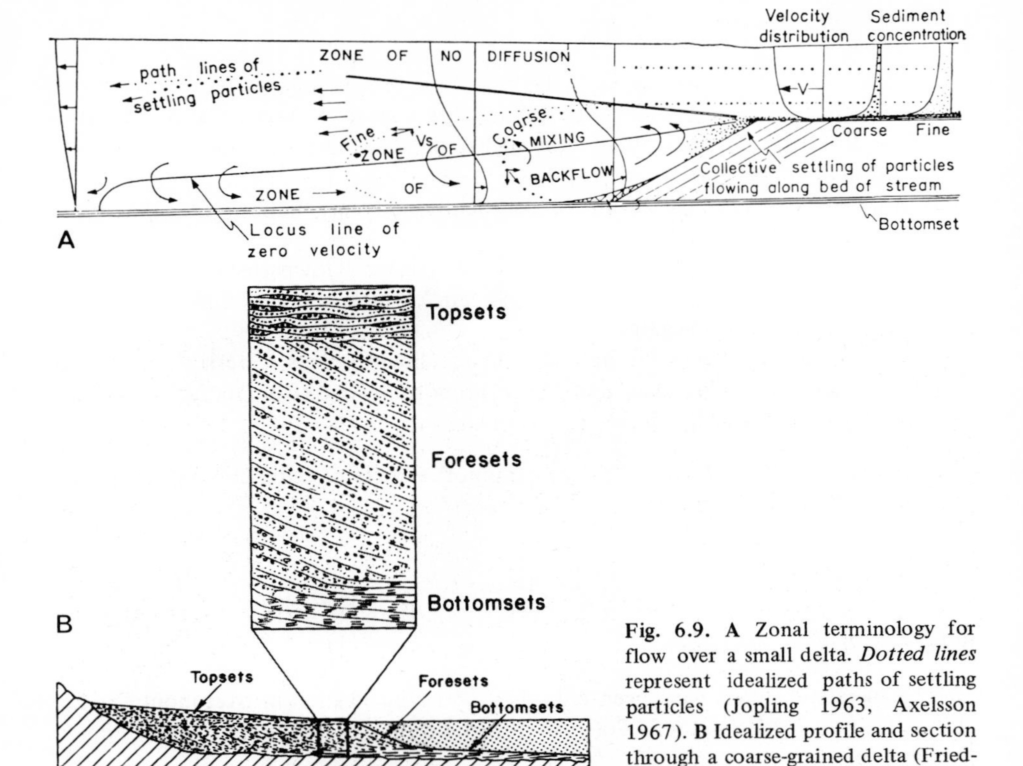

Fig. 6.9. **A** Zonal terminology for flow over a small delta. *Dotted lines* represent idealized paths of settling particles (Jopling 1963, Axelsson 1967). **B** Idealized profile and section through a coarse-grained delta (Friedman and Sanders 1978)

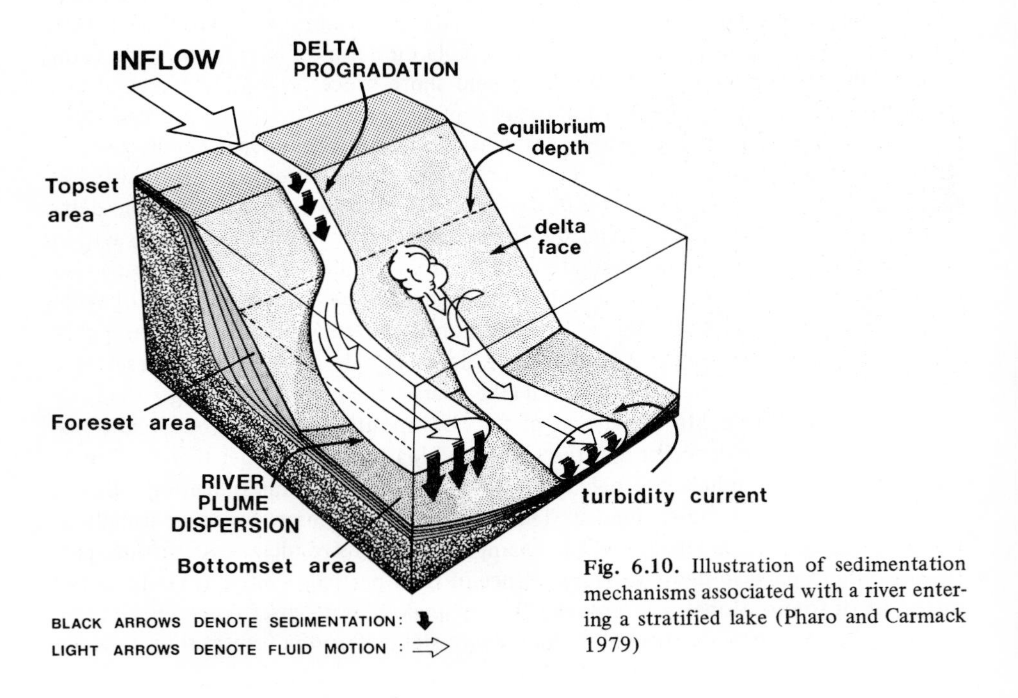

Fig. 6.10. Illustration of sedimentation mechanisms associated with a river entering a stratified lake (Pharo and Carmack 1979)

mass movements are comparatively frequent. This is illustrated in Fig. 6.10. Mass movements will be discussed in the chapter on lake bottom dynamics.

The delta advance can be very rapid in lakes receiving large loads of material. This is eminently clear from Fig. 6.11, which illustrates the fast migration of the delta front in Lillooet Lake, British Columbia. This is a lake of an alpine landscape, which is covered to 6% by glaciers.

Fig. 6.11. The advance of the delta front in Lillooet Lake, British Columbia (Gilbert 1975)

6.2.1.2 *River Plume Sedimentation*

River plume dispersion, i.e., the flow and sedimentation of comparatively fine particles by the coherent plume of river water moving into a lake, does not behave like a jet. The river plume is instead influenced by the rotation of the earth (i.e., the Coriolis force), which deflects the plume along the right-hand shore of the lake (on the northern hemisphere) in the direction of flow. The spread of the plume depends on morphometrical features (bluffs, islands, etc.) and season of the year; depending on the density differences between the lake water and the river water, the plume may spread as a *surface* current, an intermediate depth *interflow* or a near-bottom *underflow*. The very complex interrelationships between lake stratification, river plume dispersion, lake bioproduction and seasonal variability create the interpretational framework for the structures of lake bottom deposits. This is a science in itself, which will only be briefly discussed in Chapter 7 (see, e.g., Reineck and Singh 1975, Friedman and Sanders 1978, Sturm 1979).

The grain size of inorganic materials and the rate of deposition generally decrease logarithmically with the distance from the mouth of the tributary within areas dominated by river action. This is illustrated with empirical data from Lake Ekoln, Sweden, in Fig. 6.12B. Subsequently, we will utilize data from Lake Ekoln as a reference to obtain a comprehensive framework for the given examples of lake sedimentological principles. Due to the Coriolis force, outflows generally lie to the right of the midline of a lake or bay. This is illustrated in Fig. 6.12C for the northern part of Lake Ekoln. The water entering Graneberg Bay flows towards the western side of the bay and then further south. The coarse particles will make up a higher proportion in the sediments near the mouth and on the western side than further out and on the eastern side of the basin (Figs. 6.12D, 6.13C). In addition, the greatest deposition of material (Figs. 6.12C, 6.13B) and pollutants, like mercury (Fig. 6.13D), follow the same general pattern. Consequently, there exists a close relationship between dominating hydrological flow pattern, rate of deposition and physical, chemical and contaminational character of the sediments within river-mouth areas. The areal and vertical variations in sediment character are, generally, greater in river mouth areas than in areas dominated by wind/wave action. This natural difference in character and variability may be used as one criterion to establish the borderline between these two sedimentological environments. Using the water content of surficial sediments (W_{0-1}) as a key parameter, the following approach can be adopted to illustrate the concepts:

1. It is well established (see Chap. 7.3.1.2) that a close relationship exists between the water depth (D) and the water content of surficial sediments (W_{0-1}); W_{0-1} increases with D, the variability in W_{0-1} is large in shallow waters and decreases with increasing depth.

2. Lake Ekoln has been divided into four sub-areas: A_1 at a distance of 0–0.75 km from the river mouth, A_2 at 0.75–1.5 km, A_3 at 1.5–3.0 km and A_4 at greater distances than 3 km. Within each sub-basin the correlation (correlation coefficient, r) between the water depth (D) and the water content (W_{0-1}) has been determined. The result is given in Fig. 6.14A. It can be seen that the variability in W_{0-1} for the given water depths is great close to the river mouth, i.e., the r-value increases steadily with distance from the mouth to a maximum at about 3 km.

3. Mean water contents ($\overline{W}_{0-1}$) in eight sections at various distances from the mouth of River Fyris have been calculated and compared with model data for Lake Ekoln. Figure 6.14B illustrates the model. The model data are obtained by putting the values of mean relative depth for every section (D_P) into the formula given in Fig. 6.14B. The characteristic water content (W_K) for Lake Ekoln is 83.5% and the

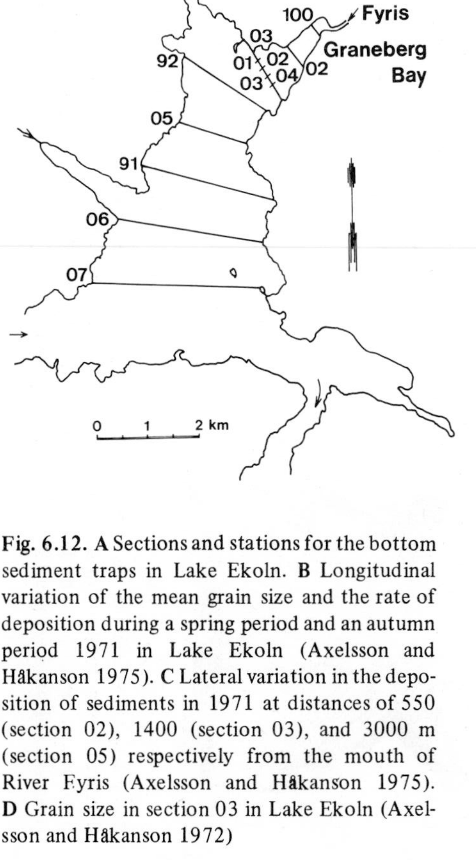

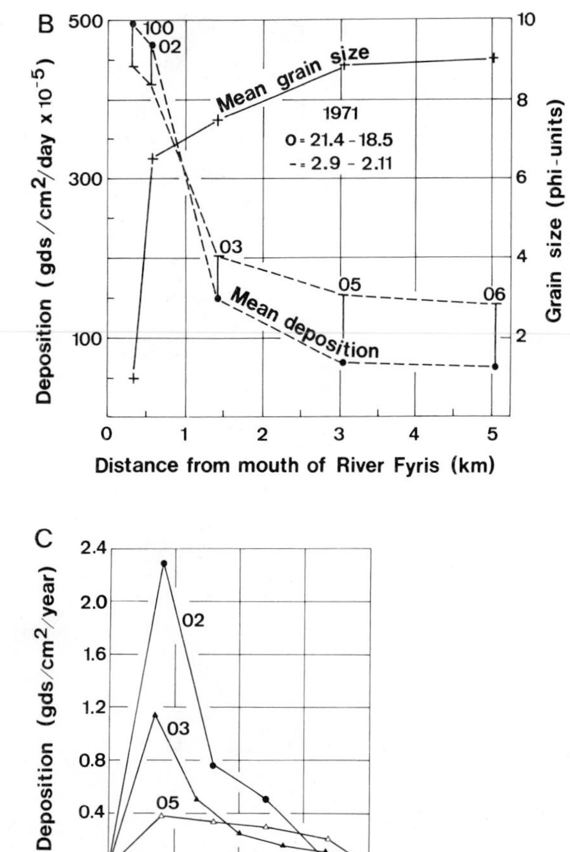

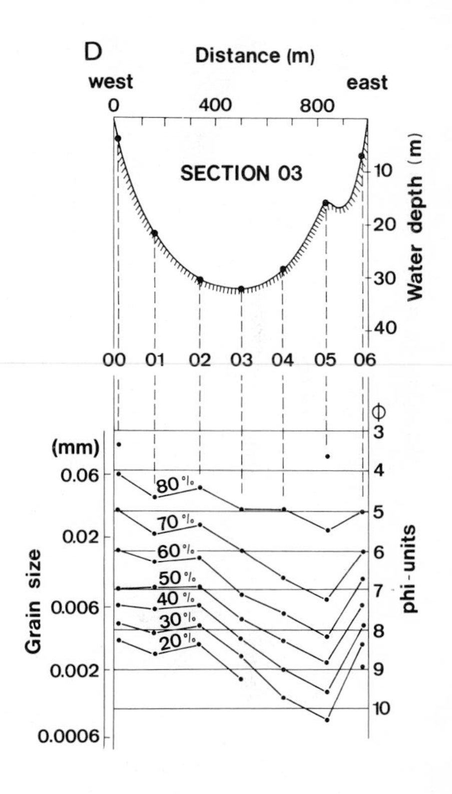

Fig. 6.12. **A** Sections and stations for the bottom sediment traps in Lake Ekoln. **B** Longitudinal variation of the mean grain size and the rate of deposition during a spring period and an autumn period 1971 in Lake Ekoln (Axelsson and Håkanson 1975). **C** Lateral variation in the deposition of sediments in 1971 at distances of 550 (section 02), 1400 (section 03), and 3000 m (section 05) respectively from the mouth of River Fyris (Axelsson and Håkanson 1975). **D** Grain size in section 03 in Lake Ekoln (Axelsson and Håkanson 1972)

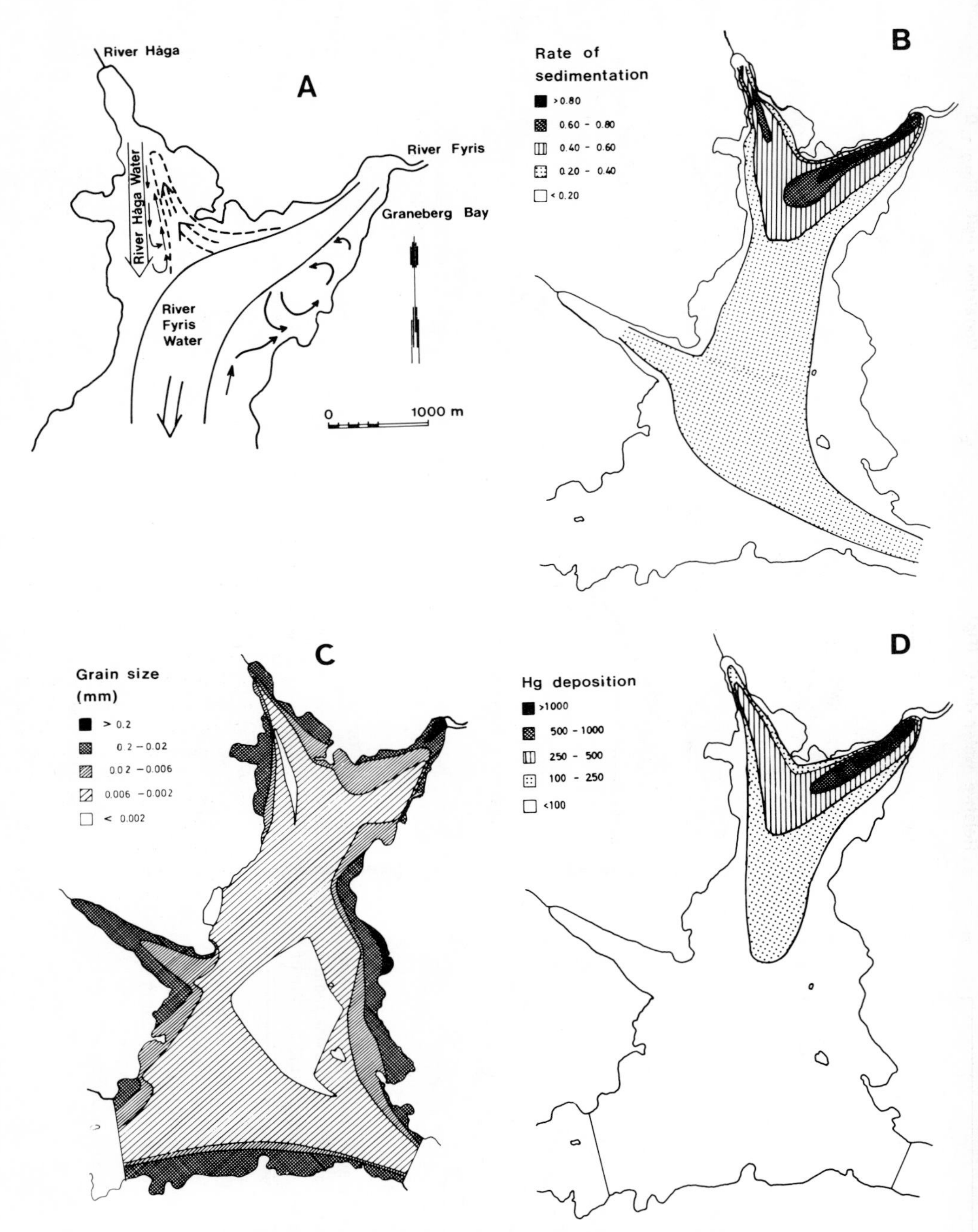

Fig. 6.13. A The dominating hydrological flow pattern in Lake Ekoln. **B** The areal variation in the rate of sedimentation (in g ds cm^{-2} yr^{-1}). **C** The areal distribution of the median grain size in surficial sediments (0–1 cm). **D** The areal variation in the deposition of mercury in ng cm^{-2} yr^{-1}. Mean values from 1971 and 1972 (Axelsson and Håkanson 1972, 1975)

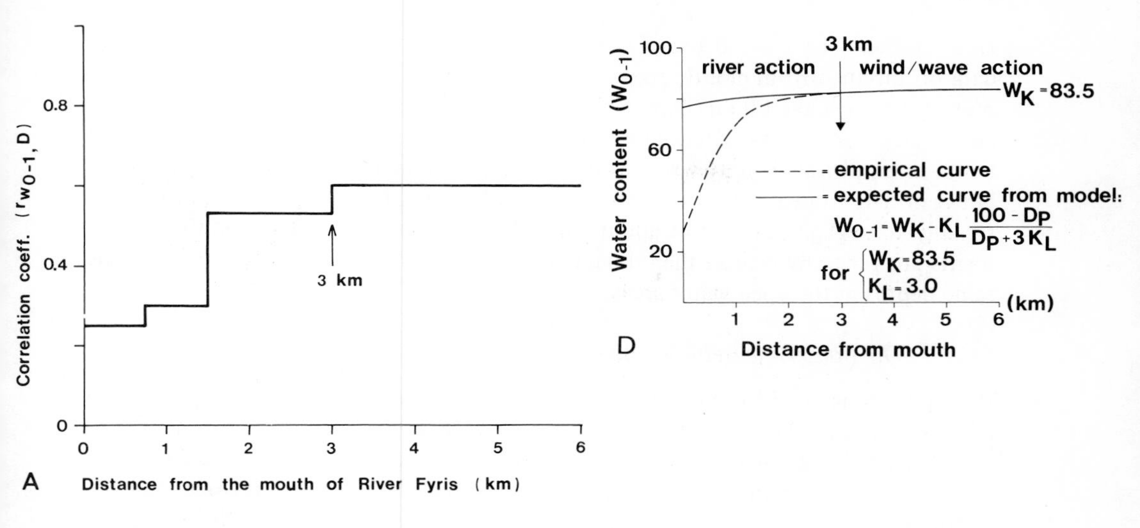

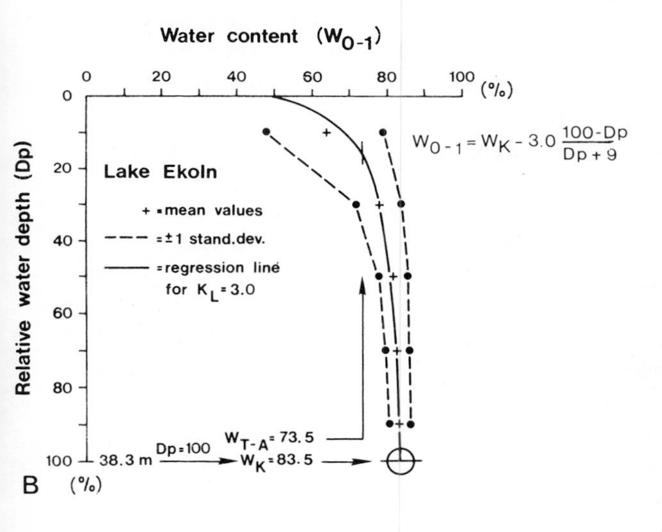

Fig. 6.14. A The variation in the correlation between the water content and the water depth within areas at different distances from the mouth of River Fyris in Lake Ekoln (Håkanson 1977a). **B** The relationship between the water content of surficial sediments (W_{0-1}) and the relative water depth (D_P) in Lake Ekoln (Håkanson 1981c). **C** Mean depths ($\bar{D}$), mean depths as a percentage of the maximum depth of Lake Ekoln (D_P), mean water contents ($\bar{W}_{0-1}$) and estimated water contents for the water depths in question (W_{0-1}) for eight sections at different distances from the mouth of River Fyris

Section	Distance	Water depth		Water content		Difference
		$\bar{D}$	Dp	$\bar{W}_{0-1}$	W_{0-1}	
	km	m	%			
02B	0	10	26.1	27.5	77.2	-49.7
02	0.55	14.3	37.5	79.8	79.4	+0.4
03	1.40	23.2	60.5	77.3	81.8	-4.5
92	1.85	23.6	61.7	79.9	81.9	-2.0
05	3.00	26.3	68.8	83.0	82.3	+0.7
91	3.90	24.2	63.3	82.8	82.0	+0.8
06	5.00	27.2	80	84.1	82.8	+1.3
07	6.00	21.4	56.0	82.4	81.5	+0.9

C

lake constant (K_L) is 3.0 (see Chap. 7.3.1.2). The table given as Fig. 6.14C gives data on distance from mouth, mean depth in meters and as a percentage of the maximum depth of Lake Ekoln (D_P), empirically determined mean water content (W_{0-1}) and water content as determined for the given model (W_{0-1}). The comparison between empirical data and model data is depicted in Fig. 6.14D. The difference between determined and expected values is great close to the river mouth but decreases steadily to negligible values at about 3 km. That is, close to the river mouth the sediments are generally coarser than should be expected, compared to sediments from the same depths in the open water areas.

6.2.1.3 The Borderline Between River Action and Wind/Wave Action

It may be concluded from the previous discussion concerning the borderline between river action and wind/wave action in Lake Ekoln that the position of this limit (Lu) will depend on, among many things, the water discharge of the tributary (Q), the stretch of the river flow in the lake, which depends on the Coriolis force, and the width and depth conditions of the basin. Subsequently, we will introduce a simple model which accounts for the major causal relationship regulating the position of Lu. This problem is related to the problem of defining and determining the mixing zone (see Brock Neely 1982). The following assumptions and simplifications of the complex reality will be utilized (see Fig. 6.15):

- assuming a conical shape of the river-mouth area, where river action dominates the sedimentological processes;
- assuming that the stretch of the river flow in the lake is deflected by the Coriolis force to the right (in the northern hemisphere) and that the maximum velocity (the trajectory) is along the dotted line at a distance of B/3 from the shore to the right of the flowing water;
- assuming that the 2/3 of the area is *directly* dominated by the river plume and 1/3 of the area is *indirectly* dominated by the river plume, due to the compensating inflow currents from the lake which would not appear if the river had not existed;
- assuming steady-state conditions, i.e., that the water transport in the river mouth (Q_0) is equal to the water transport at the distance (1) from the mouth;

we obtain:

$$Q_0 = u_0 \cdot A_0 = u(1) \cdot A(1), \tag{6.14}$$

where Q_0 = tributary water discharge $[L^3\ T^{-1}]$;
u_0 = tributary flow velocity $[LT^{-1}]$;
A_0 = cross-sectional area of river $[L^2]$;
$u(1)$ = flow velocity of river water at distance 1 from the mouth (along the stretch) $[LT^{-1}]$;
$A(1)$ = cross-sectional area of river plume at distance 1 from the mouth $[L^2]$.

But A(1) can be expressed in two ways in terms of morphometric parameters:

$$A(1) = C_1 \cdot B \cdot D \tag{6.15}$$

and

$$A(1) = C_2 \cdot 1^2 \cdot \text{tg}\alpha. \tag{6.16}$$

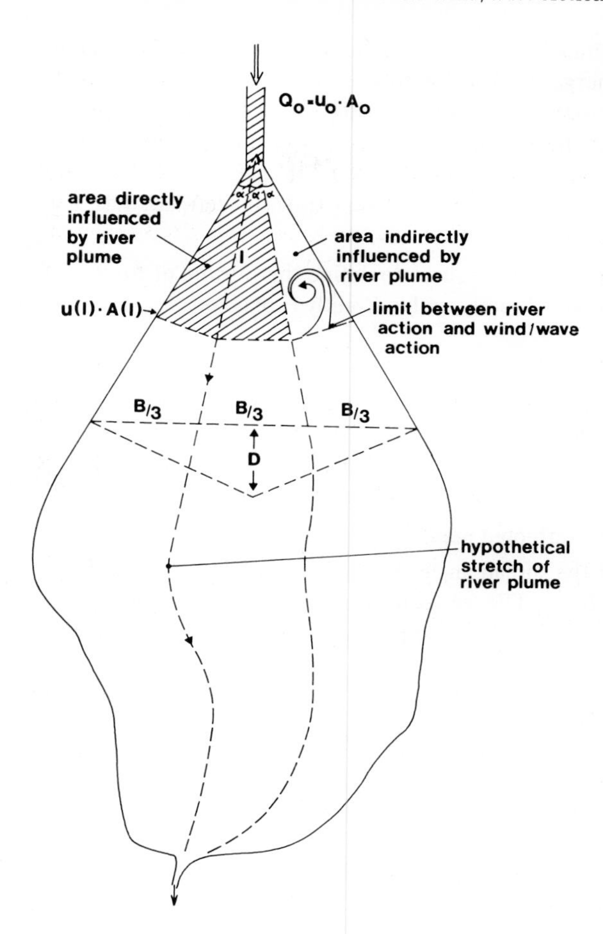

Fig. 6.15. Schematical illustration of concepts related to the limit between areas dominated by river action and by wind/wave action

That is:

$$A^2(1) = C_3 \cdot B \cdot D \cdot 1^2 \cdot tg\alpha \quad (6.17)$$

or

$$A(1) = C_4 \cdot 1 \cdot \sqrt{B \cdot D \cdot tg\alpha}\,, \quad (6.18)$$

where B = a width measure [L];
D = a depth measure [L];
α = a spread angle of the river plume (degrees);
1 = a length measure along the stretch of the river plume [L];
C_i, for i = 1, 2, 3, 4 = empirical constants linked to the definitions of the given morphometric parameters.

If Eq. (6.18) is put into Eq. (6.14), we get:

$$Q_0 = u(1) \cdot C_4 \cdot 1 \cdot \sqrt{B \cdot D \cdot tg\alpha} \quad (6.19)$$

or

$$u(1) = \frac{C_5 \cdot Q_0}{1 \cdot \sqrt{B \cdot D \cdot tg\alpha}}\,. \quad (6.20)$$

This is a general equation describing how the flow velocity of the river plume is positively related to the water discharge (Q_0) and negatively or reversely related to the length, width, depth and the horizontal (not vertical) spread component of the plume. Equation (6.2) describes the complex causal relationship determining the limit between river action and wind/wave action in a very simple way. To be useful in practice the given variables [B, D, α, u(1)] must be explicitly defined; 1 is, e.g., defined in Fig. 6.15.

- For practical reasons, it may often be adequate to define the B-value of Eq. (6.20) as the mean width of the affected (directly and indirectly) area, i.e.:

$$B = \bar{B}(1) = \frac{a(1)}{1}. \tag{6.21}$$

- Analogously, the D-value is given as:

$$D = \bar{D}(1) = \frac{V(1)}{a(1)}, \tag{6.22}$$

  where V(1) = the volume of the affected area.
- The horizontal component of the spread angle (α) can be defined, for example, by the tangential lines in Fig. 6.16. The vertical component is neglected in this approach. α, B and D depend on the definition of the affected area, i.e., the position of the requested limit (Lu).
- The mean high water discharge (MHQ), denoted Q_0^m , can, e.g., be used to represent the influence from the tributary.

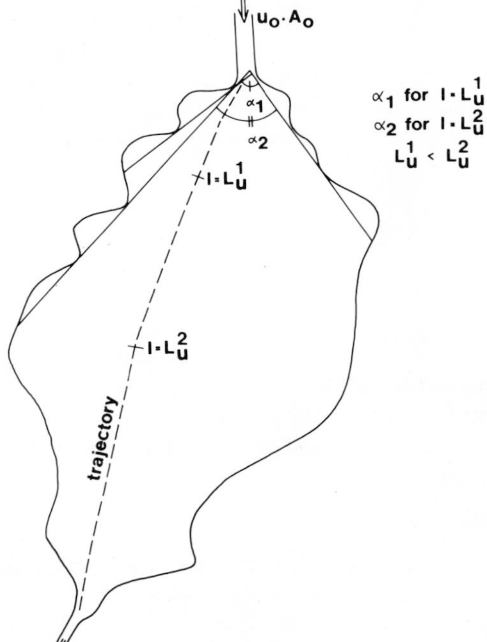

Fig. 6.16. Illustration of one way to define the spread angle (α) from the river mouth

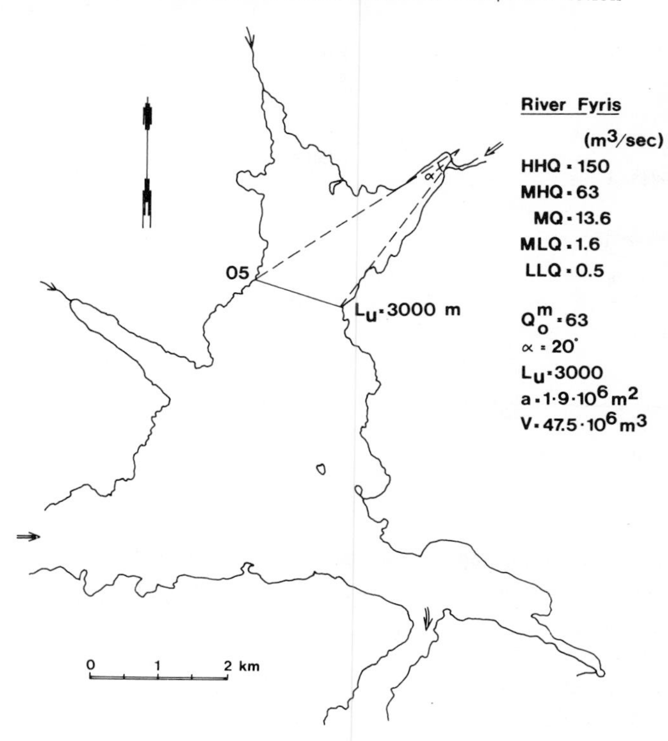

Fig. 6.17. Map of the river mouth area of River Fyris and certain hydrological data (e.g., *HHQ* highest high water discharge) relevant for determining the area influenced by river action

– Finally, the requested sedimentological limit between areas dominated by river action and wind/wave action (Lu) can be defined by the variability and character of the surficial sediments; in the area directly influenced by the river plume, the sediments are generally coarser than in the open water areas; in the indirectly influenced area the sediments would normally be finer than expected at given water depths. The given limit (Lu) can be visualized to be where the velocity of the tributary water, u(1), attains such a low value that it has lost its character in relation to the lake water, i.e., u(1) = u(Lu).

Studies in Lake Mendota (Haines and Bryson 1961) have shown that the residual surface velocity, i.e., the surface current velocity in the absence of wind, is about 5 cm s^{-1}. The velocity (u) of waves is given by (see Smith and Sinclair 1972):

$$u = \frac{\pi \cdot H}{T} \cdot e^{-2 \cdot \pi \cdot D/\lambda}, \tag{6.23}$$

where H = wave height [L];
T = wave period [T];
D = water depth [L];
λ = wave length [L].

The wave height (H) depends (see Chap. 7.2.3) on the duration and velocity of the wind and the fetch (a measure of the area affected by wind/wave action). Velocities higher than 10 cm s^{-1} are needed to move unconsolidated fine particles (see Fig. 6.18). Such velocities can only prevail at comparatively high wind speeds, high fetch values and shallow water depths (see Chap. 7.1). Subsequently, we use a flow velocity

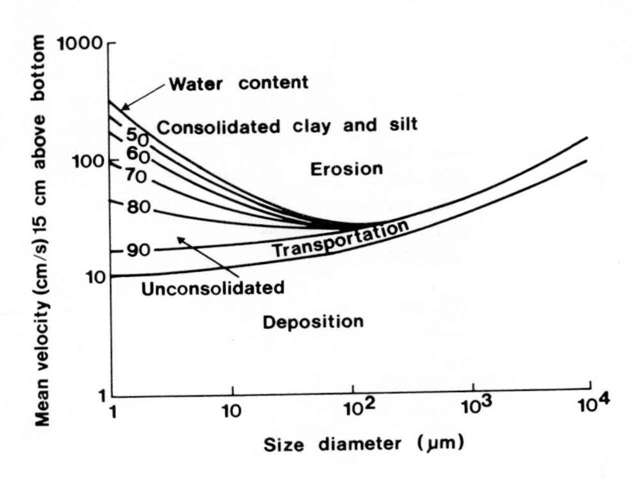

Fig. 6.18. Erosion, transportation and deposition (= accumulation) velocities for different grain sizes. Possible values for various stages of consolidation, as given by the water content, are indicated (Postma 1967)

of 10 cm s^{-1} as a measure of the requested critical value (Lu), where river action ceases to dominate the processes of sedimentation. Thus, from these presuppositions, Eq. (6.20) may be altered to:

$$\mathrm{Lu} = \frac{C_6 \cdot Q_0^m}{10 \cdot \sqrt{\bar{B}(\mathrm{Lu}) \cdot \bar{D}(\mathrm{Lu}) \cdot \mathrm{tg}\,[\alpha(\mathrm{Lu})]}} \,. \tag{6.24}$$

From the available set of empirical data from Swedish lakes, the sedimentological-morphometrical constant, C_6, can be estimated to be about 35,000.

For Lake Ekoln, where Lu = 3000 m, $Q_0^m = 63\ m^3\ s^{-1}$, $\alpha = 20°$, $a = 1.9 \cdot 10^6\ m^2$ and $V = 47.5 \cdot 10^6\ m^3$ (see Fig. 6.17), the equation becomes:

$$u(3000) = \frac{35{,}000 \cdot 63}{3000 \cdot \sqrt{630 \cdot 25 \cdot 0.36}} = 9.8.$$

This value of 9.8 is close to the ideal of 10.

When Lu is not known, it may be estimated accordingly, utilizing data from Kamloops Lake, British Columbia (see Pharo and Carmack 1979).

$Q_0^m = 1500\ m^3\ s^{-1}$, $\alpha = 10°$, $a = 52.1 \cdot 10^6\ m^2$ (for the entire lake),
$V = 3.7 \cdot 10^9\ m^3$ (entire lake). This gives for 1 = 25,000 (entire lake):

$$u(25{,}000) = \frac{35{,}000 \cdot 1500}{25{,}000 \cdot \sqrt{2080 \cdot 71 \cdot 0.18}} = 12.9.$$

This implies that the whole of Kamloops Lake (see Fig. 6.19) is dominated by river action from a sedimentological point of view. It should be emphasized that this model only describes the major causal relationships in the area dominated by river action; it should be used with due considerations to topographical obstacles and with care, since further empirical data and field studies may alter the constant, improve the accuracy of the approach, reveal limitations, and account for the vertical perspective.

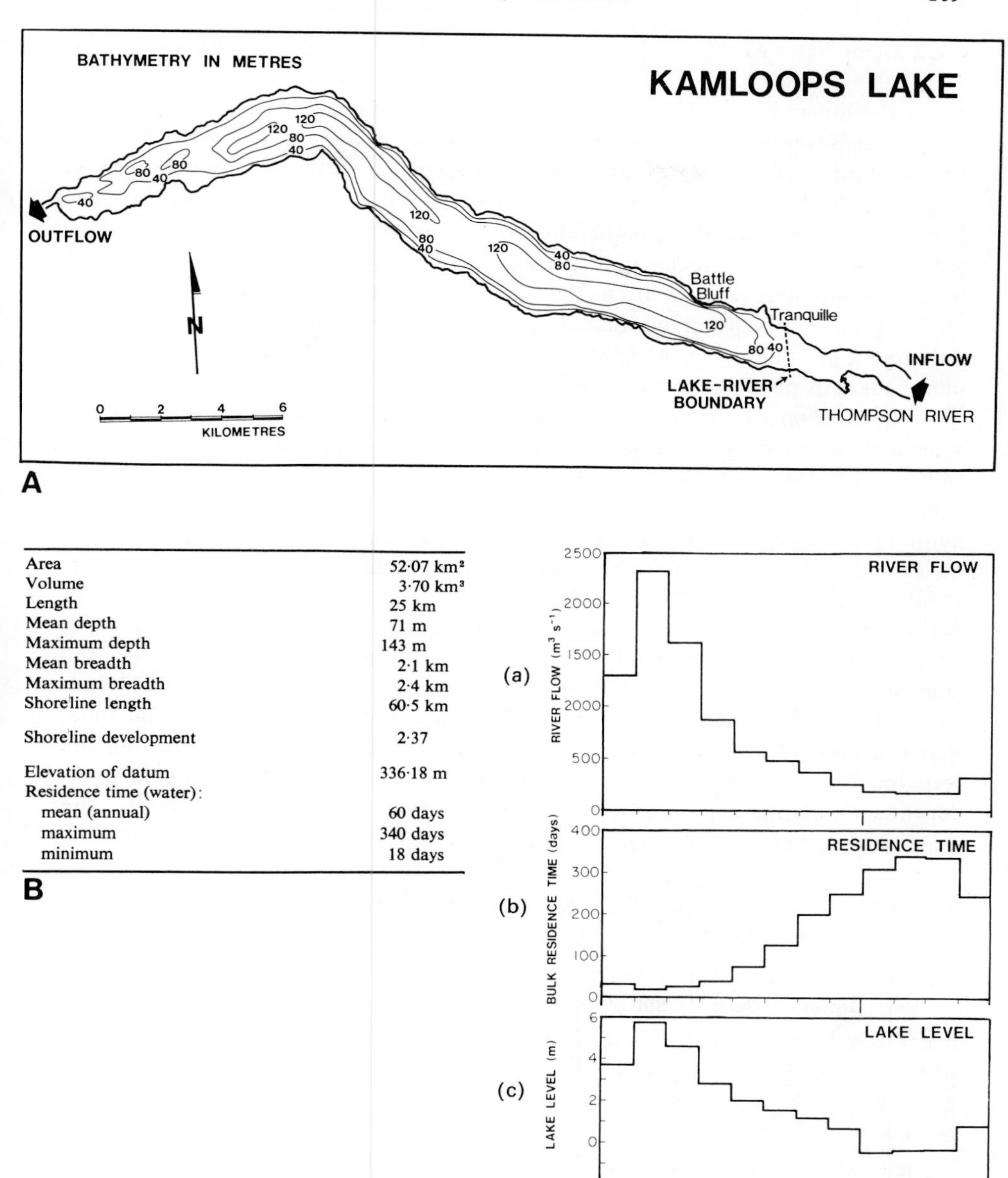

Area	52·07 km²
Volume	3·70 km³
Length	25 km
Mean depth	71 m
Maximum depth	143 m
Mean breadth	2·1 km
Maximum breadth	2·4 km
Shoreline length	60·5 km
Shoreline development	2·37
Elevation of datum	336·18 m
Residence time (water):	
mean (annual)	60 days
maximum	340 days
minimum	18 days

Fig. 6.19. A Bathymetry of Kamloops Lake, British Columbia. **B** Morphometrical data. **C** Hydrological characteristics: *a* Annual mean streamflow; *b* Residence time; *c* Lake level relative to datum elevation (336 m above sea level) (Pharo and Carmack 1979)

6.2.2 Open Water Areas

From a sedimentological horizon, it is important to distinguish between *autochthonous* materials (produced in the lake) and lakes/lake areas dominated by such deposits, i.e., primarily the open water areas, and *allochthonous* materials (from tributaries) and river-mouth areas dominated by such materials. The rate of deposition is generally higher in river-mouth areas than in open water areas, and higher in lakes dominated by allochthonous sedimentation than in, also highly productive, lakes dominated by authochthonous sedimentation; the mean annual rate of deposition is, e.g., about 20,000 g m^{-2} in oligotrophic Kamloops Lake (Pharo and Carmack 1979), while it is only 635 g m^{-2} in highly eutrophic Lake Esrom (Lastein 1976), which only receives minor amounts of allochthonous materials.

Sedimentation within open water areas is dominated by the finer particles/aggregates with Reynolds number less than 0.5. The flow around such particles is laminar and the horizontal component much larger (up to 6 times) than the vertical velocity component. Thus, the sedimentation of such particles is highly dependent on the hydrological flow pattern, on whether the lake is stratified or not, on winds and waves, etc. This means that the processes of sedimentation are complicated to understand, physically describe and empirically measure. In the fluvial environment, in rivers, there exists a simple and measurable energy parameter, the water velocity (u), which may be related directly to a simple and measurable sediment parameter, the grain size (d), yielding a curve on the critical deposition velocity [see the basic Hjulström (1935) diagram depicted in modified form in Fig. 6.20]. In the littoral zone, it is generally adequate to use the grain size as a sediment response factor, but the water energy factor is less easily characterized; here a whole set of indirect and interdependent variables must be used in an integrated form, including wave height, the velocity, direction and duration of the wind and the effective fetch. Some of these relationships are depicted in Fig. 6.21. If we leave the littoral zone for deeper waters, the picture becomes even more complicated. In deep water areas in lakes, where cohesive fine materials dominate, it may not be meaningful to utilize the grain size

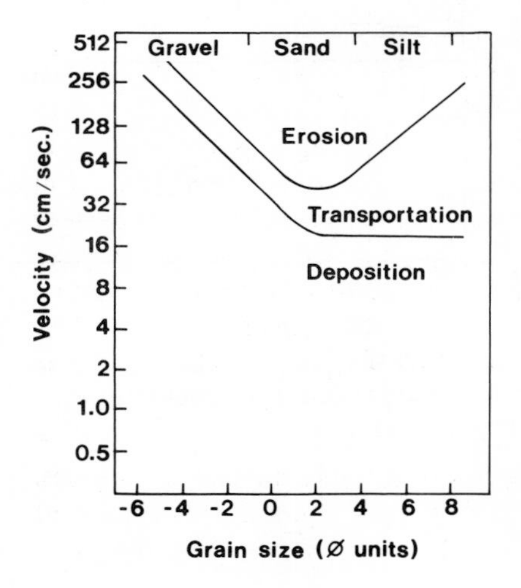

Fig. 6.20. The Hjulström (1935) curve, as modified by Sundborg (1956), illustrating the relationship between critical erosion *(upper curve)* and critical deposition *(lower curve)* velocities for various grain sizes for unidirectional flow conditions

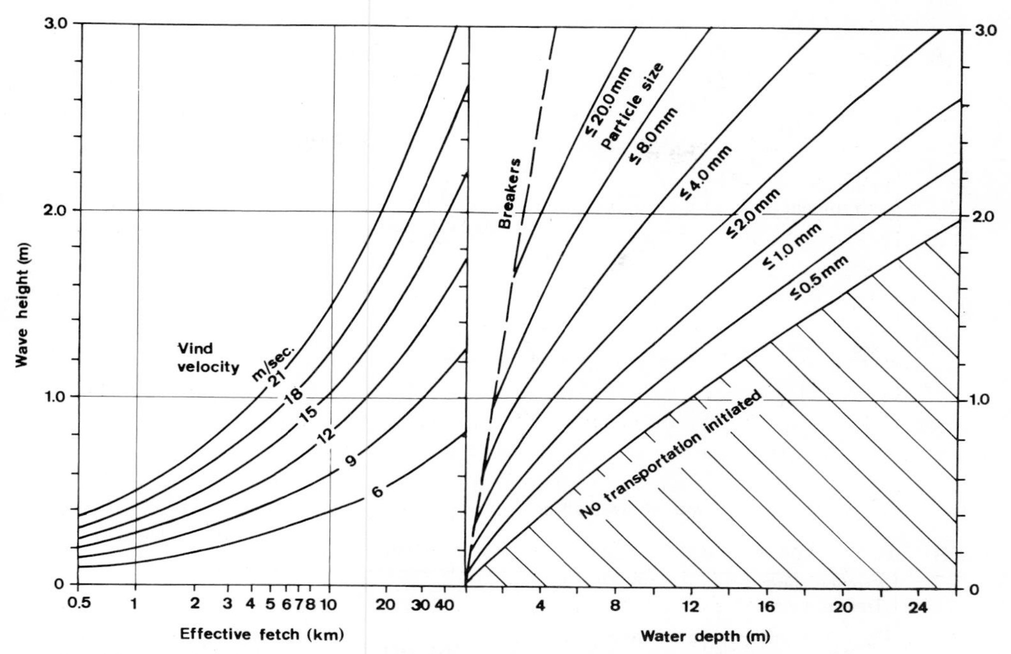

Fig. 6.21. The relationship between effective fetch (L_f) and wave height and between wave height and bottom dynamics in the shore zone. (After Norrman 1964)

of the inorganic particles or aggregates as the only or even the most accurate linkage between water- and bottom dynamics (see Chap. 7). Thus, it is often fruitful to differentiate even the basic concepts of erosion (= entrainment), transportation and accumulation (= deposition) in open water areas in lakes and in river-mouth areas, since in the latter case all types of processes can be interacting at the same time – such as the simultaneous accumulation of sand, transportation of silt and erosion of detrital particles.

It should be stressed that although promising attempts have been made, it is generally difficult and/or demanding to distinguish allochthonous and autochthonous fractions (see R.F. Wright et al. 1980) and net deposition from total deposition (i.e., net plus resuspension, see Andersen and Lastein 1981, Dominik et al. 1981, Ludlam 1981). Generally, the rate of deposition (v) increases with water depth in open water areas in lakes; from zero or negative values in shallow, high energy zones to a maximum value in the deepest part of the lake. This is illustrated in Fig. 6.22 for two different lakes for which the rate of deposition has been determined with two different methods, but the basic sedimentological pattern is still the same – a linear increase in v with water depth.

1. Bob Lake (see Evans and Rigler 1980) is a relatively small Canadian lake (area = 2.3 km^2, mean depth = 17.8 m, max. depth = 65.0 m) with several minor tributaries. The sedimentation pattern is primarily determined by lake depth and not prevalent hydrodynamical circulation patterns related to, e.g., Coriolis force or wind influences.

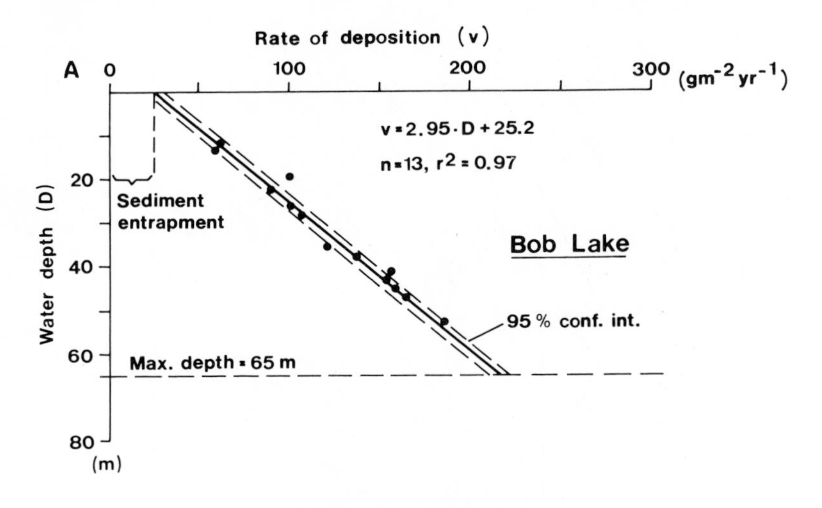

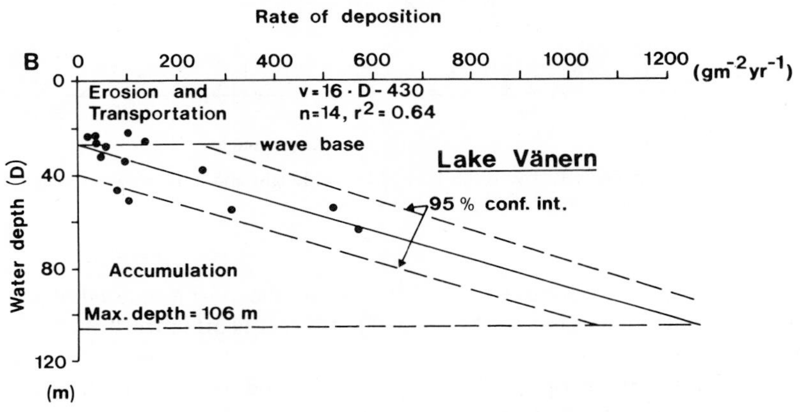

Fig. 6.22A, B. The relationship between rate of deposition and water depth: **A** In a small lake, Bob Lake, Canada (redrawn from Evans and Rigler 1980). **B** In a large lake, Lake Vänern, Sweden (Håkanson 1975)

From Fig. 6.22A, it is evident that depth alone explains 97% ($r^2 = 0.97$) of the variation between the sites in rate of deposition, which in this case was determined by lead-210 dating. If the regression line ($v = 2.95 \cdot D + 25.2$) is extrapolated to the x-axis, D = 0 gives v = 25.2 g m^{-2} yr^{-1}, which is larger than zero. Such relationships between rate of deposition and water depth are typical for small lakes, rather unaffected by wind/wave action, with an areally limited zone of hard and coarse sediments in shallow waters, where instead vegetation (reed) and/or topographically isolated enbayments can entrap suspended particles/aggregates.

2. Lake Vänern (see SNV 1978), on the other hand, is a very large lake, the fourth largest in Europe (area = 5648 km^2, mean depth = 27.0 m, max. depth = 106 m), where the sedimentation pattern in the open water areas (archipelagos excluded) is highly influenced by the hydrological flow pattern and wind/wave action. Figure 6.23A

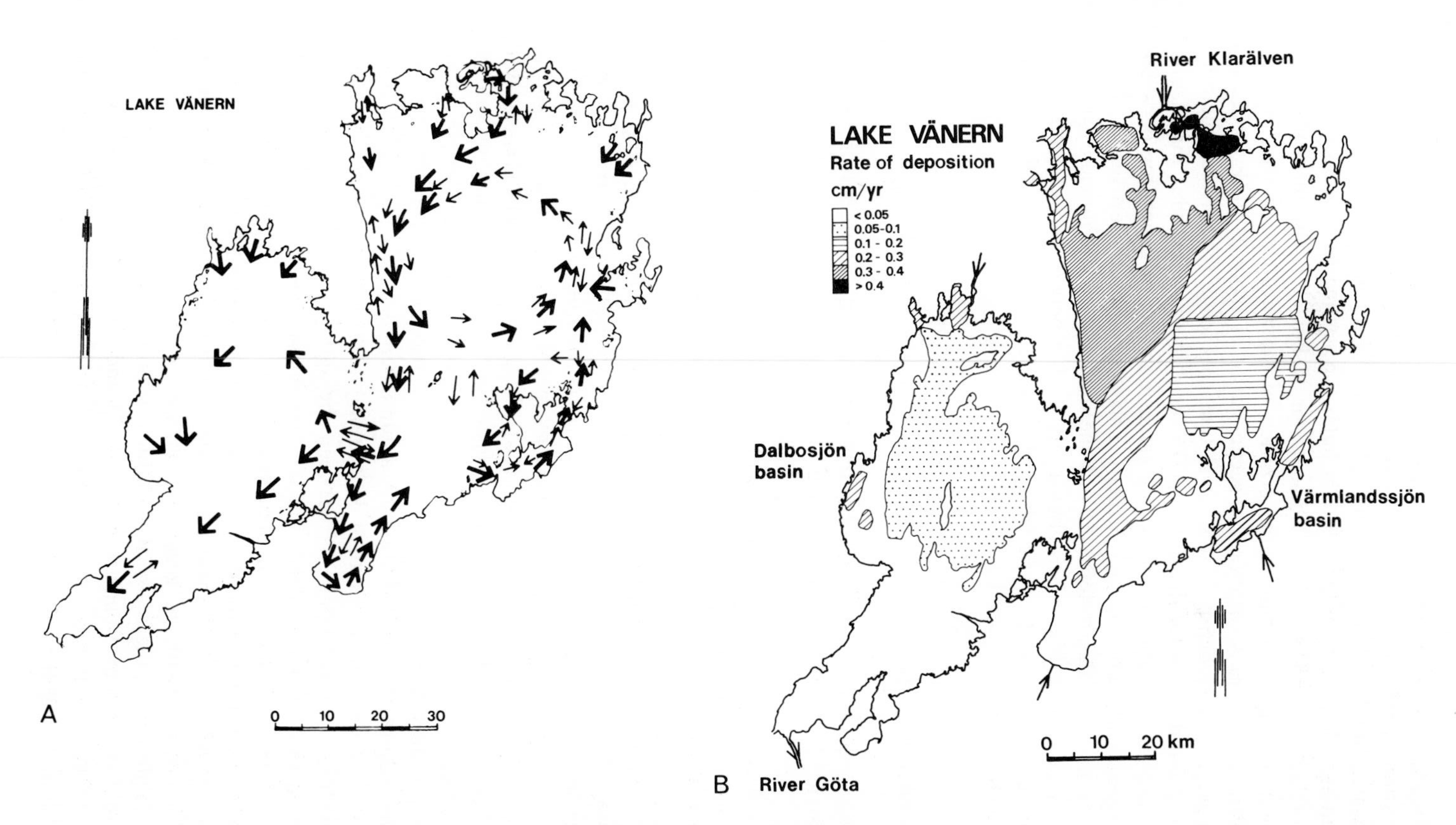

Fig. 6.23. A Schematical hydrological flow pattern in Lake Vänern. *Heavy arrows* average conditions; *thin arrows* flows of short duration (Lindell 1975). **B** The areal variation in rate or deposition in Lake Vänern (Håkanson 1978)

illustrates schematically the general circulation pattern in Lake Vänern and Fig. 6.23B the areal variation in rate of deposition. From these two figures, we can see that the dominating water circulation in each major sub-basin (Värmlandsjön and Dalbosjön) constitutes an anti-clockwise cell, which distributes the settling particles in a typical pattern, reflecting the flow of water. Sedimentation is, however, also influenced by wind/wave action, which determines the position of the wave base or the "critical limit" between areas of transportation and accumulation at about 27 m (Fig. 6.22B), and by topography/water depth – the correlation coefficient (r) between the rate of deposition (v) and the water depth (D) is 0.8, or $r^2 = 0.64$, implying that depth alone would account for about 64% of the variance in rate of deposition between sites within open water areas in Lake Vänern. This is significantly less than in Bob Lake. The data on rate of deposition in Lake Vänern have been determined by means of bottom sediment traps. Fig. 6.22B gives a relationship between water depth, rate of deposition and wave base (= "critical depth"), which is fairly typical for open water areas in comparatively large lakes.

6.2.3 Temporal Variations

Also from the viewpoint of temporal variations in rate of deposition, it is fruitful to distinguish areas dominated by river action from areas dominated by wind/wave action.

We will again utilize data from Lake Ekoln to illustrate a general temporal distribution pattern for river-mouth areas. Figure 6.24 illustrates the variation in river sediment transport (suspended load) and rate of deposition in a section (03) at 1.4 km distance from the mouth of River Fyris, well within the area dominated by river action ($<$ 3 km). During this particular year (1971), the sediment transport in River Fyris was lower than normal, but the general pattern with a marked peak in connection with the spring-flood in April is typical for rivers from this hydrological regime. The rate of deposition, in this case measured by means of bottom sediment traps, was high in connection with the spring-flood (note the slight lag phase) and especially high during September and October, when resuspension triggered by autumn storms caused a redistribution of sediments from shallow areas (erosion and transportation zones) to deep waters below the wave base or the "critical depth"; note also the low values on river sediment transport during this autumn period.

There is a typical minimum in rate of deposition just prior to the spring flood and a second much less marked minimum between the spring and the autumn peaks, which is related to the summer stagnation. The reason why the summer minimum is not nearly as deep as the winter minimum is connected with a high autochthonous production during the summer months and ample possibilities for wind/wave-induced resuspensional activities – both these factors are of minor importance during winter when the lake is covered with ice (generally from late November to late March).

Figure 6.25 illustrates the seasonal variations in rate of deposition and primary production in Lake Esrom, Denmark, during 1971 and 1972. The primary production and sedimentation (in sediment traps suspended 1.5 m above the bottom) show low values during the winter. Maximum sedimentation and production occur during

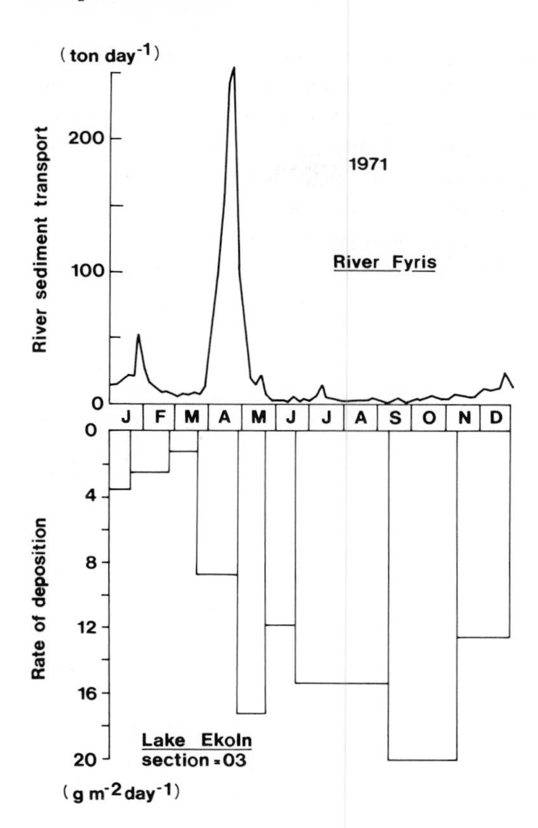

Fig. 6.24. The relationship between river sediment transport (River Fyris) and rate of deposition within the river-mouth area (section 03 at 1.4 km from the river mouth) during 1971 (Axelsson and Håkanson 1975, Axelsson 1980)

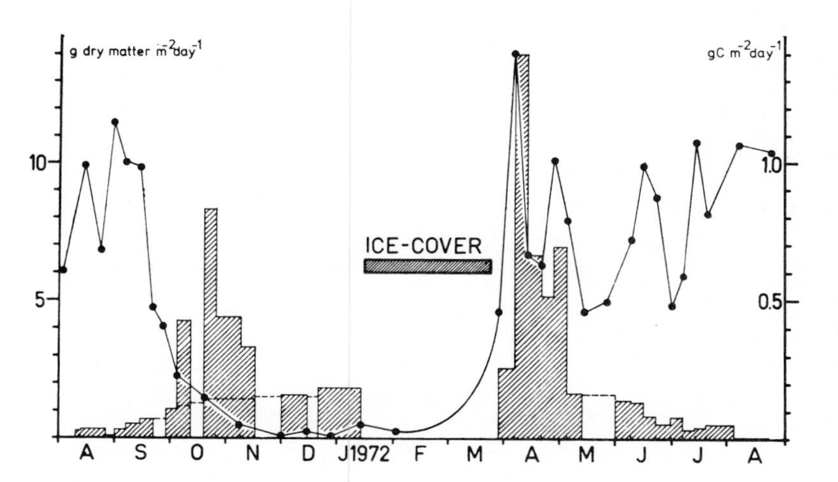

Fig. 6.25. Temporal variation in primary production and sedimentation in Lake Esrom, Denmark, during 1971 and 1972. *Solid lines* total sedimentation; *dotted lines* data corrected for resuspension and the *curves connecting dots* signify measurements of phytoplankton primary production (Lastein 1976)

and immediately following the spring bloom of algae (April–May); another peak occurs in connection with the autumn overturn in October. More than 70% of the material trapped during the autumn circulation emanates from resuspension. Lowest values of both curves occur during winter; a second minimum, especially for the rate of deposition, is apparent after the spring bloom during the summer months, when the lake is stratified. Thus, it is evident that the hydrological conditions, e.g., ice cover, stratifications, overturns and prevailing flow regime, play a fundamental role for the geographical aspects of sedimentation in lakes.

7 Lake Bottom Dynamics

7.1 Definitions

This section penetrates the relationships regulating bottom dynamics in lakes. Basic concepts will be defined in this introduction. The focus is first on the processes determining the fate of sediments after primary deposition on the lake bed, i.e., on resuspension, entrainment, turbidity, currents, then on wind/wave influences and topographical influences on bottom dynamics, and finally on methods to determine bottom dynamics.

The causal relationships in question are schematically illustrated in Fig. 7.1. Subsequently, we will adopt the following nomenclature concerning bottom dynamics in lakes (Håkanson 1977b):

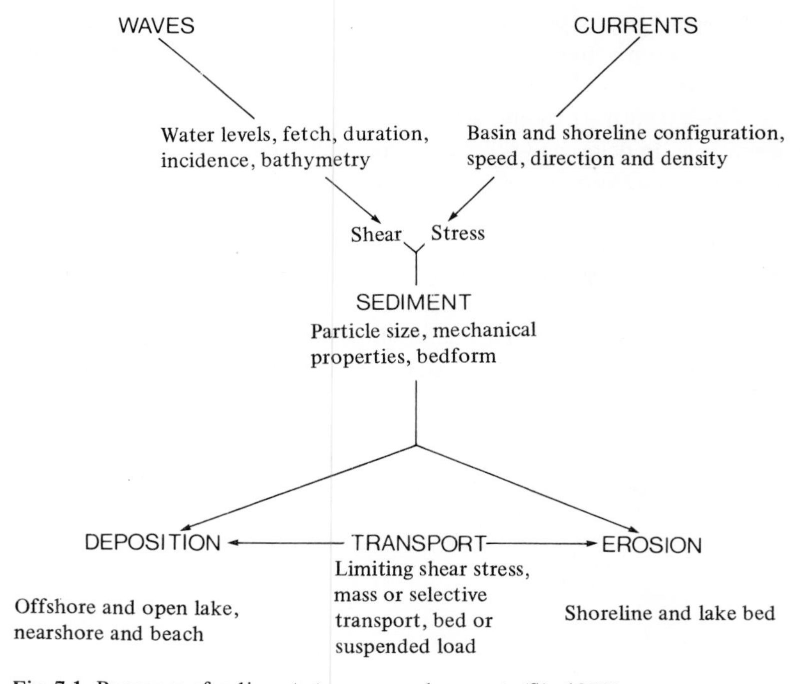

Fig. 7.1. Response of sediments to waves and currents (Sly 1978)

- *Accumulation* areas prevail where fine materials, medium silt with grain sizes less than 0.006 mm, can be deposited *continuously*.
- *Transportation* zones appear where there is a *discontinuous* deposition of fine particles/aggregates, i.e., where periods of accumulation (often rather long during winter and summer) are interrupted by periods of resuspension/winnowing/transportation (generally of short duration in connection with periods of water turnover and storms).
- *Erosion* areas prevail where there is no deposition of *fine* materials.

Areas of erosion are most frequent in shallow waters and characterized by hard or consolidated deposits – from bare rocks, gravel and sand to glacial clays. The sediments within the areas of transportation are, for natural reasons, generally very variable – from sand to loose mud. The deposits within the areas of accumulation are always comparatively loose, with high water- and organic contents and sometimes with a high content of pollutants. Figure 7.2 summarizes the most important sedimentological and bottom dynamic processes in lake. Chapter 8 will focus on bioturbation (indicated with a circle in Fig. 7.2) and the processes within the sediments.

It should be stressed that these definitions emphasize the importance of the fine materials and the distribution of fine materials, whereas the traditional definitions for the fluvial environment (see Fig. 6.20 and, e.g., Sundborg 1956), the aeolian milieu (see, e.g., Bagnold 1954) or the littoral zone (see, e.g., Norrman 1964, Muir Wood 1969) focus on sandy (non-cohesive) materials for which the grain size [d in mm or phi (ϕ)-units] may be used as a measure of capacity of entrainment. In deep water areas in lakes, where cohesive fine materials dominate, it is generally not meaningful to utilize the grain size of the inorganic particles as the only or even the most accurate link between water dynamics/velocity and bottom dynamics/entrainment (see, e.g., Terwindt 1977, J.B. Fisher et al. 1979, Fukuda and Lick 1980, McCall and J.B. Fisher 1980). It should also be noted that with these definitions we may have, for example, in deltas and other environments with high hydromechanical energy, simultaneous deposition of sand, transportation of silt and erosion of detrital particles.

Knowledge of lake bottom dynamics is essential in most sedimentological contexts, in matters related to geographical distribution and potential ecological effects of contaminants, and in many practical issues, such as choice of adequate dumping sites for dredging deposits and choice of sampling sites from aquatic pollution control. Marked relationships exist between bottom dynamic conditions, sediment physics, sediment chemistry and sediment pollution (Table 7.1), and sediment biology (Fig. 7.3). For example, pollutants appear predominantly in the loose, fine, organic-rich deposits characteristic for most areas of accumulation, while in transportation areas the physical, chemical, contaminational and biological variations may be large.

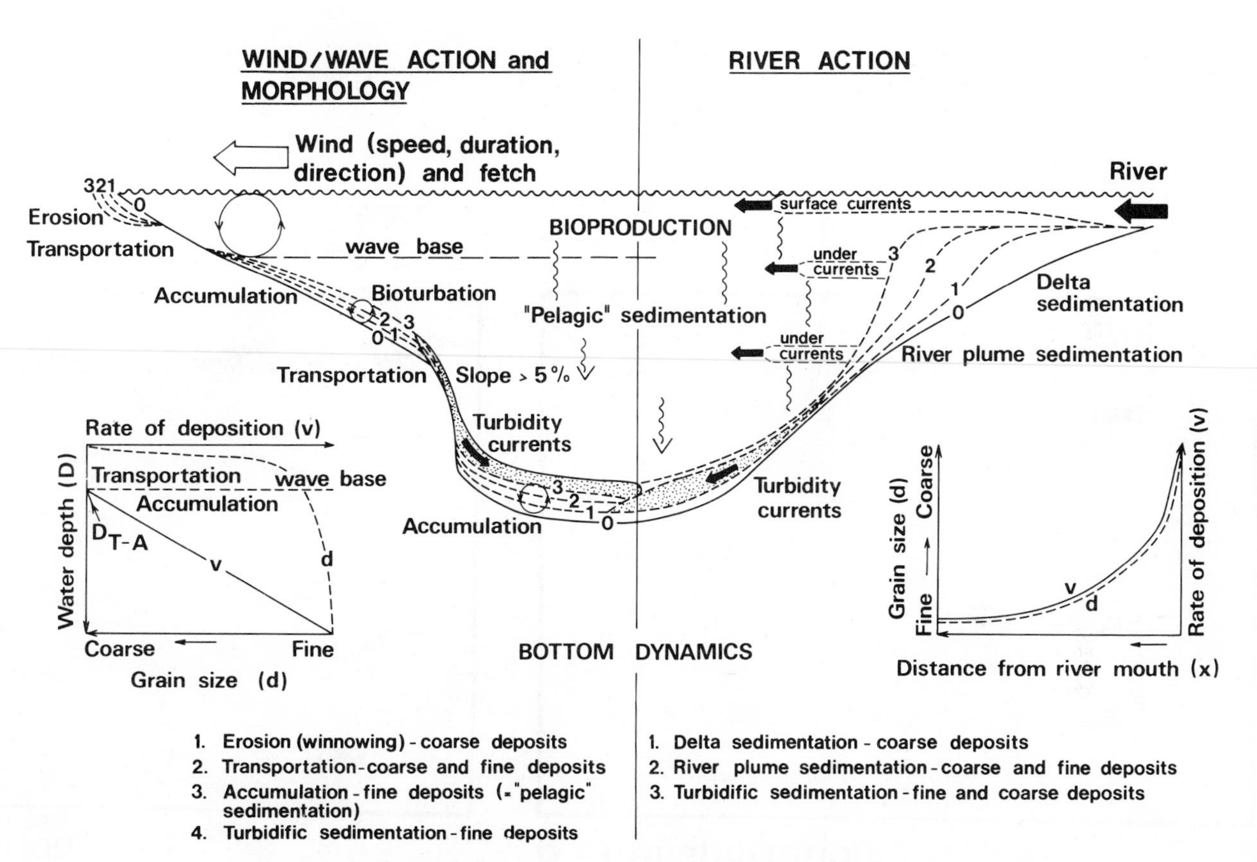

Fig. 7.2. Schematic illustration of major sedimentological and bottom dynamic processes in lakes (Håkanson 1982b)

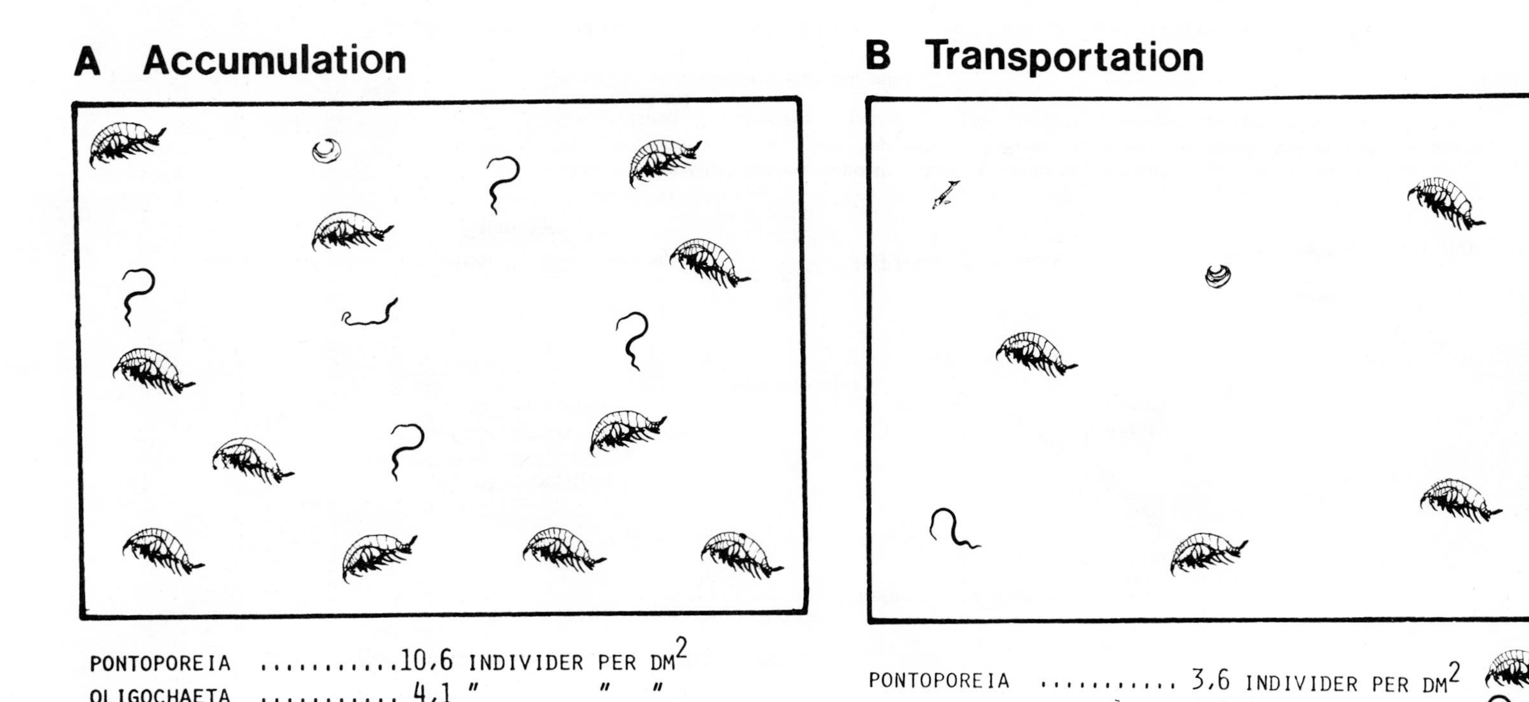

Fig. 7.3A, B. Average number of organisms on 1 dm^2 of bottom surface within representative lake areas in Lake Vänern (Värmlandssjön). **A** Accumulation areas and **B** transportation areas (Wiederholm 1978)

Table 7.1. The relationship between bottom dynamics (erosion, transportation, accumulation) and the physical and chemical character of the surficial sediments of Lilla Ullevi Bay, Lake Mälaren, Sweden. Mean values and standard deviations (in brackets) (Håkanson 1980a, raw data from Ryding and Borg 1973)

Number of analyses	Erosion 15	Transportation 10	Accumulation 14
Physical status			
Water depth. m	13.0 (5.3)	17.5 (5.4)	31.6 (8.0)
Water content. % ws	32.6 (9.0)	67.4 (9.6)	94.1 (2.3)
Bulk density. g cm^{-3}	1.72 (0.15)	1.26 (0.10)	1.03 (0.02)
Organic content. % ds	4.6 (2.2)	10.7 (4.6)	24.3 (2.5)
Chemical status			
Nutrients and indicators of nutrient status			
Nitrogen. mg g^{-1} ds	0.6 (0.4)	3.4 (1.2)	10.7 (1.5)
Phosphorus. mg g^{-1} ds	0.8 (0.4)	2.8 (2.1)	1.6 (0.5)
Carbon. mg g^{-1} ds	0.5 (0.5)	2.7 (2.0)	10.4 (1.7)
Chlorophyll. μg g^{-1} ds	5.3 (4.2)	18.5 (9.4)	167.1 (45.5)
Not contaminating, chemically mobile elements (see also P)			
Iron. mg g^{-1} ds	24.6 (10.4)	53.5 (14.4)	41.3 (3.2)
Manganese. mg g^{-1} ds	0.8 (0.8)	3.5 (2.6)	2.5 (1.5)
Contaminating elements			
Zinc. μg g^{-1} ds	41 (19)	111 (27)	189 (17)
Nickel. μg g^{-1} ds	23 (8)	40 (8)	57 (10)
Copper. μg g^{-1} ds	18 (9)	31 (13)	59 (6)

7.2 Processes of Resuspension

7.2.1 Entrainment

From Fig. 6.18, which illustrates erosion, transportation and deposition in terms of grain size, water content of sediments and mean water velocity 15 cm above bottom, one might gain the impression that grain size and water content are the only factors regulating the capacity for entrainment of sediments. This is, however, far from being the case. Figure 7.4 shows the relationship between critical entrainment stress, τ_e, and sediment water content. There is no simple relationship between the physical sediment parameters water content and/or grain size and the critical entrainment rate for *cohesive* materials (see Partheniades 1972). Thus, Fig. 6.18 can only provide order of magnitude type of information and the problem of entrainment of cohesive sediments must include a parameter expressing directly or indirectly, the "glue" properties of the deposits (McCall and J.B. Fisher 1979, Fukuda and Lick 1980). For example, tubificids may pelletize the sediments and increase the median sediment grain size and settling velocity by up to two orders of magnitude (Fig. 7.5). However, if the measurement of particle diameter is replaced by a corresponding measure, as determined from fall velocity and density (Stokes' law), possibilities to obtain more

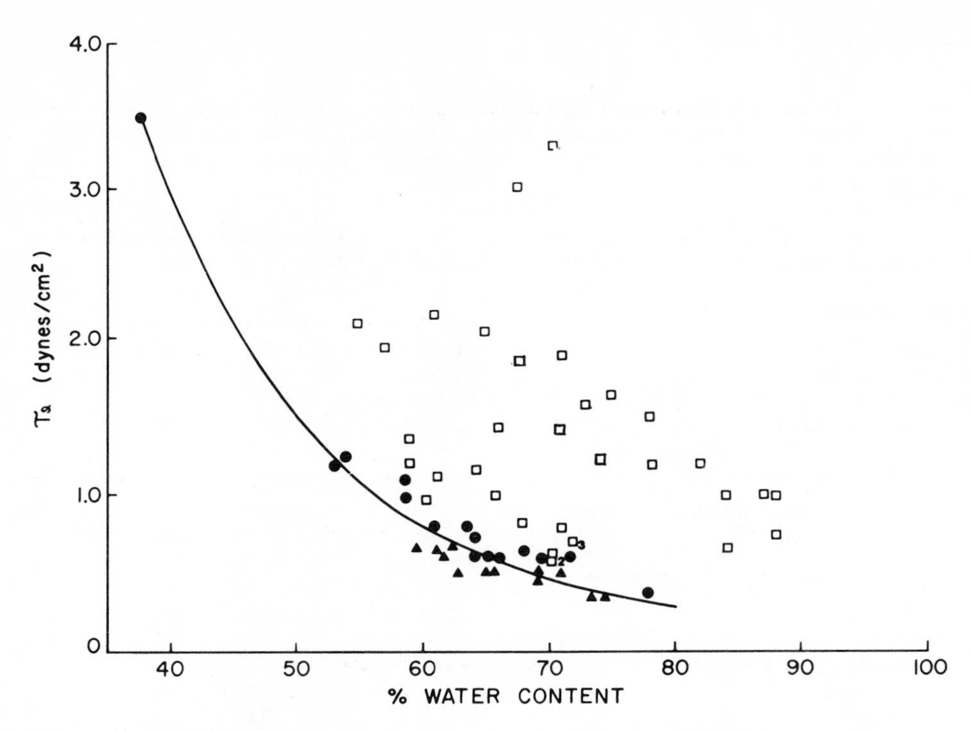

Fig. 7.4. Critical entrainment stress, τ_e, of oxidized box cores as a function of sediment water content. *Closed circles* box cores of shale based sediments. *Closed triangles* runs made in flume experiments with the entire flume covered with shale-based sediments. *Open boxes* box cores collected from locations in Lake Erie (McCall and J.B. Fisher 1980)

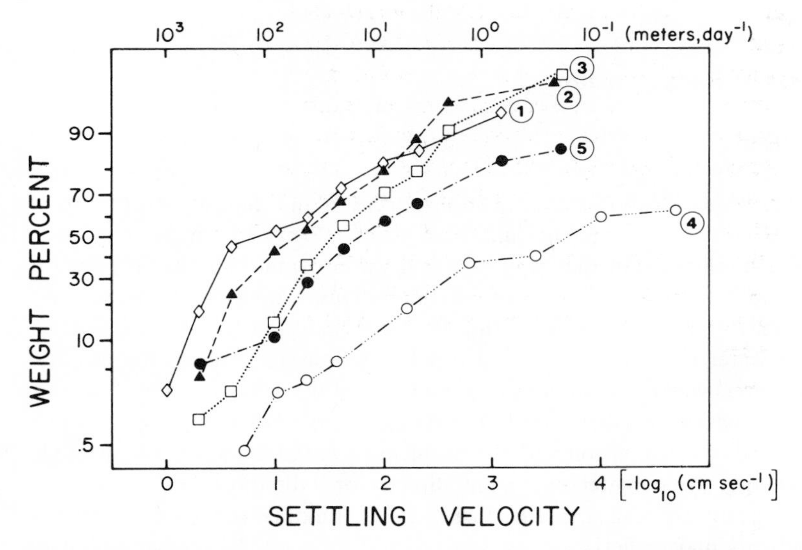

Fig. 7.5. Cumulative curves showing settling velocity distribution of sediments from western Lake Erie. *1* top 1 mm of sediments settling in lake water; *2* top 10 mm of sediments in lake water; *3* degraded pellets in lake water; *4* top 10 mm of lake sediments with blender plus dispersant treatment in distilled water; *5* top 10 mm of sediments in distilled water plus dispersant (McCall and J.B. Fisher 1980)

adequate criteria for entrainment can be obtained (see Fisher et al. 1979). The critical entrainment stress of pelletized fine sediments may be several times greater than unpelletized deposits. The opposite effect can be seen for pelletized sandy sediments, which may be more easily eroded than non-pelletized sand because the tubificids feed selectively on the finer fractions (McCall and J.B. Fischer 1979). Figure 7.6 gives a schematic illustration of the structure of sediments. It should be stressed that the complexities concerning entrainment of fine cohesive sediments in lakes, and the problems concerning the definition and measurement of the "glue" properties, has not yet been sufficiently explored. The problem of entrainment of fine sediments and the problem of resuspension, i.e., to differentiate between net and total deposition in lakes, are interrelated and important but largely unsolved. Excellent papers on the hydraulics of entrainment have been given by Smith (1975) and Fukuda and Lick (1980). Subsequently, we will only give a formula describing the principles and relationships between some crucial factors determining entrainment for *sandy* materials:

$$\tau_e = c' \cdot \rho_s \cdot (\beta \cdot d)^2 \cdot \left(\frac{du}{dz}\right)^2, \tag{7.1}$$

where τ_e = the critical stress (drag force) $[ML^{-1}\ T^{-2}]$;
c' = a constant (0.013);
ρ_s = density of particles $[ML^{-3}]$;
β = a measure of spacing between particles; a constant between 14–22;
d = particle diameter [L];
$\frac{du}{dz}$ = change of velocity (u) with distance (z) above the sediment bed $[T^{-1}]$.

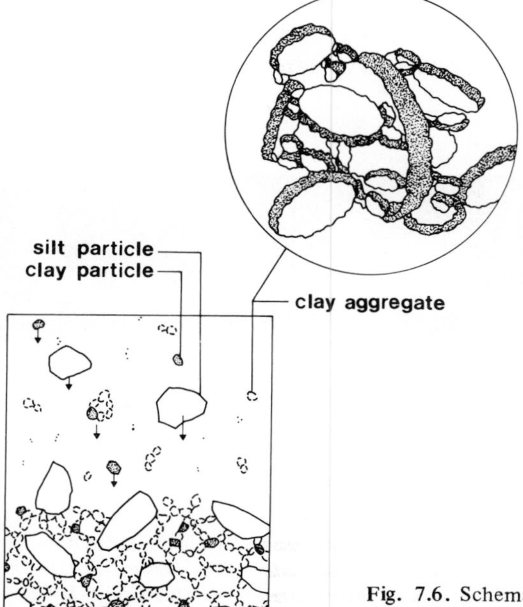

Fig. 7.6. Schematic illustration of the structure of lake sediments (Hallberg et al. 1978)

No analogous formula is, however, available for cohesive lake sediments for which the d-term of Eq. (7.1) should be replaced by a function of, at least, grain size, water content and a "glue" measure.

7.2.2 Turbidific Sedimentation

The literature on turbidific sedimentation, i.e., on turbidity currents and sub-aquatic mass movements, is very comprehensive from marine environments but scanty from lakes. Nomenclature and definitions vary, and here we will primarily follow the terminology given by Friedman and Sanders (1978) in their chapter on subaqueous gravity-displacement processes. The next section (7.3) will focus on resuspension caused by wind/wave influences.

The following types of turbidific processes can be distinguished for lakes:

1. Turbidity currents (or sediment density surges, density currents or suspension currents, see Fig. 7.7). Turbidity currents are episodic down-slope movements of sediment-laden water which redeposit material predominantly in delta areas. Pharo and Carmack (1979) distinguish this bottom process from river-plume dispersion because turbidity currents are short-lived, relocate material initially deposited in a metastable condition and have material concentrations that dominate the fluid density and hence also the down-slope drive (see Fig. 6.10). River-plume sedimentation, on the other hand, has a relatively continuous nature, involves fresh allochthonous materials and can be spread as a surface overflow, and intermediate depth interflow or a near-bottom underflow depending upon the density stratification of the lake relative to the density of the river water. Turbidity currents can attain different shapes (see, e.g., Lüthi 1980). The mechanics and geometry of turbidity currents can be studied in laboratory tank experiments (see Fig. 7.7). A turbidity current may be divided into a head and a main body (Fig. 7.8), which have different celerities or speeds;

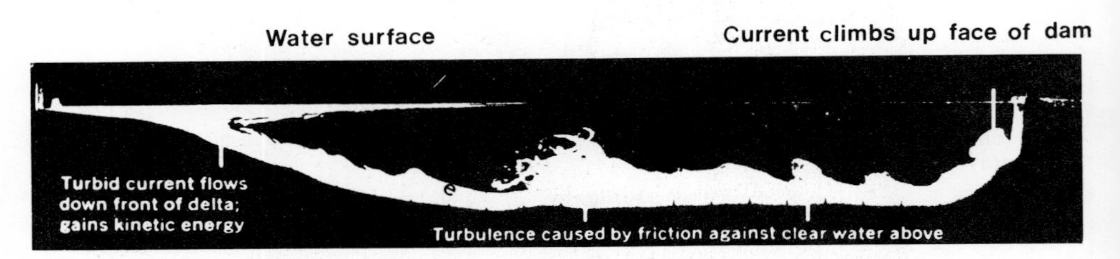

Fig. 7.7. Experimental turbidity current in glass-sided flume. *Left* where the turbidity current plunges beneath the clearer water, the water surface has been depressed (U.S. Soil Conservation Service, Cal. Tec., Friedman and Sanders 1978)

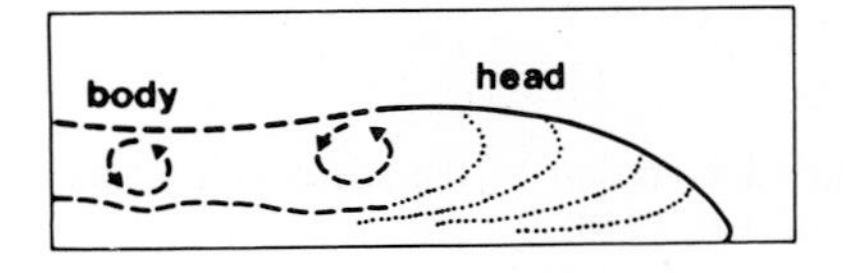

Fig. 7.8. Schematic profile of turbidity current showing head, body and flow paths *(dashed lines with arrows).* Current travelling from left to right (Middleton 1966)

the speed of the head is less than that of the body of the flow. The speed of the head may be expressed as:

$$u_2 = C_D \cdot (\Delta\rho \cdot g \cdot h_2)^{1/2}, \tag{7.2}$$

where U_2 = speed (celerity of the head) $[LT^{-1}]$;
C_D = drag coefficient (defined as the ratio of the actual measured drag force to the hypothetical force (see Smith 1975);
$\Delta\rho$ = difference in density between turbidity current and surrounding water;
g = acceleration due to gravity (981 cm s^{-2}) $[LT^{-2}]$;
h_2 = thickness of the head [L].

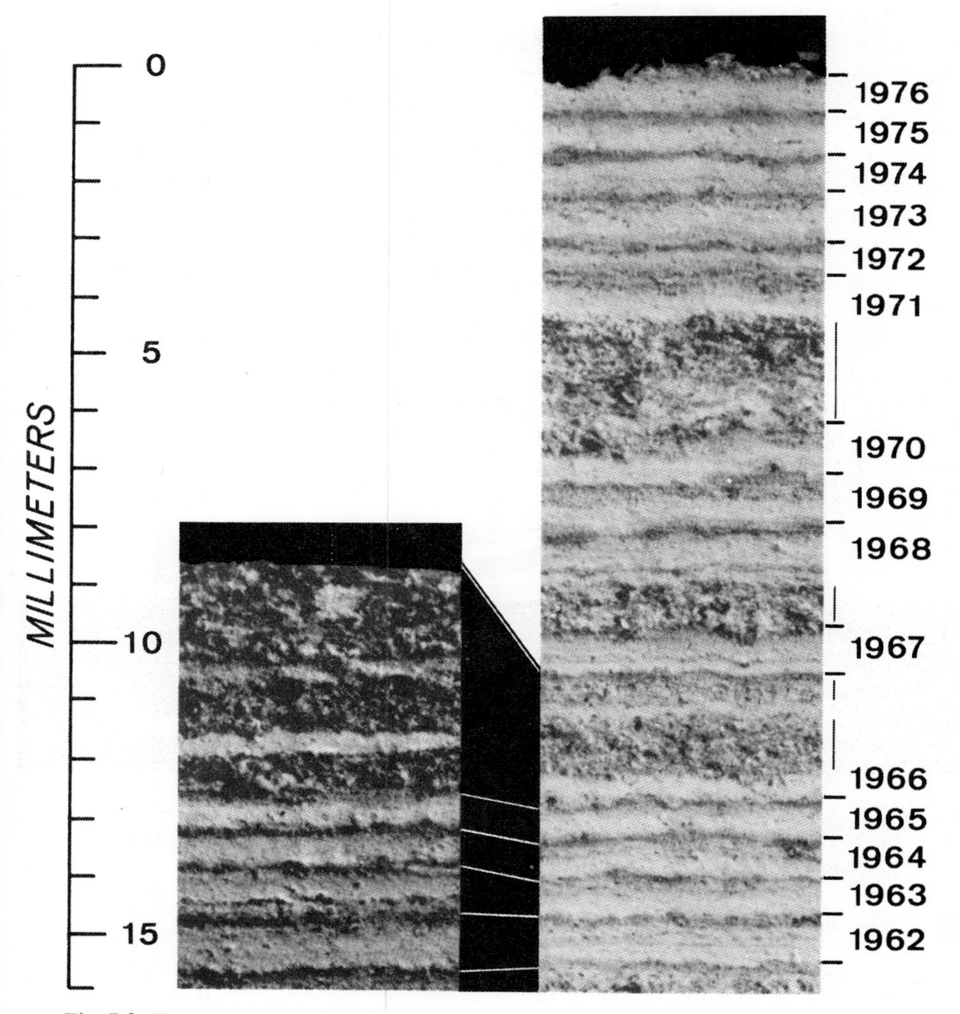

Fig. 7.9. Two vertical profiles of laminated dated sediments from Fayetteville Green Lake, New York (Ludlam 1981)

2. Subaqueous Slumps and Debris Flows. "By subaqueous slump we refer to the process involving the downslope translocation of a body of sediment along a curved surface of displacement . . ." (Quotation from Friedman and Sanders 1978). These concepts are not very well documented in limnetic environments and, probably, of rather limited interest from the perspective at this book.

Turbidity currents, on the other hand, may have a highly significant influence on lake sediments and lake sedimentological processes. This will be illustrated by two subsequent figures. The first is from Fayetteville Green Lake, New York (Ludlam 1981), which is a small (0.258 km^2) and deep (52.5 m) lake with a mean slope of 18°42′.

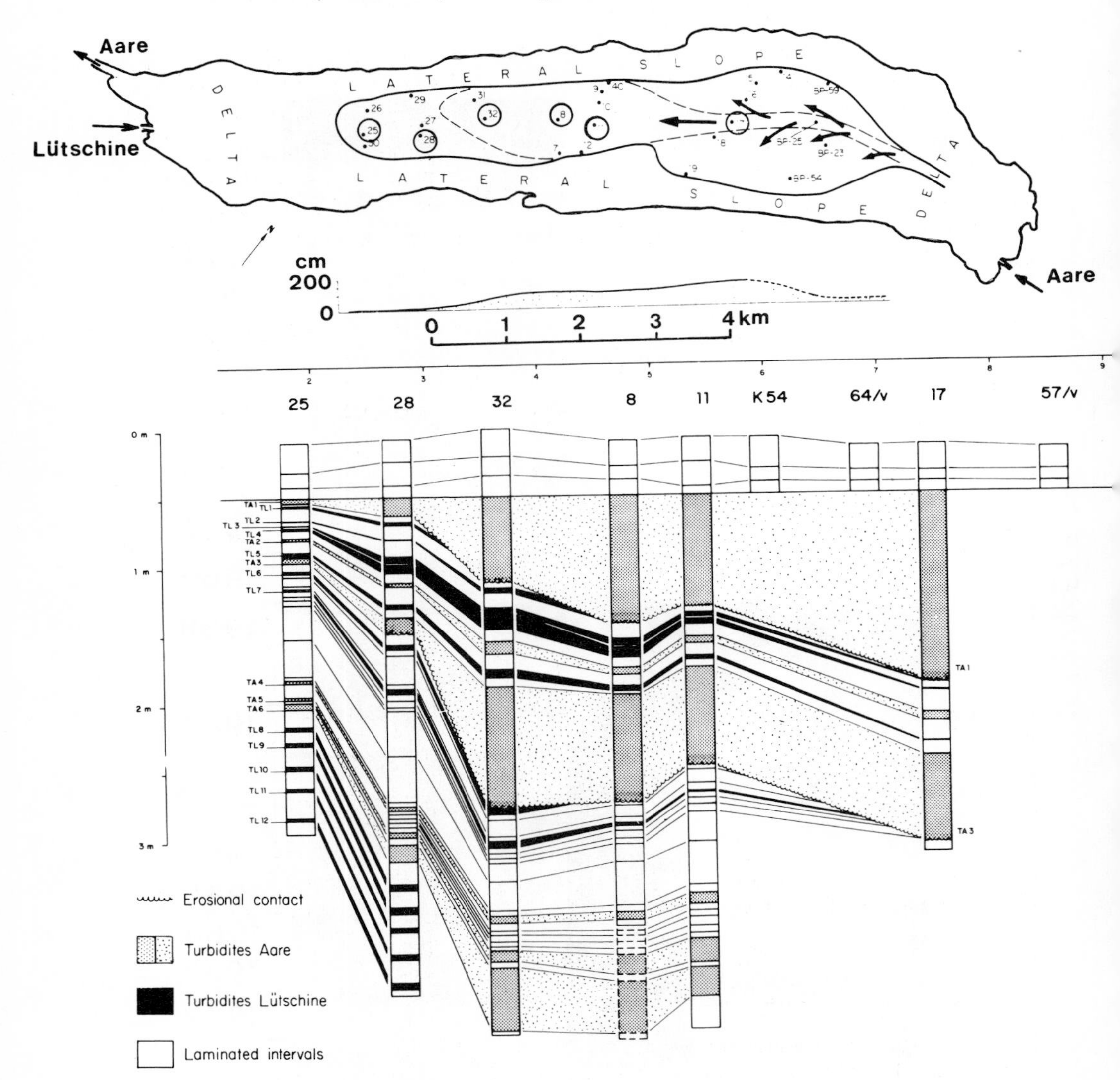

Fig. 7.10. Correlations of turbidities in a longitudinal profile of Lake Brienz, Switzerland (Sturm and Matter 1978)

The water is anaerobic below the chemocline at about 18 m. Because of the steep basin slopes, turbidity currents are frequent. This is illustrated with vertical lines on the right side of Fig. 7.9. Since oxygen is absent, there is no bottom fauna and bioturbation, which means that laminated sediments, reflecting primary sedimentological events, can be formed. The mean annual varve thickness is about 0.7 mm in dried sediment; pale lamina is deposited during spring and early summer; darker lamina during the rest of the year. Turbidities can be deposited at any season of the year and can be separated from pale, "ordinary", lamina by a greater content of organic particles from the littoral zone, a larger amount of sand and greater thickness (up to 5 cm in dry sediment).

The other example is from Lake Brienz, Switzerland (Sturm and Matter 1978), which is a fjord-like, oligotrophic lake with an area of 30 km^2 and a maximum depth of no less than 261 m. The lake sedimentation is controlled by two main tributaries – Aare River (mean water discharge, $\bar{Q}$ = 33 m^3 s^{-1}) and Lütschine River ($\bar{Q}$ = 19 m^3 s^{-1}). Both rivers have formed deltas; laminated mud and graded sand appear on the central basin plain; turbidity currents may transport great amounts of materials throughout most of the lake (Fig. 7.10). The character of the lamina depends on

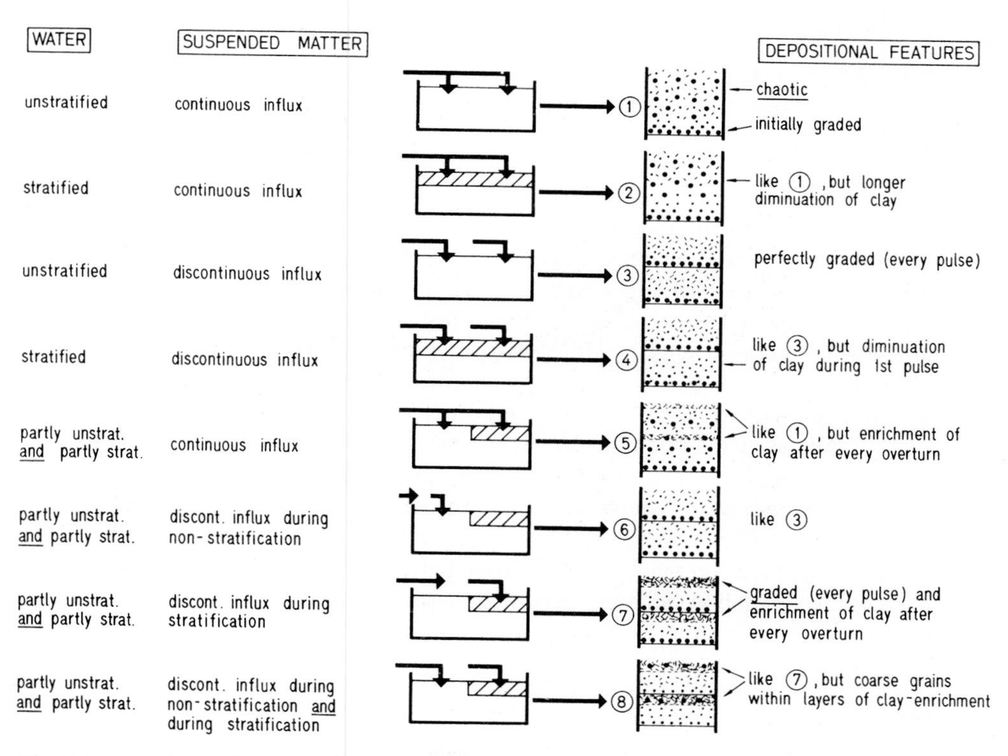

Fig. 7.11. Idealized sedimentary features as a result of two hydrological parameters (water stratification and influx of suspended matter) in an oligotrophic lake with clastic deposition. Case 7 represents the structure of an ideal clastic varve. *Heavy dots* sand; *light dots* silt; *light dashes* clay (Sturm 1979)

the season of the year, the existence of thermoclines or chemoclines, overflows, interflows and turbidity currents. A schematical key, concerning these interrelationships between lake stratification, tributary input and depositional features, is given in Fig. 7.11.

7.2.3 Wind/Wave Influences

At what depths can wind-induced waves influence the sediments in lakes? What factors determine this? The aim of this section is to try to answer these two questions. The literature on coastal processes and coastal hydraulics is extensive (see, e.g., D.W. Johnson 1919, Muir Wood 1969, Nielsen and Nielsen 1978) and the aim here is not to give a thorough review of that topic, only give a brief general background on wave theory and then focus on sediment redistribution by waves in lakes. Figure 7.12 illustrates a trace from a continuous wave recorder. For waves of uniform size, wave length, λ, period, T, celerity, c, and water depth, D, are related by the following equations:

$$\lambda = c \cdot T \tag{7.3}$$

$$c = [\tanh (2 \cdot \pi \cdot D/\lambda) \cdot g \cdot \lambda/2 \cdot \pi]^{1/2}. \tag{7.4}$$

If $D = \lambda/2$, then $\tanh (2 \cdot \pi \cdot D/\lambda) = \tanh (\lambda) = 0.996$.

So, when the depth is more than half the wave length, the wave celerity (c) is unaffected by the depth, i.e.:

$$\lambda = 1.56 \cdot T^2. \tag{7.5}$$

Wave celerity (c) is the rate of advance of the wave crests; the velocity propagation of groups of waves is denoted μg, equal to c/2. Wave groups can retain their energy, have a permanence, over large distances; individual waves may only exist for seconds. Figure 6.21 illustrated some basic concepts linking wave energy to bottom dynamics in the shore zone. Sverdrup, Munk and Bretschneider have developed a commonly utilized wave prediction technique (SMD) (see U.S. Army Coastal Engineering Research Center 1977) to obtain significant wave height, wave period and wave length (average of the one-third largest waves), when wind speed, effective fetch and water depth are known. Waves in shallow waters are smaller than waves generated in deep waters because of energy losses due to bottom friction, sediment motion and percolation.

The wave base, or the "critical" depth separating areas of transportation from areas of accumulation, and the maximum grain size that can be moved by waves,

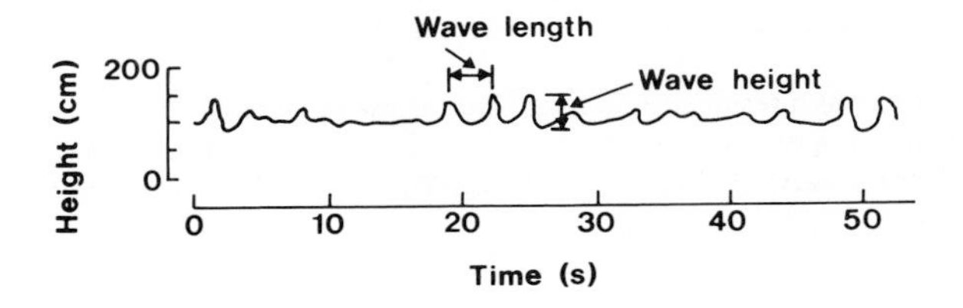

Fig. 7.12. Wave characteristics from a continuous wave recorder in Loch Leven, Scotland (Smith and Sinclair 1972)

can be determined once the size of the waves, the wave height, is determined. The effective fetch (L_f), which is a central concept in this context,may be defined according to a method introduced by the Beach Erosion Board (1972); here depicted in modified form as Fig. 7.13. The effective fetch gives a measure of the free water surface over which the winds may act upon the waves. The definition of the effective fetch accounts for several wind directions. A numerical value of L_f is rather easy to determine by means of a special transparent paper (see Fig. 7.14) put on the map from the given sample site (marked 100 in the example from Lake Vänern, Sweden, illustrated in

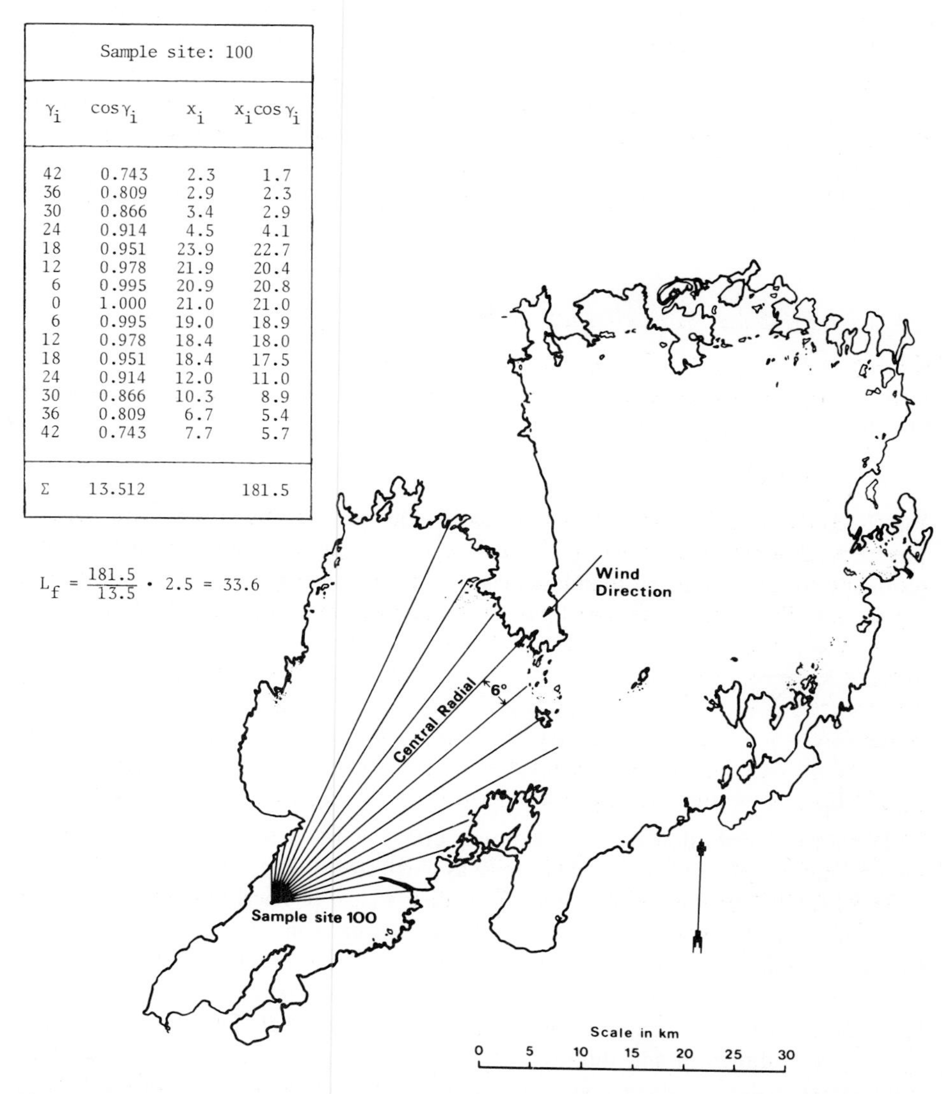

Sample site: 100

γ_i	$\cos\gamma_i$	x_i	$x_i \cos\gamma_i$
42	0.743	2.3	1.7
36	0.809	2.9	2.3
30	0.866	3.4	2.9
24	0.914	4.5	4.1
18	0.951	23.9	22.7
12	0.978	21.9	20.4
6	0.995	20.9	20.8
0	1.000	21.0	21.0
6	0.995	19.0	18.9
12	0.978	18.4	18.0
18	0.951	18.4	17.5
24	0.914	12.0	11.0
30	0.866	10.3	8.9
36	0.809	6.7	5.4
42	0.743	7.7	5.7
Σ	13.512		181.5

$$L_f = \frac{181.5}{13.5} \cdot 2.5 = 33.6$$

Fig. 7.13. Determination of effective fetch (L_f). Example from location 100 in Lake Vänern (Håkanson 1981a)

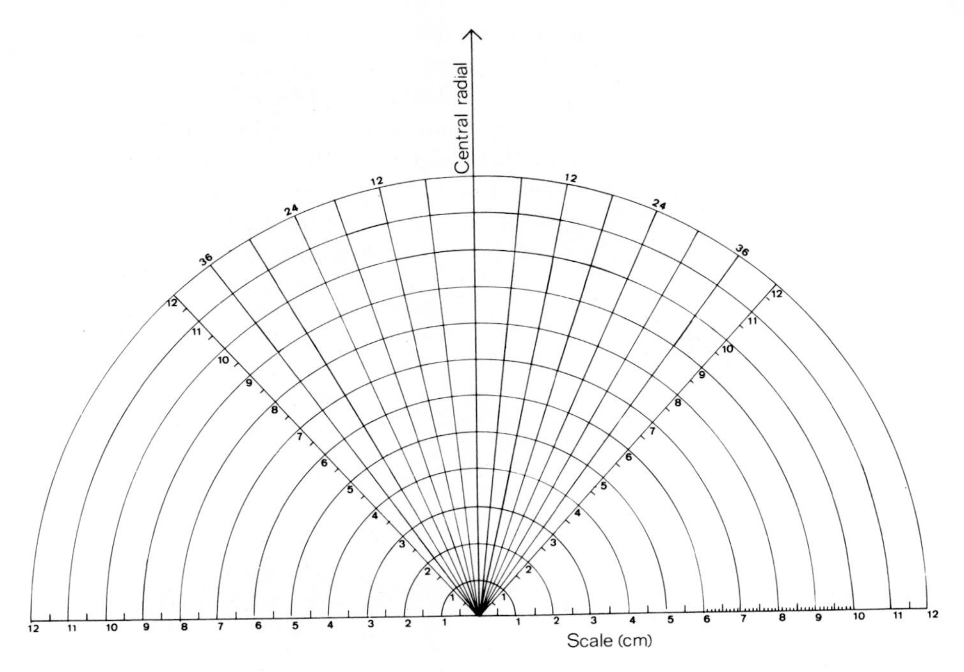

Fig. 7.14. Diagram illustrating a technique to determine the effective fetch (L_f). The diagram should be on a transparent paper (Håkanson 1981a)

Fig. 7.13). If the maximum effective fetch is required, then the transparent paper is put in the direction which gives the highest L_f-value. The distance (x in km) from the given site to land, or to an island, is measured for every deviation angle, γ_i, where $\gamma_i = \pm 6°, \pm 12°, \ldots\ldots, \pm 42°$. The L_f-value is calculated from the formula:

$$L_f = \frac{\Sigma\, x_i \cdot \cos \gamma_i}{\Sigma \cos \gamma_i} \cdot s', \tag{7.6}$$

where $\Sigma \cos \gamma_i$ = 13.5, a constant;
s' = the scale constant; $s' = 2.5$ for a map scale of 1:250 000, etc.;
L_f = the effective fetch in km.

The areal distribution of the maximum effective fetch shows a typical pattern with minimum values in central lake areas, as illustrated in Fig. 7.15 for Lake Vänern.

In lakes, there are many problems in making a clear physical linkage between the wave energy and the sediment response. In the littoral zone, it is generally adequate to use the grain size as a measure of the type of material which dominates at a given size of exposure (as illustrated in Fig. 6.21), but the water energy factor must account for a whole set of indirect and interdependent variables including the wave height, the velocity, direction and duration of the wind and the effective fetch. In deep water areas in lakes, where cohesive fine materials dominate, the picture gets even more complicated. The basic equations relating wave height (H), period (T), wind speed (w) and fetch (L_f) are (see Smith and Sinclair 1972):

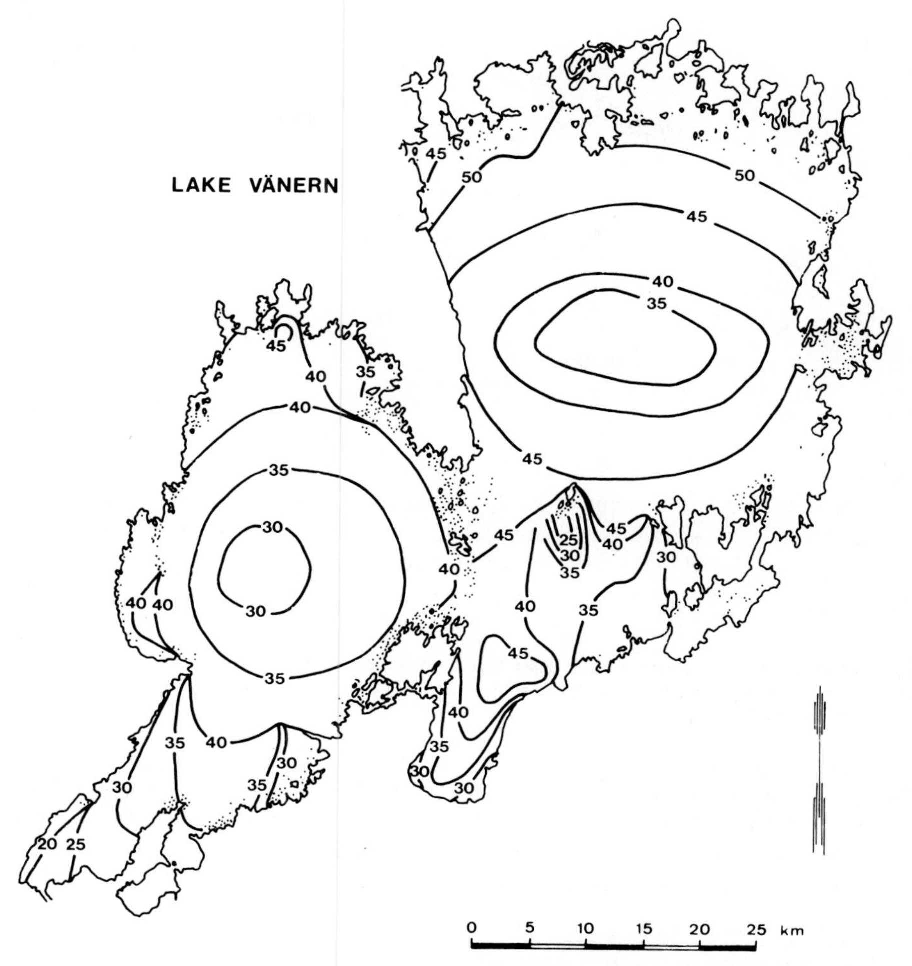

Fig. 7.15. The areal distribution of the effective fetch (L_f) in Lake Vänern (Håkanson 1981a)

$$g \cdot H/w^2 = 0.0026 \cdot [g \cdot L_f/w^2]^{0.47} \quad (7.7)$$

and

$$g \cdot T/w = 0.46 \cdot [g \cdot L_f/w^2]^{0.28}. \quad (7.8)$$

The w-values is the mean wind speed for the given area measured 8 m above the lake surface. The relationship between wave height (i.e., significant wave height), wind speed and effective fetch, as given by Eq. (7.7), is shown in Fig. 7.16. The rapid decline in fetch (and also in wave steepness, i.e., the ratio H/λ) for L_f-values below 5 km is important; it indicates that wind/wave influences in small lakes would be limited. Circumventing the problems to obtain adequate data on capacity for entrainment of cohesive sediments, instead focusing on the principal aspects, we may utilize the results given in various publications by Komar and co-authors (Komar et al. 1972,

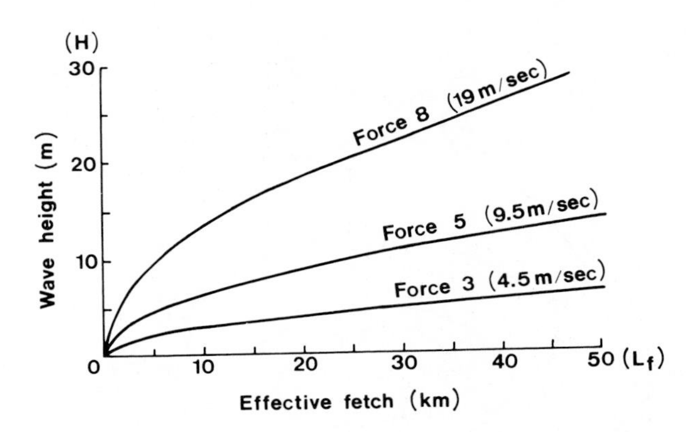

Fig. 7.16. Relationship between wave height, wind speed and effective fetch (Smith and Sinclair 1972)

Komar and Miller 1973, 1975, Komar 1976) to describe the linkage between wave orbital motion and entrainment of *sandy* sediments. The elliptical orbital motion of the water near the lake bed becomes oscillating with a horizontal displacement (l_n), given by the wave height (H), wave length (λ) and water depth (D) as:

$$l_n = H/\sinh(2 \cdot \pi \cdot D/\lambda). \tag{7.9}$$

The maximum horizontal velocity associated with this motion (u_m) is:

$$u_m = \pi \cdot l_n/T. \tag{7.10}$$

An empirical relationship between the wave motion and sediment grain size (d) moved by the passing wave, for grains smaller than 0.5 mm, are (from Komar and Miller 1973, 1975):

$$\rho \cdot u_m^2 (\rho_s - \rho) \cdot g \cdot d = C \cdot \sqrt{l_n/d}, \tag{7.11}$$

where ρ_s = the grain density $[ML^{-3}]$;
ρ = the water density 1.0 (g/cm^3, $[ML^{-3}]$);
C = an empirical constant (= 0.13 according to Sternberg and Larsen 1975).

Based on the given presuppositions, T.C. Johnson (1980) has produced some illuminating diagrams in this complicated topic – here depicted as Fig. 7.17. It should be stressed that this figure "cannot be used to determine sediment grain size moved at various depths in a deep lake as waves migrate into shallower water, because in the large fetch, shallow-water region of the graph the wave sizes generated are not as large as waves generated in deep water that migrate into shoal regions" (quotation from T.C. Johnson 1980). The diagram illustrates clearly, however, that wave activity in lakes significantly affects the sediments, even at considerable water depths. This is also demonstrated with numerical data in Table 7.2. Here the wave base is put at approximately 25% of the wave length. Various methods to determine potential bottom dynamics (erosion, transportation, accumulation) and the very important limit for the wave base, below which there is continuous deposition of fine particles/aggregates and above which there is discontinuous deposition, will be given in Section 7.3.

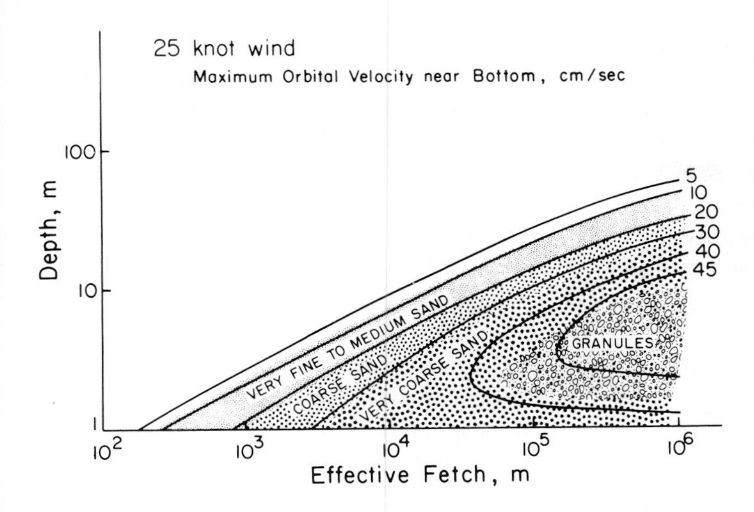

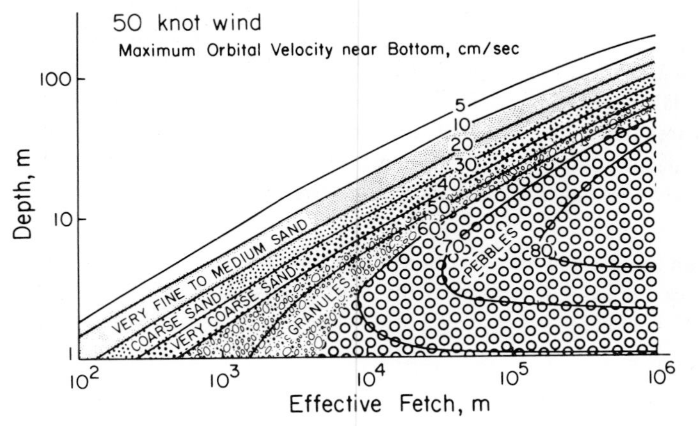

Fig. 7.17. The relationship between maximum orbital velocity at the lake floor, the sediment size moved, water depth and effective fetch at two different wind speeds (T.C. Johnson 1980)

Table 7.2. Potential wind/wave influence on bottom sediments (Sly 1978)

	Continuous wind speed				Continuous wind speed			
	74 40	56 30	37 20	28 15	74 40	56 30	37 20	28 (Kph) 15 (knots)
Fetch (km)	Significant wave height (1/3 H) (m)				Wave base (WB) – approx. 25% of wave length (m)			
600	7.3	4.9	2.6	1.6	77	53	35	24
500	7.0	4.7	2.5	1.6	68	50	31	22
400	6.4	4.4	2.4	1.5	59	45	27	20
300	5.8	4.0	2.3	1.4	50	39	22	17
200	5.0	3.4	2.0	1.3	41	31	18	15
100	3.8	2.7	1.6	1.2	29	22	14	12
50	2.9	2.1	1.3	0.9	20	15	10	8
20	2.1	1.5	1.1	0.6	12	10	6	4.5
10	1.5	1.2	1.0	<0.5	9	6.5	4.5	3
5	1.2	0.9	0.8	–	6	4.5	3	2
2	1.0	0.6	<0.5	–	3	3	1.5	–

7.2.4 Topographical Influences

It has already been emphasized that slope is an important factor in connection with turbidity currents. In this section, we will give some further examples on the impact of bottom slopes on sediments in lakes and also discuss the linkage between lake form and lake bottom dynamics.

Figure 7.18 illustrates three important concepts in lake sedimentology; that the prevailing bottom dynamics in open water areas in lakes can be expressed in terms of:

- An *energy* factor related to the wind/wave influence and the position of the wave base (= the "critical" depth).
- A lake *form* factor related to the hypsographic curve of the lake and expressing the fact that lakes with convex hypsographic curves (like lake I in Fig. 7.18) would have a larger bottom area of shallow water influences by wind/wave action than lakes with more concave hypsographic curves (like lake II in Fig. 7.18); 64% of lake I as compared to 13% of lake II lies above the wave base.
- A *slope* factor which should reflect the fact that fine deposits rarely stay permenently on slopes inclining more than 4%–5%, i.e., 4–5 m height difference at 100 m length difference. This is illustrated in Fig. 7.19 with data from a test area in Lake Vänern (see Håkanson 1977b). The position of this test area, in terms of effective fetch and water depth, is such that one would expect uniform sedimentological conditions (accumulation), if it were not for the differences in slope between the given sites. The major factor controlling the physical sediment character is thus the slope.

The slope at any given sample site may be determined directly from an echogram, provided the lane follows the same direction as the major slope axis, which can be

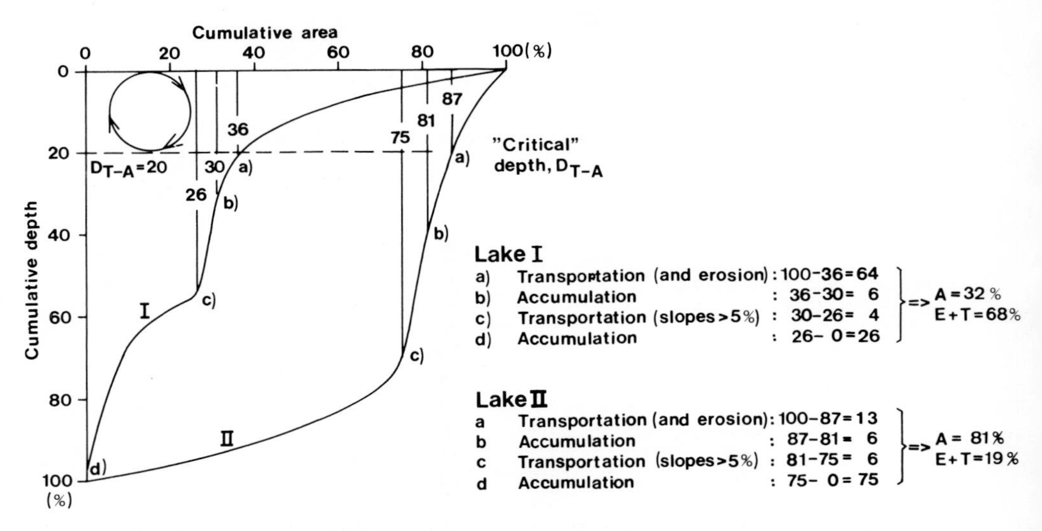

Fig. 7.18. Schematic illustration of the impact of the energy factor on the position of the wave base (or "critical depth") and how the form of the lake and the slope conditions influence the areal distribution of erosion and transportation zones (i.e., the bottom dynamics) (Håkanson 1982b)

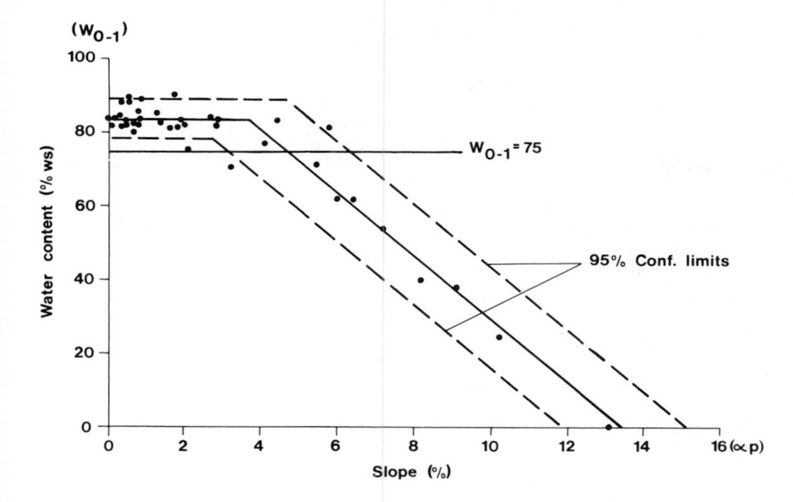

Fig. 7.19. The relationship between slope (α_p) and water content of surficial sediments (W_{0-1}) within a test area in central Lake Vänern. The position of this test area is such that one would expect uniform sedimentological conditions; the major factor controlling the physical sediment character (water content) is the slope (Håkanson 1977b)

determined from the bathymetric map. The slope is given in degrees (α^o) or preferably as a percentage (α_p) of the height versus the length. The slope between two contour lines in the bathymetric map can be determined from the following formula:

$$\alpha_p = \frac{(l_1 + l_2) \cdot l_c}{20 \cdot a''}, \tag{7.12}$$

where α_p = the slope in %;
l_1 and l_2 = the length of the two contour lines (km);
l_c = the contour-line interval (m);
a'' = the area between the two contour lines (km^2).

The mean slope ($\bar{\alpha}$) for an entire lake may be determined from, e.g., the two following formulas (see Håkanson 1974, 1981a):

$$\bar{\alpha}_p = \frac{(l_o + 2 \cdot l_t) \cdot D_{max}}{20 \cdot n \cdot a}, \tag{7.13}$$

$$tg\bar{\alpha}^o = \frac{D_{med}}{0.165 \cdot \sqrt{a}}, \tag{7.14}$$

where D_{max} = the maximum depth (m);
D_{med} = the median depth (m);
l_o = the length of the shoreline (km);
l_t = the total length of the contour lines in the bathymetric map (km);
a = the lake area (km^2);
n = the number of contour lines.

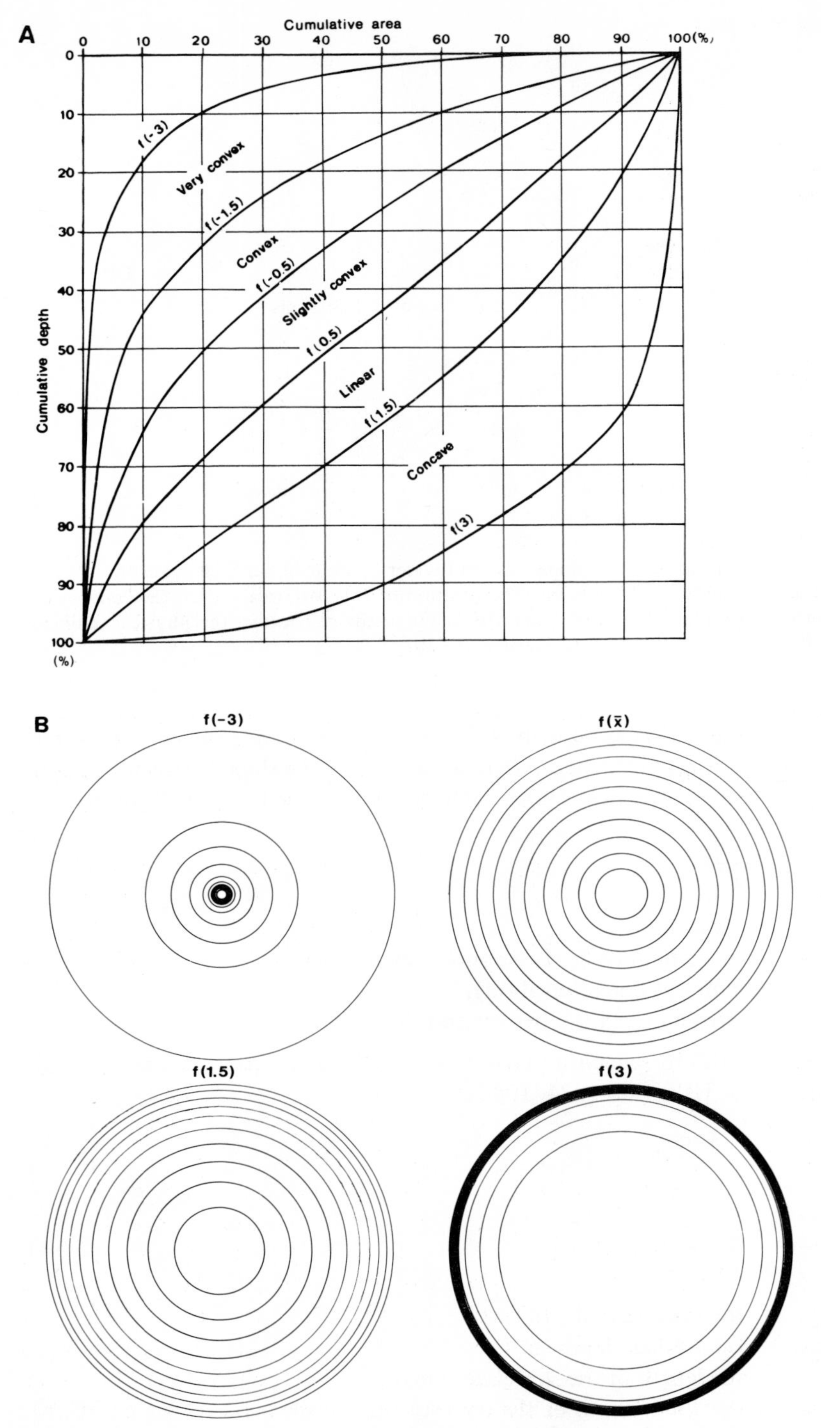

Fig. 7.20. A Terminology and class limits for the classification system of lake forms from relative hypsographic curves. **B** Schematical bathymetric interpretation of four statistical lake forms (Håkanson 1981a)

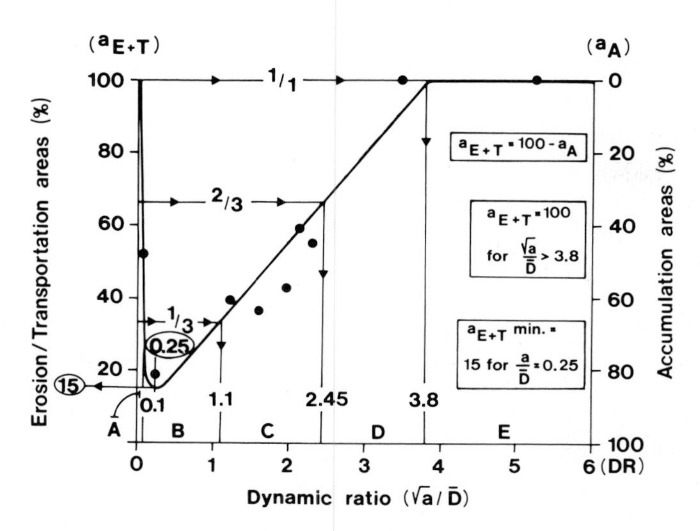

Fig. 7.21. The relationship between the percentage area of erosion and transportation (a_{E+T}) and the dynamic ratio ($\sqrt{a}/\bar{D}$). Empirical data from nine Swedish lakes illustrated with *solid rings*. The correlation coefficient between empirical data and model data is 0.97 (Håkanson 1982d)

Table 7.3. Terminology, probability and class limits for various lake forms (Håkanson 1981a)

Name	Lable	Probability	Class limits	$V_d^{-1} = \frac{D_{max}}{3 \cdot \bar{D}}$
Very convex	VCx	6.545	f (− 3)–f (− 1.5)	20–3.0
Convex	Cx	24.170	f (− 1.5)–f (− 0.5)	3.0–1.5
Slightly convex	SCx	38.300	f (− 0.5)–f (0.5)	1.5–1.0
Linear	L	24.170	f (0.5)–f (1.5)	1.0–0.75
Concave	C	6.545	f (1.5)–f (3)	0.75–0.5

Returning to the form factor, this may be expressed in terms of the relative hypsographic curve. The mean lake form, as determined from a large set of lakes (Håkanson 1977c), labelled f ($\bar{x}$), signifies that there is a 50% chance for an unknown lake to have a relative hypsographic curve above (on the convex side) or below (on the concave side) the f ($\bar{x}$)-curve. The statistical deviations corresponding to ± 0.5, ± 1.5, and ± 3.0 standard deviations, labelled f (± 0.5), etc., are represented in Fig. 7.20A. A lake with a relative hypsographic curve of the f(− 0.3)-type has one (or more) areally limited deep hole(s), but is generally very shallow (Fig. 7.20B). A lake of the f(− 3.0)-type is through-like with steep inclining walls and a very plane and dominating bottom. Terminology and class limits for the various lake forms are given in Table 7.3. A simple measure of lake form is the volume development (V_d, dimensionless), which is defined as the quotient between the lake volume ($V = a \cdot \bar{D}$) and the volume of a cone whose base area is equal to the lake area (a) and whose height is equal to the maximum depth (D_{max}), i.e.:

$$V_d = \frac{3 \cdot \bar{D}}{D_{max}} . \tag{7.15}$$

Table 7.3 illustrates the relationship between V_d and the various lake forms as determined from the relative hypsographic curves.

The areal distribution of erosion and transportation (a_{E+T}, in percent of the lake area) and accumulation ($a_A = 100 - a_{E+T}$) processes at lake bottoms may be expressed in terms of (Håkanson 1982d):

– an energy factor, given as

$$E = \sqrt{a}/D_{max} \tag{7.16}$$

– a slope factor

$$S = 41^{0.061 \cdot \bar{D}/\sqrt{a}} \tag{7.17}$$

– and a form factor (the inverse of the volume development), i.e.:

$$V_d^{-1} = D_{max}/3 \cdot \bar{D}. \tag{7.18}$$

Based on an empirical set of data on the percentage area of the lake bed subject to erosion plus transportation (a_{E+T}) and accumulation (a_A), as determined by a combination of sedimentological field work and ETA-analysis (see Sect. 7.3.2.1) from nine Swedish lakes/basins (see Table 7.4), a formula which expresses a_{E+T} (and a_A) in terms of these simple morphometric data is given as:

$$a_{E+T} = 100 - a_A = 25 \cdot (\sqrt{a}/\bar{D}) \cdot 41^{0.061 \cdot \bar{D}/\sqrt{a}} . \tag{7.19}$$

The quotient $\sqrt{a}/\bar{D}$ is called the dynamic ratio to emphasize its linkage to bottom dynamics. Formula (7.19) is illustrated in Fig. 7.21. From this figure we may emphasize:

Table 7.4. Basic morphometric data for nine studied Swedish lakes/basins. a_{E+T} = empirically determined percentage area occupied by erosion and transportation (Håkanson 1981c)

	Area (a, km²)	Max. depth (D_{max}, m)	Mean depth ($\bar{D}$, m)	Shore development (F, dim.less)	Shape factor (SF″,dim.less)	a_{E+T} (%)
Hem Bay	25.4	2.6	0.96	2.80	3.3	100 (100) [a]
Mellan Bay	40.1	3.2	1.82	1.88	1.8	100 (95.8)[a]
Dalbosjön	2066	89.0	21.0	5.06	3.4	59.0
Great-Hjälmaren	277.5	22.0	7.21	3.82	3.3	55.5
Lilla Ullevi Bay	1.9	52.0	22.0	2.74	7.4	52.5
Värmlandssjön	3582	106.0	30.0	5.62	2.8	43.0
E. Hjälmaren	35.7	20.0	4.99	3.32	4.5	39.6
S. Hjälmaren	99.1	16.0	6.21	2.41	3.5	36.4
Ekoln	18.6	38.3	19.0	2.75	3.9	19.4

[a] Data within brackets illustrate percentage area occupied by erosion (a_E)

- That the formula gives a very accurate description of the available empirical data.
- That a_{E+T} = 100% when $\sqrt{a}/\bar{D}$ is larger than 3.8. For such lakes (with comparatively large area relative to mean water depth) the resuspension activity due to wind/wave action will dominate many aspects of the limnological "character" (e.g., transparency).
- That a_{E+T} decreases when $\sqrt{a}/\bar{D}$ below 3.8 and a_{E+T} attains a certain minimum value of 15% for DR = 0.25; i.e., for lakes larger than 1 km^2 we may expect that a_{E+T} would not normally be smaller than 15%.
- That a_{E+T} increases very rapidly with decreasing DR-values below 0.25; for such lakes the slope processes increase in importance.
- That a_{E+T} = 100% for DR-values smaller than 0.052; the bottom dynamics of such lakes is totally dominated by slope processes.
- That the dynamic ratio could be used as a simple and attractive abiotic diagnostic tool in many limnological contexts. This is indicated by the following classes A–E, where:

  A) Lakes with large mean slopes. Bottom dynamics dominated by slope processes.

  B) Lakes with less than 33% areas of erosion and transportation. The resuspension activity would play a comparatively small role and the lake basin would serve as a as a rather effective "sediment trap"; $0.1 < DR < 1.1$.

  C) Lakes with a_{E+T} in the range 33%–67%. Resuspension rather important; $1.1 < DR < 2.45$.

  D) Lakes with a_{E+T} in the range 67%–100%. Resuspension important. The bottom dynamics governing distribution pattern of, e.g., pollutants. Budget calculations hazardous; $2.45 < DR < 3.8$.

  E) Lakes with $a_{E+T} > 100$. Most limnological characteristics influenced by resuspension.

It should be stressed that formula (7.19) is only tested for lakes in the size range from a = 1.9 km^2 to a = 3583 km^2. It is probable that the formula *cannot* be applied for small lakes (with $a < 1$ km^2), since wind/wave influences according to Fig. 7.16 will be limited for such lakes; or for very large lakes ($a > 5000$ km^2), which may have DR-values > 3.8 but still be deep enough to have areas of accumulation. The conceptual framework of the energy-topography Eq. (7.19), and the fact that the square root of the lake area is used as a measure of potential energy, focus the attention to single basins and circular lakes. Would the applicability depend on shore irregularity and lake shape? The answer, as given from available data, is – no. And Fig. 7.22 gives some schematical graphical interpretations illustrating that the percentage area of erosion and transportation, a_{E+T}, may very well be independent of lake topographical irregularities. The shore irregularity, or shore development, F, is defined as (see Chap. 3.2.2):

$$F = l_0/2 \cdot \sqrt{\pi \cdot A}\,. \tag{7.20}$$

The F-value illustrates the relationship between the actual length of the shoreline (l_0 in km) and the length of the circumference of a circle with an area equal to the lake area (A in km^2; A stands for the total lake area, a for the water surface). The shape factor, SF″, is defined as the quotient between the maximum effective length (L_e) and the mean width ($\bar{B}$) (see Håkanson 1981a), i.e.:

$$SF'' = L_e/\bar{B}\,. \tag{7.21}$$

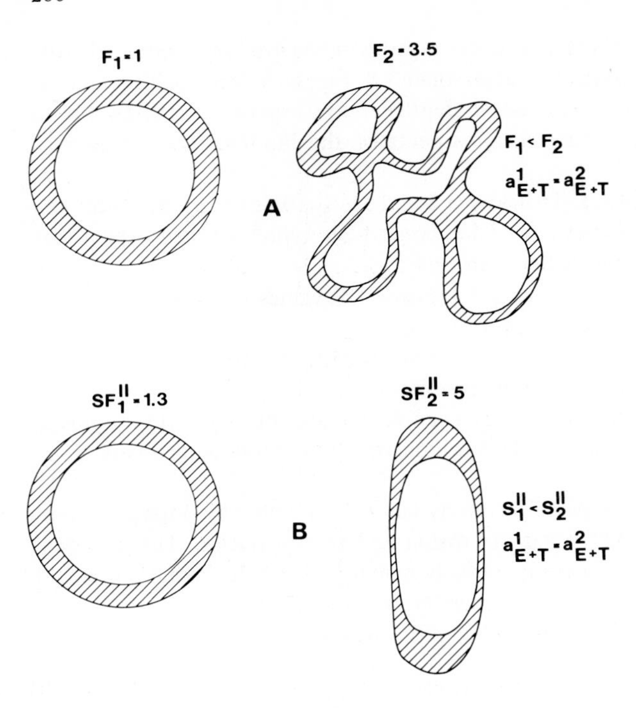

Fig. 7.22A, B. The percentage area of the lake bed subject to erosion and transportation (a_{E+T}) may very well be independent of shore irregularity *(F), upper figures,* and lake shape *(SF''), lower figures* (Håkanson 1982d)

7.3 Methods to Determine Prevailing Bottom Dynamics

Since there are, at present, no physical models to describe the relationship between the capacity for entrainment of cohesive and partly cohesive fine sediments and the appropriate water energy factor, it is necessary to accept empirical or semi-empirical models describing these complicated casual relationships. The aim of this section is to discuss four such methods to determine the potential bottom dynamic situation in lakes and the "critical" limit between areas of transportation and accumulation.

A) Lake-specific methods
   1. based on indirect, theoretical data,
   2. based on field observations.
B) Site-specific models
   1. based on indirect, theoretical data,
   2. based on field observations.

Lake-specific methods mean that the technique yield results applicable for a given *whole lake.* Indirect, theoretical data imply that the methods do not require collection of field data; all determinations are based on readily available bathymetric charts and the use of established relationships (which, of course, are based on empirical data determined elsewhere).

There are several reasons why this limit is called the "critical" limit and why it is worth special attention:

1. The conditions within the areas dominated by erosion and transportation are variable and difficult, if not impossible, to predict and cannot be described by a general equation of continuity; the opposite is valid for deposits from accumulation areas.
2. The distribution of the fine materials (particles and aggregates) within the areas of accumulation have a special meaning in environmental/pollution studies; most types of pollutants show high affinity for small inorganic and organic "carrier" particles/aggregates (see Förstner and Wittmann 1982).
3. Dredging deposits should always be disposed of within areas of accumulation (see, e.g., Blomqvist 1982).

7.3.1 Lake-Specific Methods

7.3.1.1 The Energy-Topography Formula (A1, see Håkanson 1982d)

The basic idea behind this formula [Eq. (7.19)] has already been discussed. The method requires knowledge of two morphometric standard parameters: the lake area (a, in km^2) and the mean depth ($\bar{D}$ in m). First, the percentage area of erosion and transportation, a_{E+T}, is determined from Eq. (7.19). Assuming a = 100 km^2 and $\bar{D}$ = 6.5 m, the formula gives:

$$a_{E+T} = 25 \cdot (\sqrt{100}/6.5) \cdot 41^{0.061 \cdot 6.5/\sqrt{100}} = 45 \text{ (in \%)}.$$

When this value is determined, the "critical" depth, $D_{T\text{-}A}$, may be obtained directly from the relative hypsographic curve, as illustrated in Fig. 7.23. The obtained value in this example is $D_{T\text{-}A}$ = 17% (of the maximum depth, D_{max}, which is supposed to be known). Assume D_{max} = 50 m, then $D_{T\text{-}A}$ is at 0.17 · 50 = 8.5 m. Areas beneath this depth would, on the average for the entire lake, be classified as areas of accumulation. It is evident that particular *sites,* e.g., on steep slopes in deep waters or in river-mouth areas, could not be classified according to this formula which yields lake-specific values.

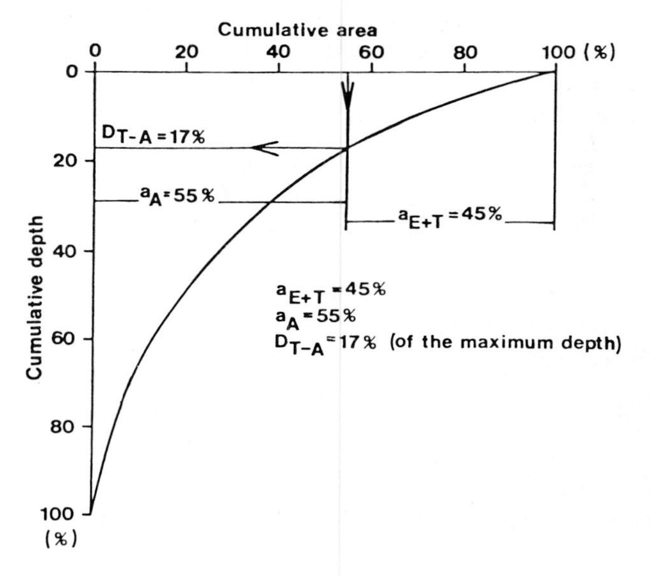

Fig. 7.23. Relative hypsographic curve for a hypothetical lake. The area occupied by erosion and transportation is assumed to be 45% of the total lake area. The "critical" water depth, $D_{T\text{-}A}$, is obtained at a depth of 17% of the maximum depth

7.3.1.2 The Characteristic Water Content Model (A2, see Håkanson 1981c)

This model has already been utilized in the discussion on river-plume sedimentation (see Chap. 6.1.2 and Fig. 6.15B), and mentioned in Chapter 4.1.8 on areal variations of physical parameters. Here we will discuss the principles behind the model more thoroughly and demonstrate its applicability to determine the "critical" limit, $D_{T\text{-}A}$.

The model is based on the following premises concerning the distribution of physical parameters (water content, bulk density, grain size, porosity, etc.) in surficial lake sediments:

- the distribution depends on water depth;
- the spread around the mean is generally large in shallow waters, where erosion, transportation, as well as accumulation of fine particles/aggregates may appear;
- the spread decreases with increasing water depth;
- the spread is relatively small at all water depths within the area of accumulation.

This allows the determination of a characteristic value of physical sediment parameters (as well as chemical and biological sediment parameters showing high correlation with physical parameters, e.g., non-contamination elements) at the maximum depth (D_{max} or $D_p = 100$, where D_p = the relative depth). This model will only be valid in lakes with areas of accumulation, i.e., the dynamic ratio, $DR = \sqrt{a}/\bar{D}$, must be smaller than 3.8 or larger than 0.052.

The general formula is:

$$Ph_{0-1} = Ph_k - K_L \cdot [(100 - D_p) / (D_p + K_L \cdot K_{Ph})], \tag{7.22}$$

where Ph_{0-1} = a given physical sediment parameter from surficial sediments (0–1 cm);
Ph_k = the characteristic content of the given parameter in the lake/basin;
K_L = an empirical lake constant, which can be utilized for any given sediment parameter for a given lake:
K_{Ph} = a parameter constant, which can be used for all lakes for any given sediment parameter;
D_p = the relative water depth.

The model, which is only tested and applicable for lakes larger than 1 km^2, is illustrated in Fig. 7.24 for the water content (i.e., $Ph_{0-1} = W_{0-1}$), which can be considered as a key parameter also in this context. The lake constant (K_L) is defined from the distribution of the water content; the K_L-value is simply defined as the empirical constant yielding the best correlation (by the standard method of least squares) with an empirical set of data (W_{emp}). The technique is illustrated in Table 7.5 with empirical data from various water depth intervals (D_p) in Lake Vänern, Sweden. The best fit, the least error, $[\Delta W^2 = (\bar{W}_{emp} - W_{theor})^2]$, is obtained for $K_L = 7.0$, which is the lake contant for this particular lake. The water content is also used in determining the requested "critical" limit, $D_{T\text{-}A}$. The parameter constant (K_{Ph}) for the water content (K_w) is 3.0. Hence, for the water content, Eq. (7.22) may be rewritten as:

$$W_{0-1} = W_K - K_L \cdot [(100 - D_p) / (D_p + 3.0 \cdot K_L)]. \tag{7.23}$$

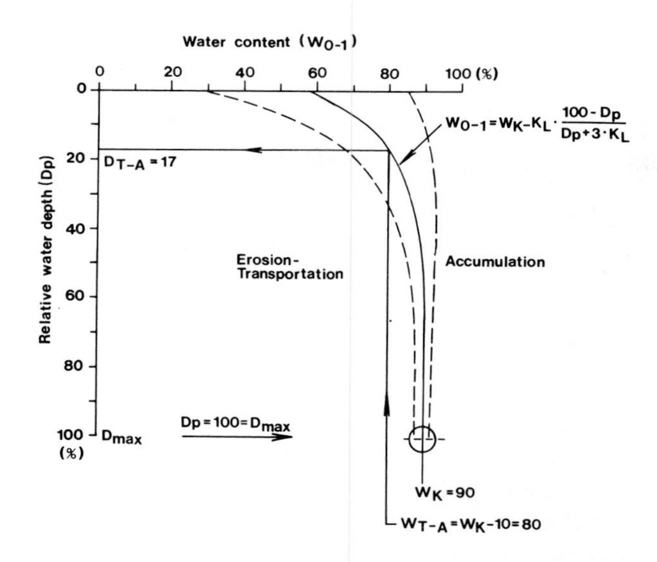

Fig. 7.24. Definition of the characteristic water content, W_K, the "critial" water content, W_{T-A}, and the "critical" water depth, D_{T-A}, separating areas of transportation from areas of accumulation (Håkanson 1977b)

Table 7.5. Test to establish the best fit regression line and the lake constant, K_L, for the data from Lake Vänern (Håkanson 1981c)

	D_p	$\overline{W}_{emp}$	Tested K_L values for $K_W = 3.0$					
	10.0	65.1	10	5	8	7.5	7	6.5
	30.0	80.6	77.3	81.2	78.6	79.0	79.4	79.8
	49.0	82.2	82.5	85.0	83.4	83.7	83.9	84.2
	68.0	85.5	85.7	87.1	86.2	86.3	86.5	86.6
	90.5	88.8	88.9	88.9	88.9	88.0	88.9	88.9
W_K		89.0						
W_{T-A}		79.0						
ΔW^2			11.03	10.77	5.94	5.46	5.34	5.86

D_p is the relative water depth (%), W_{emp} is the empirically determined mean water content, W_{theor} is the model prediction, W_K is the characteristic water content for Lake Vänern, W_{T-A} is the critical water content for Lake Vänern, and $\Delta W = W_{emp} - W_{theor}$

It is evident that for $D_p = 0$ the formula becomes $W_{0-1} = W_K - 100/3$; and for $D_p = 100$ it gets $W_{0-1} = W_K$.

The characteristic water content, W_K, will provide better information of what is lake-typical than any other sediment measure, e.g., statistical measures like the mean value for samples from the whole lake area ($\overline{W}$) or the mean value from accumulation areas ($\overline{W}_A$); the $\overline{W}$-value is obviously less adequate, since this value also includes data from erosion and transportation bottoms, which do not differentiate between various lakes; the $\overline{W}_A$-value would be an acceptable measure, but a somewhat less attractive one, since it is basically a statistical measure related to the sampling system whereas the W_K-value is related (at least in principle) to the causal relationships determining the character of the sediments.

To establish first the "critical" water content, $W_{T\text{-}A}$, and then the requested "critical" water depth, $D_{T\text{-}A}$, the following simple approach may be used:

$$W_{T\text{-}A} = W_K - 10. \tag{7.24}$$

The rationale behind this relationship is as follows:

- The $W_{T\text{-}A}$ should follow W_K; in, e.g., dystrophic lakes, and other lakes with very loose sediments ($W_K > 95\%$), also the deposits within the transportation zones tend to be comparatively loose; in lakes receiving large amounts of inorganic allochthonous materials, the W_K-value may very well be in the order 75%–80%, and this would also influence the sediments within the zones of transportation, which would have a water content in the range of about 60%–70%.
- For practical reasons, and for reasons of simplicity, it is advantageous to utilize a constant to differentiate $W_{T\text{-}A}$ and W_K.
- This constant must not be too low, e.g., lower than 5, since there generally exists a certain spread also around W_{0-1}-values, within the area of accumulation.
- It must be not be too high, e.g., higher than 15, since this would yield a low resolution, i.e., take us far into the wider, upper part of the "crooked funnel" to areas where erosion and transportation prevail.
- The constant should lie at about two standard deviations from the W_K-value.
- The constant 10 meets these requirements.

Thus, with this definition of the "critical" water content, $W_{T\text{-}A}$, the corresponding "critical" water depth, $D_{T\text{-}A}$, can be determined directly from Eq. (7.23) in the following manner (see Fig. 7.24). Assuming $W_K = 90$,

- then $W_{T\text{-}A} = 80$. Assuming $K_L = 3.2$, we obtain:

$$80 = 90 - 3.2\,[(100 - D_{T\text{-}A}) / (D_{T\text{-}A} + 3.0 \cdot 3.2)].$$

- That is:

$$D_{T\text{-}A} = 17$$

Thus, the $D_{T\text{-}A}$-value is 17% of the maximum depth, which is assumed to be known.

7.3.2 Site-Specific Methods

Bottom dynamic maps, like the one illustrated in Fig. 7.25 for the open water areas in Lake Vänern, can be constructed from site-specific data of the prevailing or potential bottom dynamic situation. The areal distribution of the different dynamic types must, as has been pointed out, by necessity, vary from one lake to another. In Lake Vänern the figures are as follows: Erosion 30%; transportation 19% (of which 3% constitute slopes); and accumulation 51%.

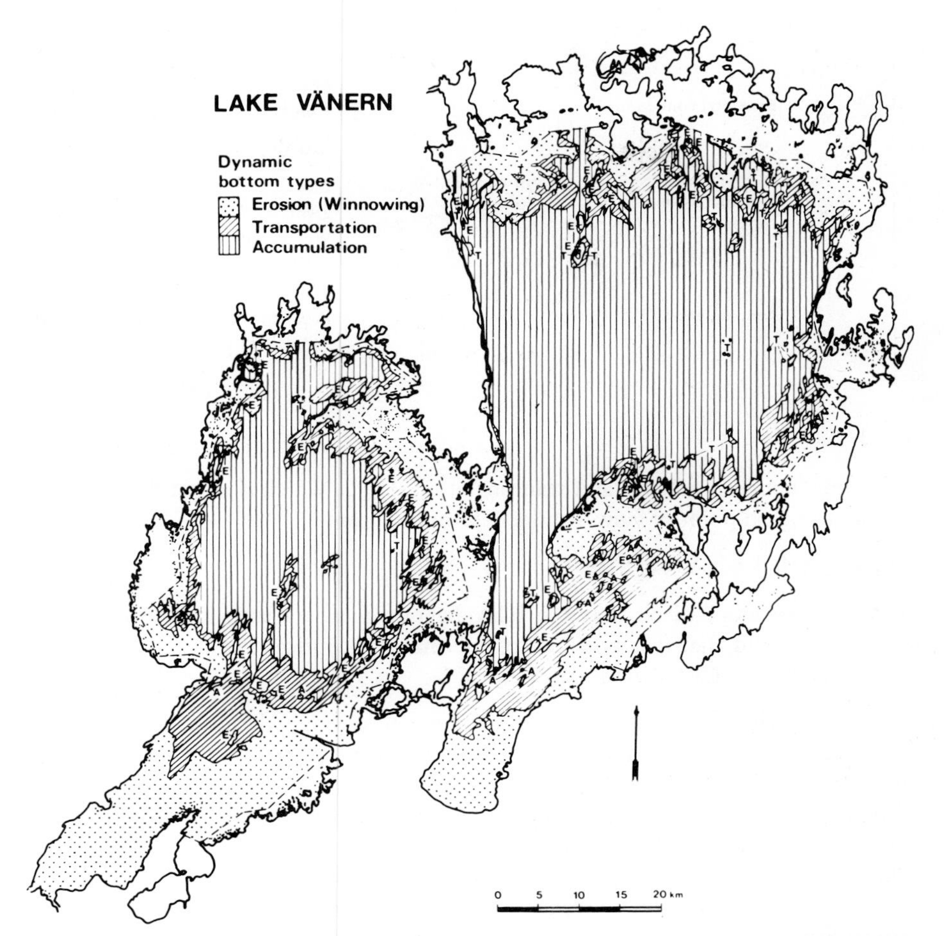

Fig. 7.25. The areal distribution of dynamic bottom types in the open water areas of Lake Vänern, Sweden (Håkanson 1977b)

7.3.2.1 The ETA-Diagram (B1, see Håkanson 1977b, 1981b)

This diagram (Fig. 7.26) illustrates the effective fetch, as an energy measure on the x-axis, and the water depth, as a measure of the influence of wind/wave action on the bottom, on the y-axis. Also, in this case, the water content of surficial sediments, W_{0-1}, may be used as a key physical indicator of the sedimentary conditions. If the effective fetch, L_f, is comparatively small and/or the water depth, D, large, areas of accumulation will prevail. For any given site in a lake, the "critical" water depth, $D_{T\text{-}A}$, may be determined from the equation:

$$D_{T\text{-}A} = (45.7 \cdot L_f) / (L_f + 21.4), \qquad (7.25)$$

where $D_{T\text{-}A}$ = the "critical" water depth in m;
L_f = the effective fetch (i.e., the potential maximum effective fetch) in km.

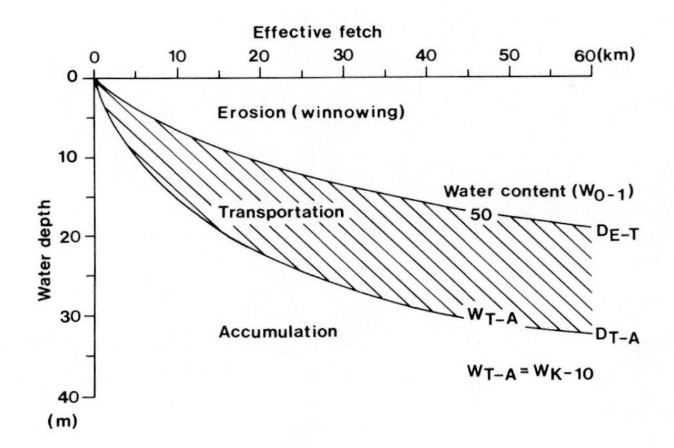

Fig. 7.26. The ETA-diagram. The rough general distinction between erosion and transportation is based upon the water content W_{0-1} = 50%. The "critical" limit, between areas of transportation and accumulation, may be given by $W_{T\text{-}A} = W_k - 10$, where W_k is the characteristic water content and $W_{T\text{-}A}$ is the "critical" water content of surficial lake sediments (0–1 cm) (Håkanson 1981b)

The limit between erosion and transportation, $D_{E\text{-}T}$, which is rather diffuse from a sedimentological point of view, can be determined from:

$$D_{E\text{-}T} = (30.4 \cdot L_f) / (L_f + 34.2). \tag{7.26}$$

It should be emphasized that the ETA-diagram means a simplication of a very complex issue, and that it is possible to find many exceptional cases to the general rule, as given by the ETA-diagram. For example:

1. In areas where the L_f-value has no meaning per se, e.g., where winds only blow from certain directions and/or in areas where higher or lower wind velocities prevail;
2. Close to river mouths, where sediment movement is controlled by current action rather than by wind/wave action;
3. On subaquatic slopes, where no fine material is deposited;
4. In areas where the water movement may be influenced by various bathymetric controls;
5. In areas influenced by high turbidity current activity;
6. In shallow areas where plants can trap fine material;
7. In areas affected by subaquatic ground water discharge;
8. In very small lakes ($a < 1\ km^2$) where wind/wave impact on bottom dynamics is limited.

7.3.2.2 The Cone Apparatus (B2, see also Chap. 3.4)

Here we shall demonstrate how the cone technique can be used to determine the potential bottom dynamics at given sites in lakes. The ultimate goal would be to actually measure, for example, water velocity and entrainment in situ, but unfortunately we are far from that goal today. This means that more indirect techniques must be adopted. The cone technique is an example of such an indirect, crude but practical method. To determine the potential bottom dynamic situation by means of the cone method would be very simple indeed if an adequate calibration had been made against a large set of lakes with different sediment types. For the present, there exists no such wide range set of empirical data, only a limited number from a few lakes. Consequently, the results presented here must be regarded as preliminary and qualitative rather than quantitative. Moreover, these results do not emanate from the in situ apparatus illustrated in Fig. 3.25 but rather from the two-cone apparatus depicted in Fig. 7.27. The function of this instrument may be described by the following steps (see Håkanson 1982b):

1. The sampler, which is a modified Ekman-sampler (Fig. 3.4), is first lowered to the lake bottom.
2. The sampler is retrieved and put in its cradle. The trigger assembly is then removed. After the water has been siphoned off, the sediments are accessible from the top of the box.
3. The cone apparatus is put on the box, and the top of the cones adjusted to the sediment surface (zero adjustment).
4. The two cones are allowed to penetrate the sediments for 5 s (released by a button, stopped by another), whereafter the penetration depth of the cones, L_1 and L_2, can be read on the two scaled axis. For example, $L_1 = 3$ cm and $L_2 = 7$ cm would imply a transportation site; $L_1 = 9$ cm and $L_2 = 18$ cm would imply silty clay (accumulation); and $L_1 = 15$ cm and L_2 = max. would mean very loose sediments (accumulation).

The tested cones have the following characteristics:

	Cone angle	Static load (g)	Height of cone (cm)
L_1	90°	250	3.0
L_2	30°	250	7.5

To calibrate the instrument, the bottom dynamic conditions characterizing a particular site have been determined by three different, complementary methods:

Method 1. The W_K-model (see Sect. 7.3.1.2)
This is the least adequate method, yielding only a rough indication of the prevailing bottom dynamics. This method requires empirical data of the water content of surficial sediments, W_{0-1}, from various sample sites at different water depths within a lake. From such data, the characteristic water content, W_K, may be determined from Eq. (7.23) and the "critical" water content, $W_{T\text{-}A}$, from Eq. (7.24). Thus, if the measured water content at a sample site, W_{0-1}, is less than the "critical" $W_{T\text{-}A}$, this would indicate that the sample emanates from a transportation area. $W_{0-1} > W_{T\text{-}A}$ indicates an area of accumulation.

Method 2
This method is based on the fact that the vertical variation of the water content in a sediment core from an accumulation area generally, but certainly not always, is smaller than the vertical variation within zones of transportation; discontinuous deposition of fine and coarse particles will create discontinuous layers and a high probability for variable sediments within zones dominated by transport processes.

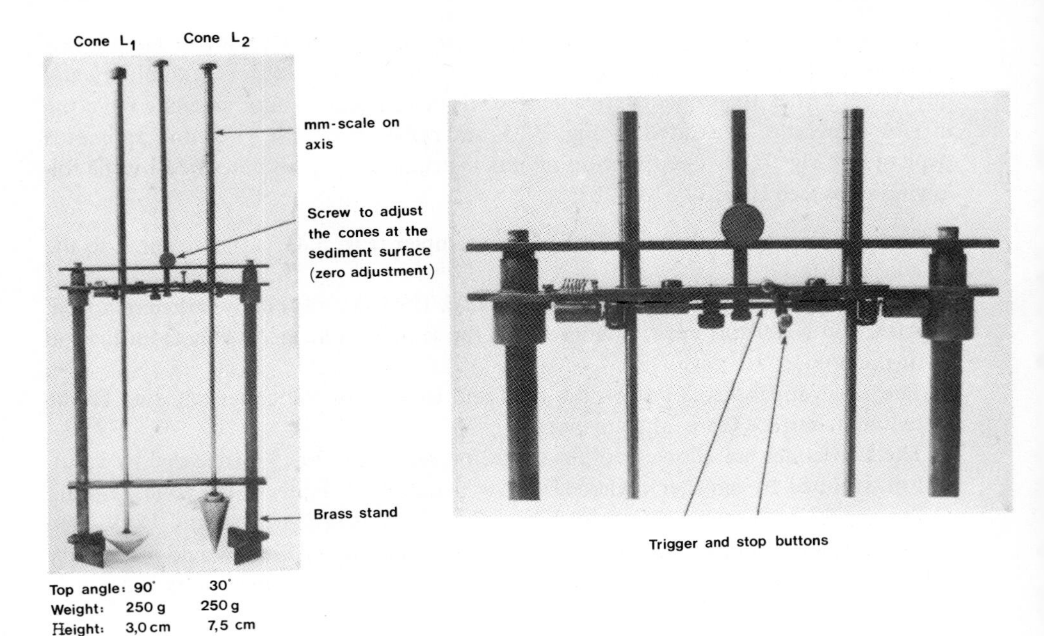

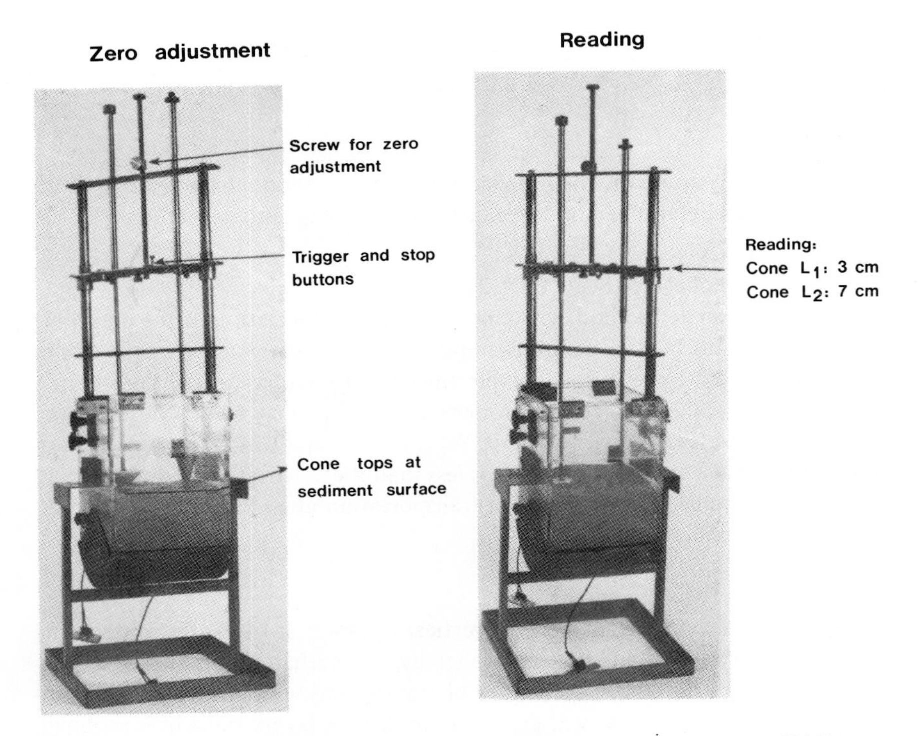

Fig. 7.27. The two-cone apparatus and the modified Ekman-sampler (Håkanson 1982b)

The vertical decrease in water content may be described by the following simple general formula (see Chap. 4.1.9):

$$W(x) = W_{0-1} + K_s \cdot \ln(2x), \tag{7.27}$$

where $W(x)$ = the mean water content at the sediment depth x ± 0.5 cm;
W_{0-1} = the water content at the uppermost centimeter, i.e., the 0–1 cm value;
K_s = the sediment constant, i.e., the empirical constant which provides the best fit (the method of least squares) against the empirical data.

Method 2 is illustrated in Fig. 7.28 with data from one transportation site and one accumulation site in Lake Ekoln. In this example, we obtain the following regression line for the sediments from the transportation site:

$$W(x) = 71.2 - 6.0 \cdot \ln(2x).$$

The correlation coefficient between the empirical data (solid rings) and the model data (the curve) is only r = 0.45.

The fit between the empirical data (open rings) and the model data is significantly better for the accumulation area, r = 0.64.

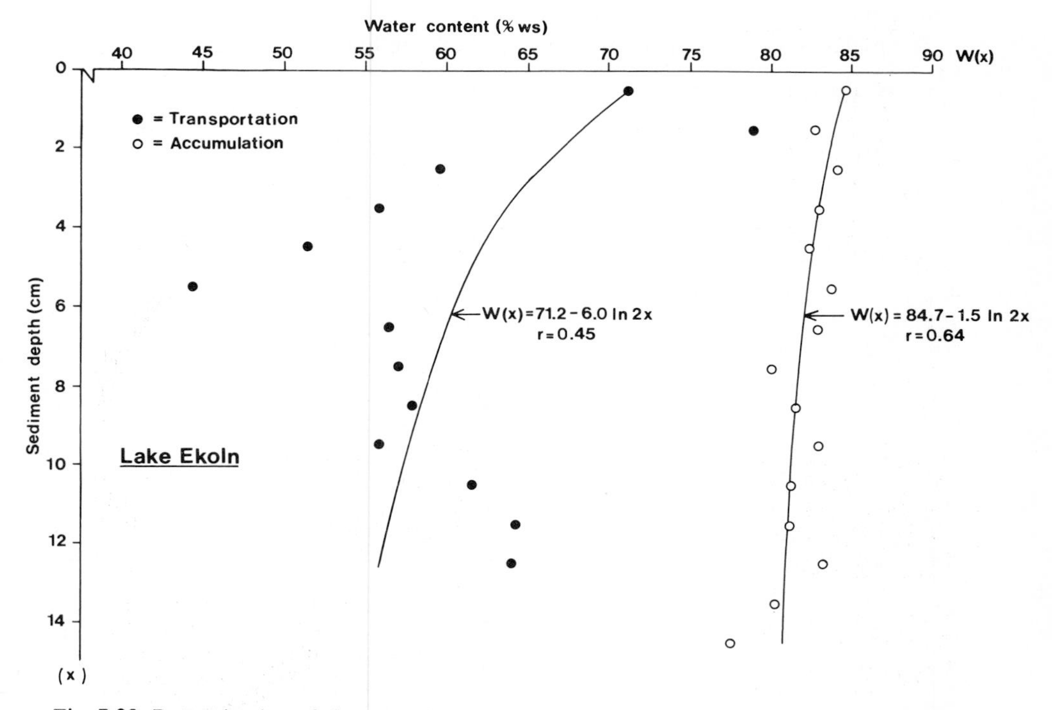

Fig. 7.28. Determination of the potential bottom dynamic type according to method 2 (Håkanson 1982b)

The regression line is:

$$W_{0-1} = 84.7 - 1.5 \cdot \ln(2x).$$

Thus, if the obtained r-value is low, that would *indicate* a transportation site. The relationship between t (i.e., t from the standard Student's t distribution), r (the correlation coefficient) and n (the number of samples) is given by:

$$t = r \cdot \sqrt{n-2}/\sqrt{1-r^2}. \qquad (7.28) \qquad (7.28)$$

The t-values are given in standard tables (see Appendix 1).

Method 3. The ETA-diagram. See Fig. 7.26
If all three methods indicate that a given site should be classified as a transportation site, then this has been depicted with a solid ring in the preliminary calibration wave, Fig. 7.29; "down" triangles are used to indicate sites which cannot be determined unambiguously as belonging to either transportation or accumulation areas; "up" triangles are used for unanimously determined sites from accumulation areas; and filled triangles for data from zones of erosion. From Fig. 7.29 it may be noted that all "down" triangles lie on a line which separates transportation areas from accumulation areas. Of all the 35 sites investigated hitherto, there are 19 "pure" A, 6 on the "critical" limit, 5 "pure" T, 1 on the "diffuse" limit between erosion and transportation and 4 "pure" cases totally dominated by sand and/or coarser materials with a water content (W_{0-1}) of 50% or less (i.e., erosion).

The preliminary "critical" limitation line has the following equation:

$$L_1 = 8.3 - 1.5 \cdot L_2/L_1. \qquad (7.29)$$

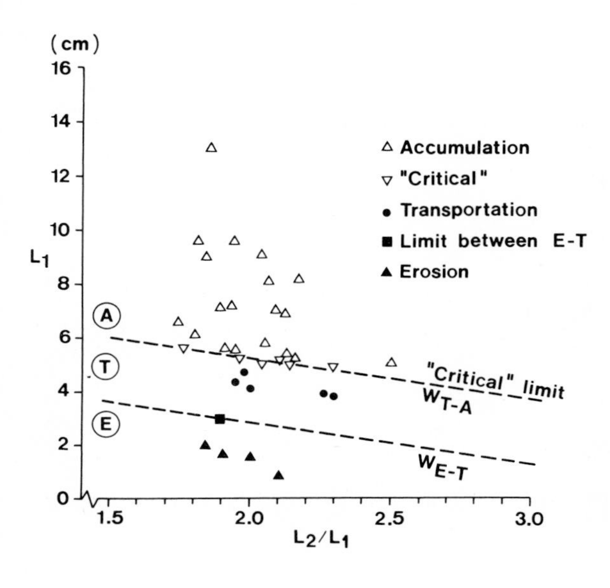

Fig. 7.29. Preliminary calibration curves for the two-cone apparatus based on empirical data from 35 sites in different Swedish lakes. Samples above the "critical" limit are within areas of accumulation, samples below this limit emanate from zones of transportation and/or erosion

The line separating zones of erosion from zones of transportation has the following equation:

$$L_1 = 2.5 - 1.5 \cdot L_2/L_1 . \tag{7.30}$$

Table 7.6 summarizes the empirical tests and the utilized methods to calibrate the cone apparatus with data from a transportation site in Lake Ekoln. It is probable that these results from the two-cone instrument can be correlated, but not directly applied, to the in situ apparatus; the two tips of the two-cone instrument are adjusted at the

Table 7.6. The physical character of lake sediments as described by the water content (W), the organic content (IG) and the cone penetration depths (L_1 and L_2)

Lake Ekoln-transportation

Sediment depth cm	W % ws	IG % ds		L_1	L_2
0– 1	71.2	11.0			
1– 2	78.9	15.9			
2– 3	59.6	7.2			
3– 4	55.8	5.3			
4– 5	52.8	5.1			
5– 6	44.4	3.5			
6– 7	56.4	6.7			
7– 8	57.0	7.1			
8– 9	57.9	7.5			
9–10	61.6	9.6	$\bar{L}$:	4.33	8.49
10–11	61.6	9.6	s_L:	0.26	0.55
11–12	64.3	10.7	N:	10	10
12–13	64.0	10.9			

Method 1: $W_K = 83.5 \rightarrow W_{T\text{-}A} = 73.5$
$W_{0-1} = 71.2 < W_{T\text{-}A} \rightarrow$ *transportation*

Method 2: $W(x) = K_S \cdot \ln(2x) + W_{0-1}$
$\left.\begin{array}{l} W_{0-1} = 71.2 \\ K_S = -6.0 \end{array}\right\} \rightarrow r = 0.45$
t-test yields t = 1.67. Correlation is not significant at the 90%-level. This indicates *transportation*

Method 3: The ETA-diagram
$L_f = 1.8 => D_{T\text{-}A} = 3.6$ m
$D_{T\text{-}A} = 3.6 \text{ m} > D = 1.5 \text{ m} \rightarrow$ *transportation*

Three different methods to estimate the potential bottom dynamic situation at a given site:

Method 1 is based on the characteristic water content, W_K (which is 83.5 in Lake Ekoln) and the corresponding "critical" water content, $W_{T\text{-}A} = W_K - 10$. If the water content, W_{0-1}, at a given sediment site is less than the "critical" water content, $W_{T\text{-}A}$, then this would indicate a transportation site

Method 2 is illustrated in Fig. 7.27. It is based on the fact that the physical character of the sediments generally varies much more within areas of transportation than within areas of accumulation

Method 3 implies that the potential bottom dynamic type may be directly estimated from the ETA-diagram (i.e., from knowledge of effective fetch and water depth)

sediment-water interface, whereas the three tips of the in situ apparatus are not triggered until the whole instrument is stopped, which may be at about 1–3 cm down in the sediments on very loose bottoms. These results, however, illustrate some major causal relationships and important principles regarding cone penetration and potential bottom dynamics. Once the bottom dynamics for given sites has been established, the bottom dynamic map may be drawn according to the illustration in Fig. 7.30.

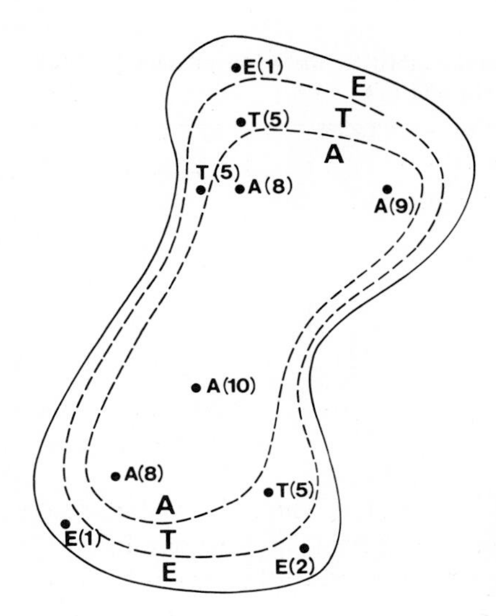

Fig. 7.30. Map illustrating the drawing of isolines from sites for which the bottom dynamics *(E, T, A)* have been determined by the two-cone apparatus (L_1-values in *brackets*)

8 Sediment Dynamics and Sediment Age

The focus in this chapter is on the conditions within the sediments (internal structures and bioturbation), including causal relationships determining the age of recent sediments and methods of age determination. For a more thorough and theoretical approach to the processes (bioturbation, diffusion, advection) within sediments, the book *Early Diagenesis* by Berner (1980) can be recommended.

8.1 Laminated Sediments

Where and why do laminated lake sediments appear? Figure 7.12 gave a schematic interpretational code on depositional features in lakes relative to the hydrological/sedimentological regime.

Laminated deposits predominate in anaerobic sediments, where the bioturbulent mixing is negligible due to extinction of the bottom fauna (Fig. 8.1), and in river-mouth areas, where the rate of deposition of minerogenic allochthonous materials is high, i.e., where the substrate is meagre and the bottom fauna too poor to level out primary structures. Lamination may also exist for other reasons, e.g.:

- In Lake Superior, Dell (1973) found that highly calcareous layers were formed during summer, when the rate of sedimentation was high and the carbonates rapidly buried, whereas during winter, when the rate of sedimentation was low, the carbonates were partially dissolved and a low-carbonate winter layer formed.
- As a result of rhythmic deposition of turbidites or resuspended materials from shallow areas (see, e.g., Ludlam 1974);
- As a response to fluctuating bioturbation (secondary lamination, see Håkanson and Källström 1978).

The formation of anaerobic conditions in the sediments is favoured in deep, stagnant, stratified, highly productive lakes. The formation of gas bubbles, e.g., methane and hydrogen sulfide, may cause considerable disruption of laminated deposits.

Apparently homogeneous sediments may reveal lamination if dried (Edmondson 1975), examined by X-ray (Axelson and Händel 1972, H.H. Roberts 1972) or thermoluminescence (Wintle and Huntley 1980). Subsequently, we will give examples of what laminated sediments look like and how they are distributed in lakes.

The first example emanates from Lillooet Lake, Canada, and Fig. 8.2 illustrates the variation, the decrease, in mean annual varve thickness with distance from the

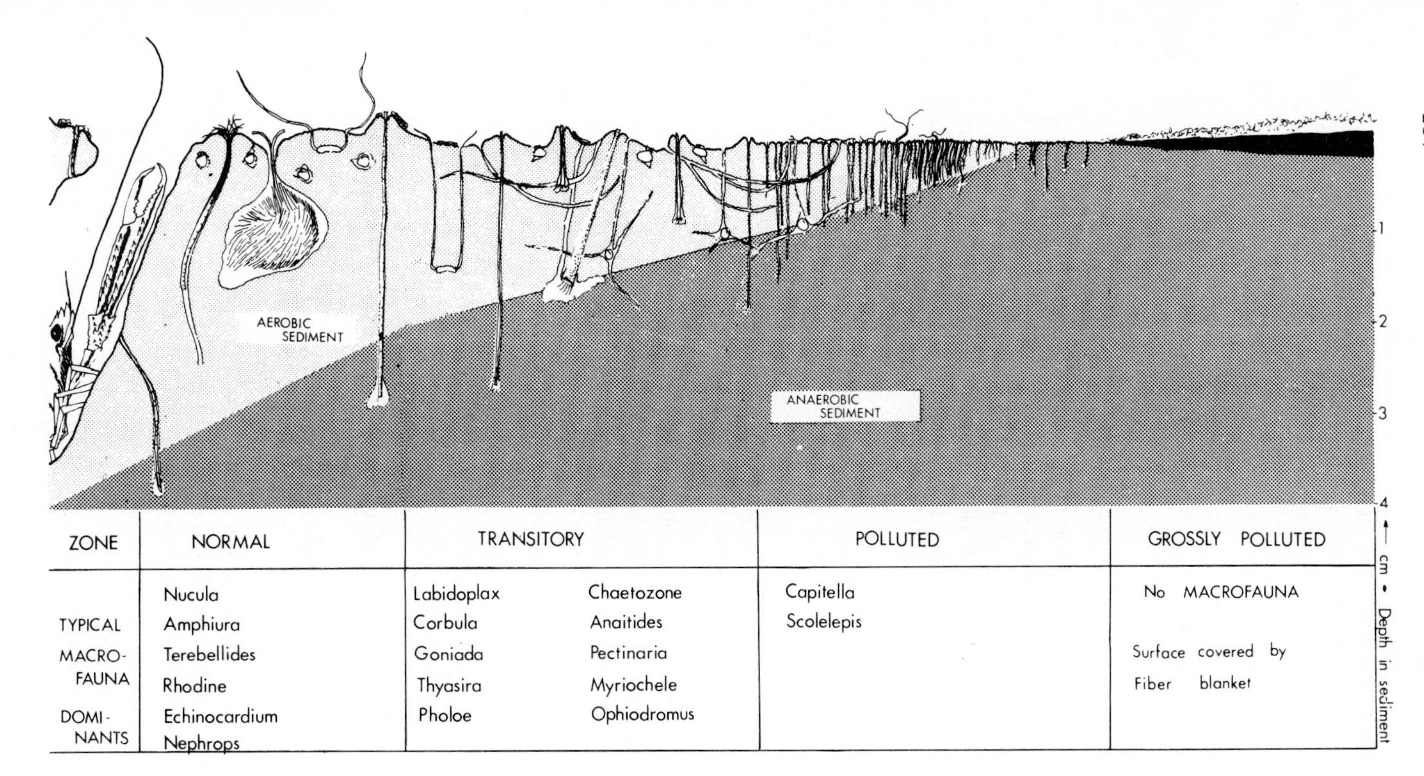

Fig. 8.1. Bioturbation and oxic conditions in a marine environment (Pearson and Rosenberg 1976)

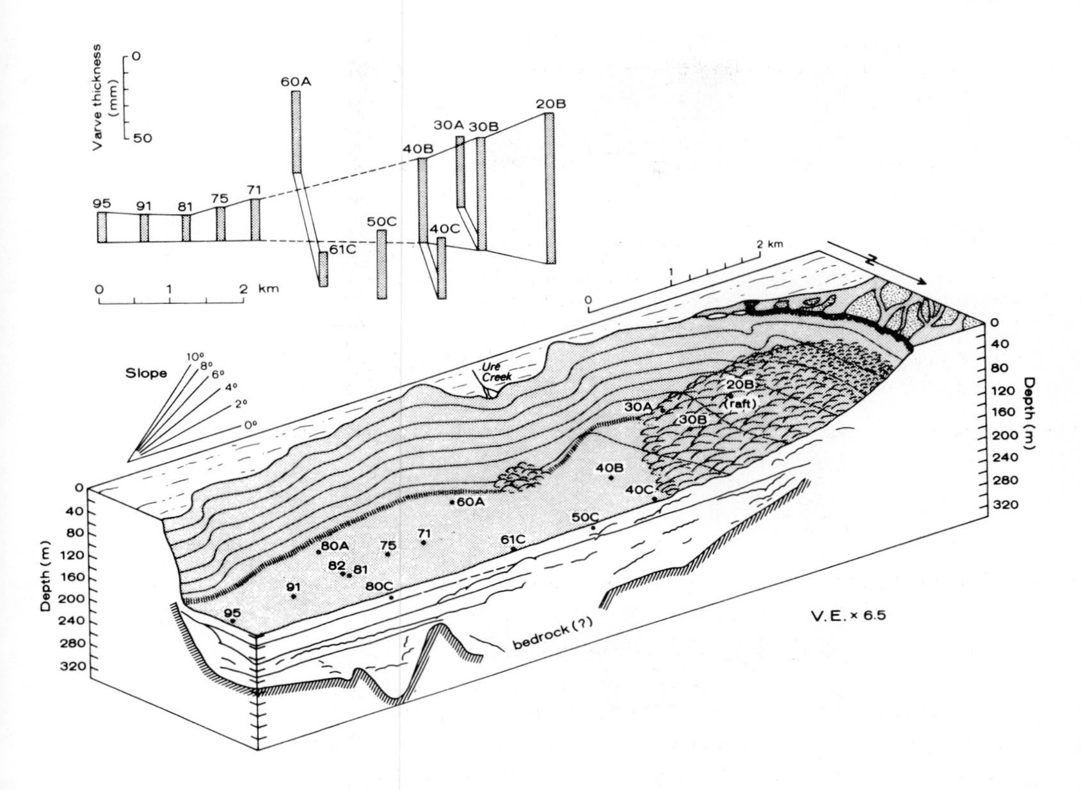

Fig. 8.2. *Upper diagram* mean varve thickness in Lillooet Lake sediments at different distances from the river mouth; *lower diagram* block sketch of the northwestern part of Lillooet Lake showing lake bathymetry and sample sites (Gilbert 1975)

tributary (cf. Fig. 6.11). Thicker, coarser and lighter lamina are generally created in connection with periods of high water discharge; thinner, finer and darker layers during more stagnant conditions.

The second example comes, once again, from Lake Brienz, Switzerland, where varves on the slopes are generally more regular and show less frequent microlamination than varves on the basin plain (Fig. 8.3), which is highly influenced by turbidity current activity. A summary of the different depositional environments in Lake Brienz and important controlling factors are given in Fig. 8.4.

Renberg (1981b) has published a very informative microphoto (Fig. 8.5) of varved lake sediments from northern Sweden. Mineral grains (labelled A in Fig. 8.5) are generally deposited during periods of high water discharge (spring flood); organic materials (B), like dead algae and animals, dominate the deposition during the productive season (summer and autumn); fine, darkish particles/aggregates are laid down (C) during winter.

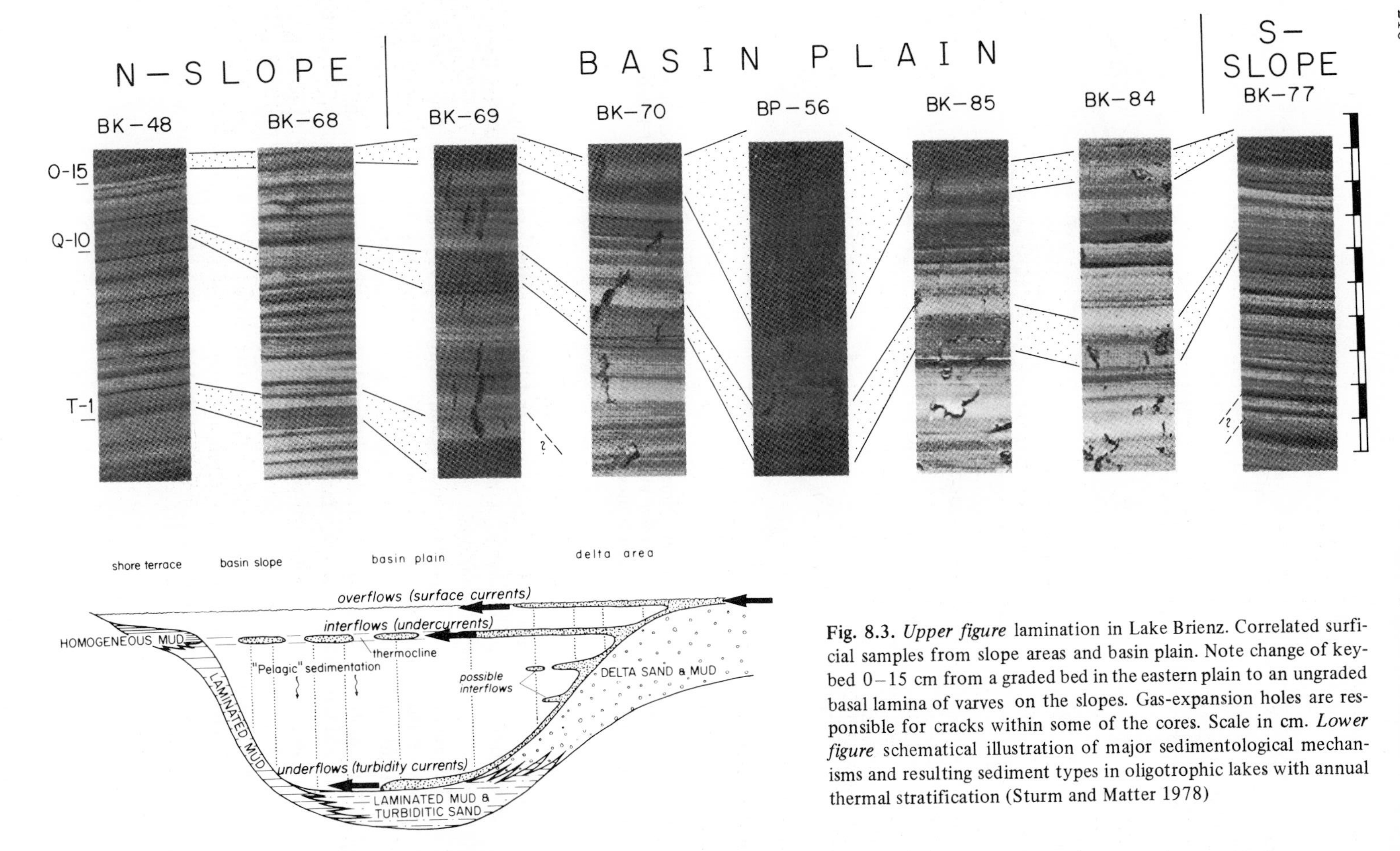

Fig. 8.3. *Upper figure* lamination in Lake Brienz. Correlated surficial samples from slope areas and basin plain. Note change of key-bed 0–15 cm from a graded bed in the eastern plain to an ungraded basal lamina of varves on the slopes. Gas-expansion holes are responsible for cracks within some of the cores. Scale in cm. *Lower figure* schematical illustration of major sedimentological mechanisms and resulting sediment types in oligotrophic lakes with annual thermal stratification (Sturm and Matter 1978)

	DELTA AREA	CENTRAL BASIN PLAIN	LATERAL SLOPE
AREAL DISTRIBUTION (in % of lake floor)	38 %	23%	40 %
RELIEF	rough, uneven	smooth, flat	smooth, steep
INCLINATION	8°- 20° (foreset)	0° - 1°	30° - 40°
SEDIMENT SOURCE	proximal	proximal ⇔ distal	distal
ENERGY LEVEL	high	low, occ. high	low
ENVIRONMENT	erosional = depositional	erosional ≪ depositional	depositional
MODE OF TRANSPORT	turbidity current	turbidity curr. / undercurrent	undercurrent
CHANNELS	common	not distinct	absent
SEDIMENTATION RATE (last 75 years)	40 - 50 mm/y	5.5 - 6.8 mm/y	2.5 - 3.3 mm/y
GAS EXPANSION HOLES	common	common	± absent
GRAIN SIZE DISTRIBUTION	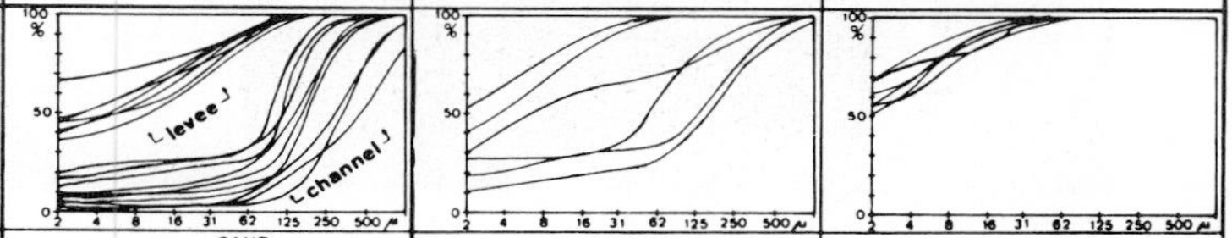 levee; channel		
CLAY/SILT/SAND - RATIO	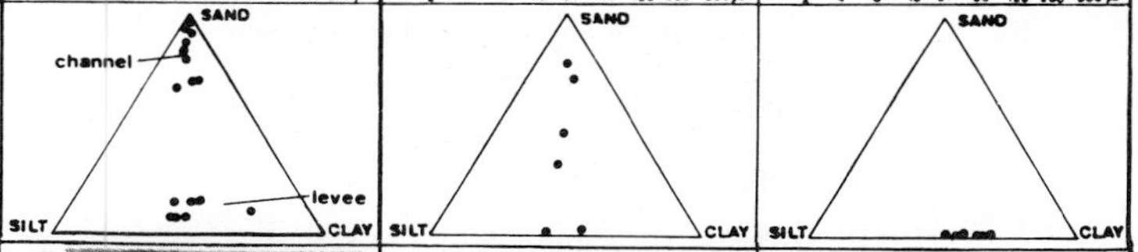 SAND; SILT; CLAY; channel; levee	SAND; SILT; CLAY	SAND; SILT; CLAY
SEDIMENTARY STRUCTURES	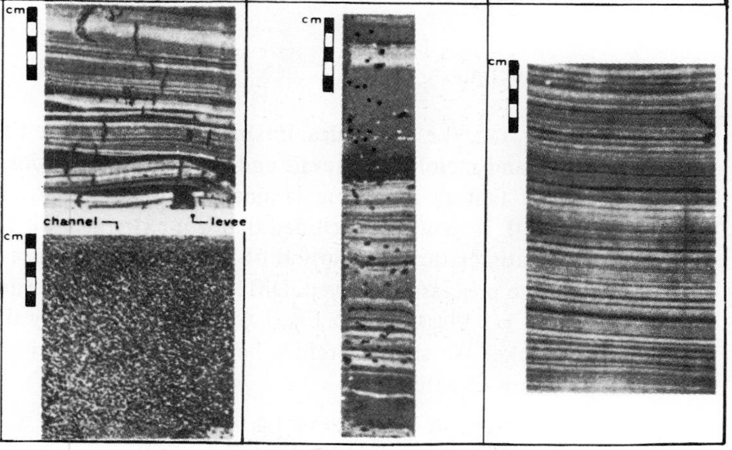 cm; channel; levee	cm	cm

Fig. 8.4. Summary of different depositional environments and their corresponding sedimentological structures in Lake Brienz (Sturm 1975)

Fig. 8.5. Microphotograph of a thin section of varved lake sediment (Renberg 1981b)

8.2 Bioturbation

8.2.1 Introduction

Bioturbation, i.e., the mechanical mixing of sediments due to bottom foraging fishes, macrofauna and meiofauna prevail under aerobic conditions. Thus, it is favoured by the opposite factors regulating lamination. According to Lindeström (1979) and Berner (1980), it may be concluded that it is extremely difficult to gain exact knowledge and mathematical description of bioturbation because of the variety and complexity of the processes involved. Different benthic organisms display different mixing activities (as illustrated in Fig. 8.6), and are unevenly distributed (in space and time) in a lake. We shall therefore begin this section with a brief presentation of different forms of patchiness.

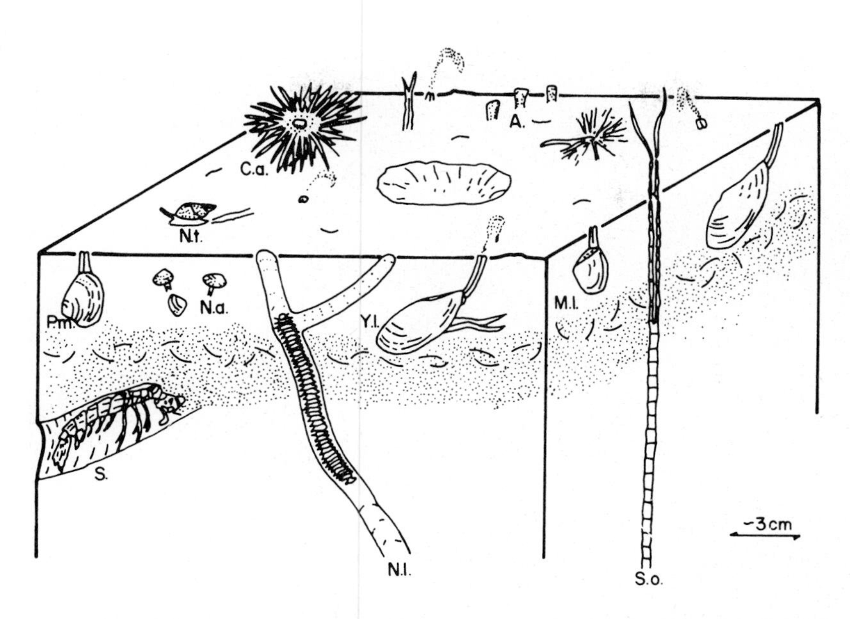

Fig. 8.6. Schematic illustration of the bioturbulent activity of different types of marine bottom animals. *C.a. Ceriantheopsis americanus* (infaunal anemone); *A. Ampelisca* sp. (deposit-feeding amphipod); *N.t. Nassarius trivitatus* (deposit-feeding gastropod); *S. Squilla* (deep-burrowing shrimp); *N.a. Nucula annulata* (deep-feeding bivalve); *Y.l. Yoldia limatula* (deep-feeding bivalve); *P.M. Pitar morrhuana* (suspension-feeding bivalve); *M.l. Mulinia lateralis* (suspension-feeding bivalve); *N.i. Nephtys incisa* (very mobile polychaete); *S.o. Spiochaetopterus oculatus* (deep-burrowing, sedentary polychaete) (Berner 1980, from Aller 1977, 1980)

8.2.2 Patchiness

8.2.2.1 Areal Patchiness

Figure 8.7 illustrates the areal distribution of four important and common groups of bottom fauna (large and small chironomids, *Pisidium* and oligochaetes) in Lake Latnjajaure, northern Sweden. The numbers represent mean values from 1967, 1968 and 1970 (from Nauwerck 1981). In this oligotrophic, cold, high-mountain lake, the majority of small chironomids live in comparatively shallow areas influenced by tributary brooks, whereas the large chironomids prefer the deep water regions. The benthic communities can be very sensitive to changes in the abiotic life frames (e.g., O_2-conditions, pH and Eh). A minor shift in any of these internal governing factors may upset the normal conditions, causing a drastic effect on the composition and number of the benthic community; deep-holes may become anoxic and the bottom fauna extinct; hydrological flow patterns, rates and patterns of sedimentation and composition of sediments may vary. This will imply both areal and temporal patchiness, which is illustrated in Fig. 8.8.

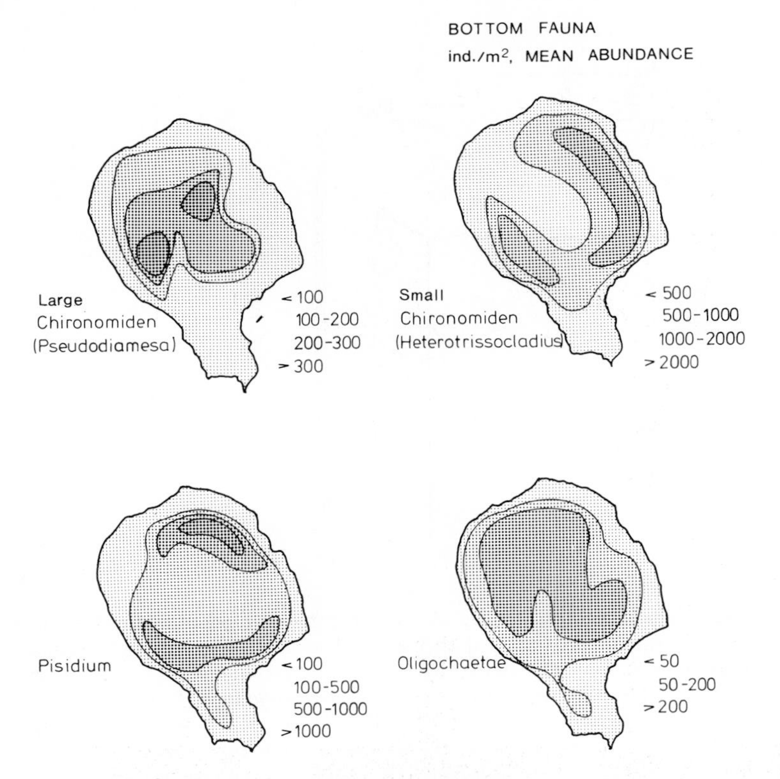

Fig. 8.7. Areal distribution of four important types of bottom fauna in Lake Latnjajaure, northern Sweden (Nauwerck 1981)

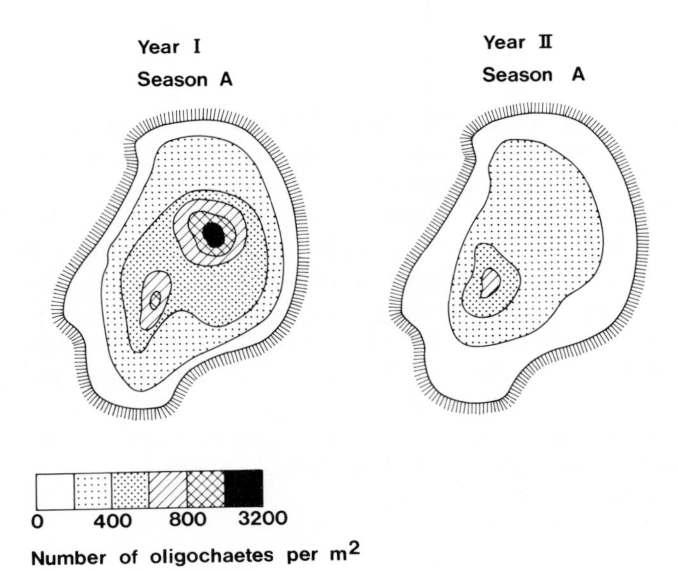

Fig. 8.8. Schematic illustration of areal and temporal patchiness of a given benthic organism (oligochaetes) in a given hypothetical lake

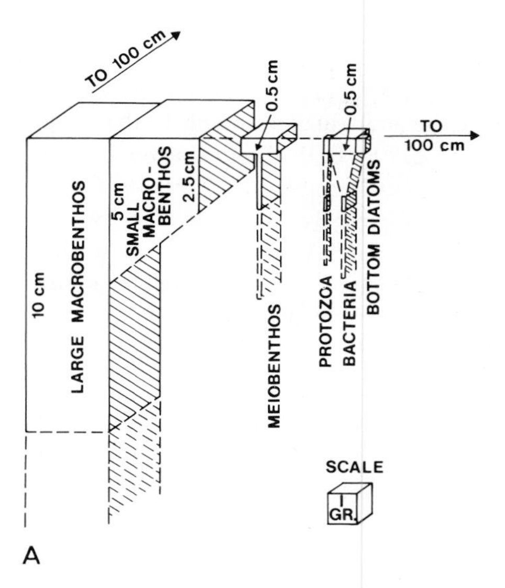

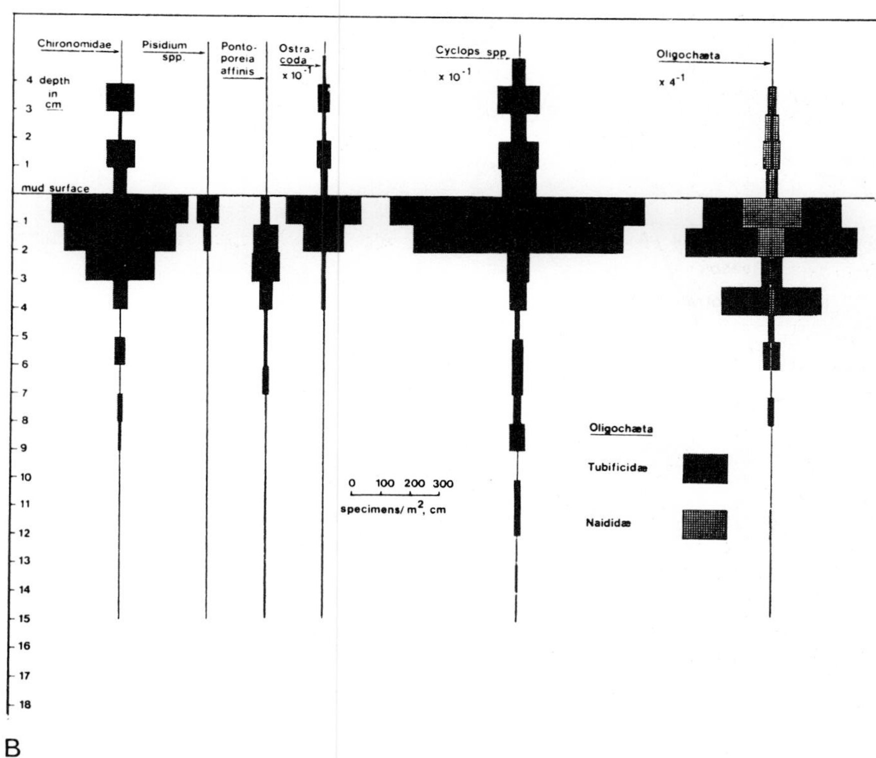

Fig. 8.9A, B. Vertical variation of bottom fauna in lake sediments. **A** A schematization from a sandy-silty bottom habitat (Mare 1942, as given in Rhoads 1974). **B** Data on six species in Hovgårds Bay, Lake Mälaren, Sweden (Milbrink 1973)

8.2.2.2 *Vertical Patchiness*

Figure 8.9 illustrates the fact that various types of bottom-dwelling species occupy different vertical niches in the sediments; the general rule of thumb is that large species have a large vertical register and vice versa. Fig. 8.9B shows the vertical distribution more specifically for six different species: Chironomids, *Pisidium, Pontoporeia,* ostracodes, *Cyclops* and oligochaetes in Hovgårds Bay, Lake Mälaren, Sweden (from Milbrink 1973). It is evident that the vertical patchiness depends on the areal distribution, since various species dwell at various water depths, as well as of the temporal distribution, since all species have a given yearly cycle of life.

8.2.2.3 *Temporal Patchiness*

This is illustrated in Fig. 8.10. Most species appear in low numbers, and a corresponding low bioturbulent activity, during winter, and in high numbers during summer. This means that the potential biotransport would fluctuate with the season of the year.

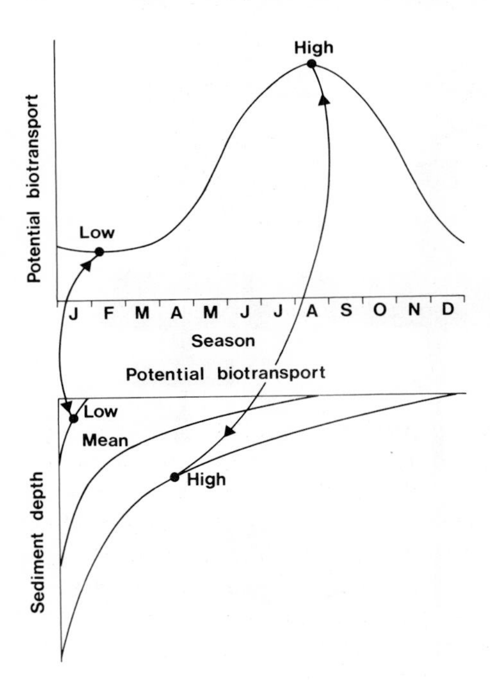

Fig. 8.10. Schematic illustration of seasonal variations in abundance (i.e., potential bioturbation) in benthic communities and its effect on the biotransport of materials in lake sediments

8.2.2.4 *Species-Specific Patchiness*

Biotransport of materials in lake sediments may be separated into at least three major categories (Fig. 8.11).

Type I, exemplified by tubificids (see Davis 1974), which transport material from a certain feeding depth for delivery as faeces on the sediment surface. Integrated over time and space, the upward component, b_u (x), should be much larger than the downward component, b_d (x).

Type II, exemplified by *Pontoporeia,* which may be responsible for an intense bioturbation at the sediment-water interface. In this case the upward and downward biotransport would be of the same order of magnitude.

Type III, a more or less diffuse residual factor, i.e., the biotransport that may not be categorized as types I or II. This would be the net integrated effect of, e.g., chironomids, which presumably would cause a net downward transport of materials, bottom foraging fish, which may create intense bioturbation in the upper part of the sediment column, bivalves (e.g., *Pisidium*), which would give rise to very intense biotransport along the paths of movement, and gas bubbles, which, most likely, only cause upward transport. A probable result of all these activities is schematically illustrated by the spiral ($b_u > b_d$) in Fig. 8.11. Very few empirical data exist to substantiate this type of biotransport.

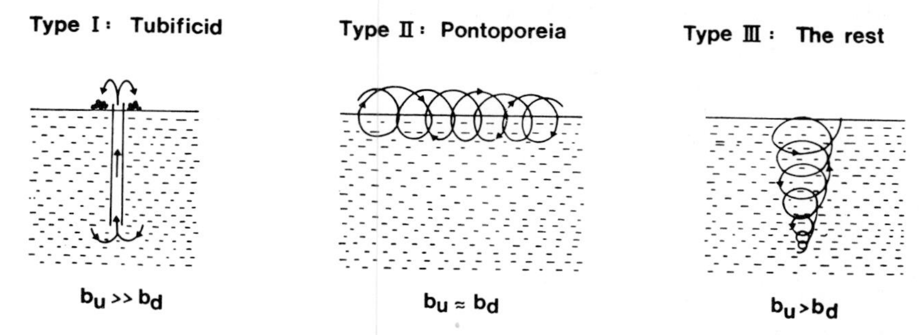

Fig. 8.11. Schematic illustration of three different modes of biotransport

It should be stressed that neither models nor reliable empirical data are presently available which account for these processes and patterns of patchiness in a quantitative way. However, for tubificids, J.B. Fisher et al. (1980), have conducted very elegant laboratory experiments, by means of radioactive tracer (Cs-137) combined with a marker horizon, to study and describe the mixing process by these particular worms. Their study illustrates that the activity of tubificids alone may give rise to significant mixing, that the reworking is largely directional and cannot be adequately described by eddy diffusion, but is better approximated as an advective process, and that experimental data may be used to explain results obtained from empirical investigations (e.g., Davis 1974). Figure 8.12 illustrates the results after 15 days of reworking when the active layer was installed at various depths; the mixing gradually decreases from the sediment-water interface to the bioturbation limit. It may be concluded that:

- Worms (polychaetes and oligochaetes) and bivalves (e.g., *Pisidium*) may burrow several tens of centimeters into the sediments in digesting sedimentary particles. These burrows may lodge organisms and enhance vertical exchange of pore water, a process referred to as irrigation.

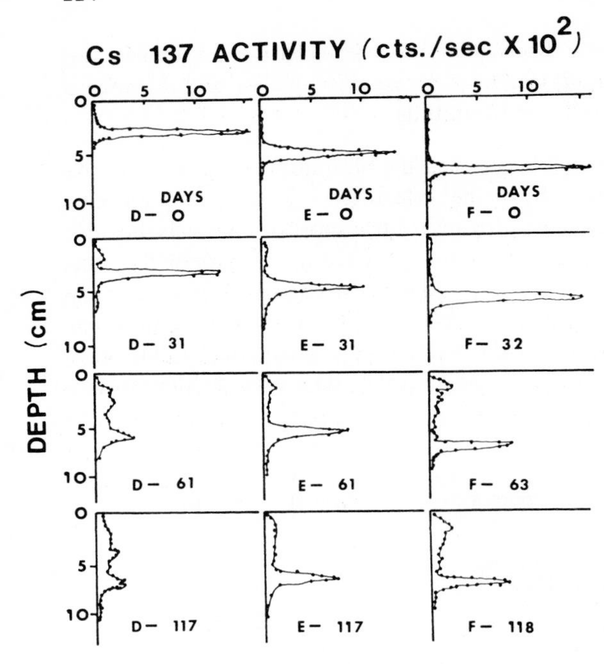

Fig. 8.12. The biotransport caused by *Tubifex tubifex* in a laboratory experiment where the worms were incubated at different levels and the particle reworking rate determined from Cs-137 activity (J.B. Fisher et al. 1980)

- Certain crustaceans (e.g., *Pontoporeia*) may cause very intense mixing in the upper centimeter of the sediment column, since these organisms can be present in great abundance during certain periods in restricted areas.
- Insect larvae (e.g., chironomids) and bottom-foraging fishes (e.g., bream, roach, perch, eel, whitefish) may cause bioturbation, if present in sufficient numbers.
- Meiofauna, microorganisms and gases generally play a minor role for the bioturbation of materials under aerobic conditions.
- The conditions favouring gas formation are eutrophic environments and high temperatures, i.e., shallow, soft bottoms during summer.

8.2.3 Modelling of Bioturbation/Biotransport

The fact that lake sediments represent a dynamic and complex environment, and that bioturbation displays considerable areal, vertical, temporal and species-specific patchiness, motivates the use of mathematical modelling in bioturbation studies. However, it should be stressed that for the present it must be concluded that no dynamic models, which account for all aspects of patchiness and rank the underlying mechanisms of bioturbation in terms of importance, are available or even close at hand. It is, furthermore, probable that such dynamic models would be extremely complex, difficult to verify and of limited practical value. Therefore, much of the research has focused on rather simple models which would tell "most of the story, but not give the entire picture". The simplest approach (see, e.g., Berger and Heath 1968, Guinasso and Schink 1975), the "one-box model", is to assume uniform bioturbation in the whole active layer down to the bioturbation limits (b_z), beneath which the bioturbulent mixing would be zero. This is, of course, an oversimplification, since the mixing rate evidently varies very much within the biologically active layer; from high bioturbation near the sediment surface to zero at b_z.

It is important to distinguish particle bioturbation from fluid bioturbation because these two processes often take place at different rates due to irrigation (see Aller 1977, Berner 1980). This is schematically illustrated in Fig. 8.13, where dark and light "balls" mark two different types of solid particles and the space in between represents interstitial water. Example A represents an inhomogeneous distribution of particle types and a homogeneous distribution of porosity; bioturbation results in complete homogenization. Example B represents an inhomogeneous distribution of porosity and a homogeneous distribution of solid particles; bioturbation means complete homogenization. Example C shows inhomogeneous distribution of both porosity and particle types; bioturbation causes homogenization. The mathematics of these processes has been thoroughly discussed by Berner (1980) and will not be dealt with in this context.

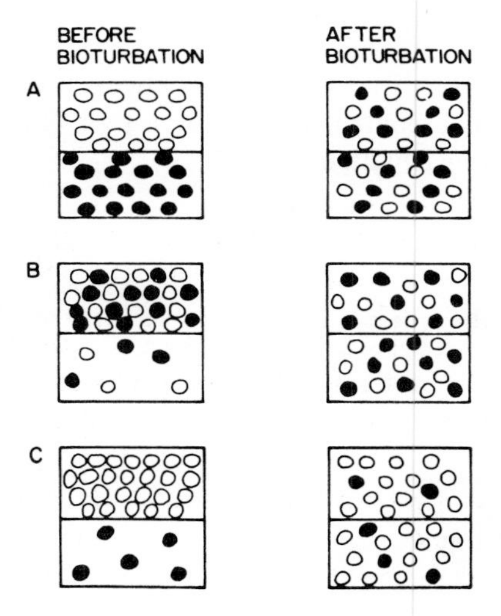

Fig. 8.13A–C. Schematic illustration of bioturbation of solids and pore water. *Dark* and *light balls* two different types of solid particles. The *space in between the balls* interstitial water. Note in all three cases that bioturbation yields homogenization of balls and water (Berner 1980)

8.2.3.1 A Dynamic Model

From what has been mentioned, it is evident that the modelling of bioturbation implies an act of balance between too much complexity and too much simplification. Subsequently, we will discuss a dynamic model which is meant to describe only the most important causal relationships regulating the distribution of solid particles and hence also the age of the sediments. The concepts accounted for in the model are illustrated in Fig. 8.14, i.e.:

- the sediment depth (x in cm); in this context we will only utilize one cm thick layers, i.e., $x_i = 1, 2, 3, 4, \ldots$;
- the rate of sedimentation (v in cm yr^{-1});

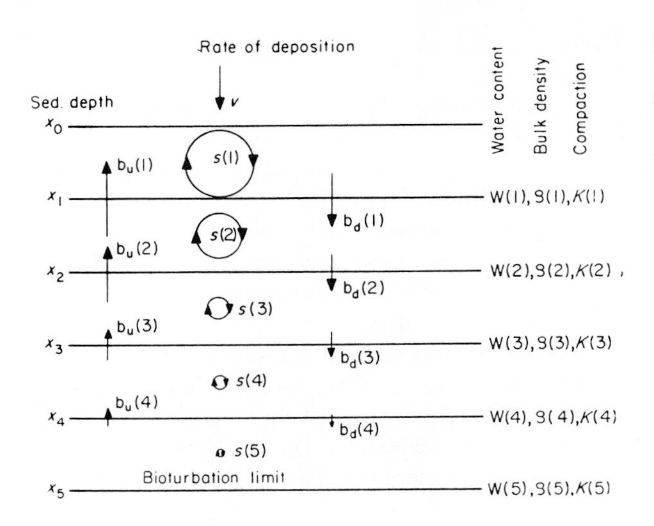

Fig. 8.14. Schematic illustration of the bioturbation model given by Håkanson and Källström (1978)

- the water content [W(x) in % wet substance = ws]. The vertical distribution of the water content may be described by Eq. (4.3);
- the bulk density [ρ(x) in % dry substance = ds]. The relationship between the bulk density, the water content and the organic content is given by Eq. (4.7).;
- the degree of compaction, K(x), as defined by the equation:

$$K(x) = \frac{(100 - W_{0-1}) \cdot \rho_{0-1}}{[100 - W(x)] \cdot \rho(x)}, \tag{8.1}$$

where W_{0-1} = W(0.5) = the water content of the sediments at the interval 0–1 cm.

- The biotransport [b(x) in g ws cm^{-2} yr^{-1}]. The biotransport can be divided into an upward component, $b_u(x)$, and a downward component, $b_d(x)$, where

$$b(x) = b_u(x) - b_d(x). \tag{8.2}$$

$b_u(x)$ is generally larger than $b_d(x)$, which implies that the net biotransport of materials, b(x), is generally larger than zero at all sediment depths (x) in the active layer. The numerical relationship between b(x), $b_u(x)$ and $b_d(x)$ is assumed to be regulated by the compaction; if the compaction gradient is high, downward transportation would be more difficult than if the gradient is low. This may be given by:

$$\frac{b_u(x)}{b(x)} = \frac{K(x + \Delta x) - K(x)}{K(x) - K(x - \Delta x)}. \tag{8.3}$$

- The substrate decomposition [s(x) in g ws cm^{-2} yr^{-1}], which should describe the possible loss of material within the layer due to, e.g., loss of heat, kinetic energy and/or gas exchange with the interstitial water. s(x) is here assumed to be small compared to, and proportional to, the net biotransport, b(x).
- The bioturbation limit (b_z in cm), i.e., the level below which b(x) is zero.

These are the factors to be accounted for in the model, which should describe how steady-state conditions (a given density distribution) can prevail in the sediments.

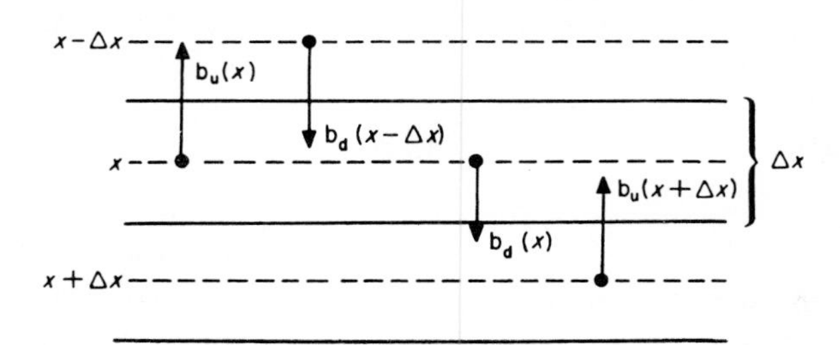

Fig. 8.15. Notations

Consider a layer at level x of thickness Δx (see Fig. 8.15). We would like to know that happens during the time Δt.

1. The amount of material in this layer is $\rho(x) \cdot \Delta x$;
2. During the time Δt, the following has happened:
   - $b_u(x) \cdot \Delta t$ and $b_d(x) \cdot \Delta t$ have flowed out (up and down) into neighbouring layers;
   - $b_d(x - \Delta x) \cdot \Delta t$ and $b_u(x + \Delta x) \cdot \Delta t$ have flowed into the layer;
   - $s(x) \cdot \Delta t \cdot \Delta x$ is decomposed;
   - the layer has sunk by $v \cdot \Delta t$ cm;
   - the thickness has decreased to $\Delta x' = \Delta x \cdot K(x + v \cdot \Delta t)/K(x)$.
3. After the time Δt, the amount in the layer is $\rho(x + v \cdot \Delta t) \cdot \Delta x'$.
4. To maintain a steady state, we get:
   $\rho(x) \cdot \Delta x - b_u(x) \cdot \Delta t - b_d(x) \cdot \Delta t + b_u(x + \Delta x) \cdot \Delta t + b_d(x - \Delta x) \cdot \Delta t - s(x) \cdot \Delta x \cdot \Delta t$ equal to
   $\rho(x + v \cdot \Delta t) \cdot \Delta x \cdot K(x + v \cdot \Delta t) / K(x)$

Using elementary calculus, i.e., dividing by $\Delta x \cdot \Delta t$ and letting $\Delta x \to 0$ and $\Delta t \to 0$, we obtain the equation:

$$b_u'(x) - b_d'(x) - s(x) = v \cdot \left[\frac{K'(x)}{K(x)} \cdot \rho(x) + \rho'(x)\right], \tag{8.4}$$

where the prime denotes derivate with respect to x, i.e., $\rho'(x) = d\rho/dx$. By putting Eq. (8.2) into Eq. (8.4), and substituting $s(x)$ by $C \cdot b(x)$, we get:

$$b'(x) - C \cdot b(x) = v \cdot \left[\frac{K'(x)}{K(x)} \cdot \rho(x) + \rho'(x)\right], \tag{8.5}$$

where C is a substrate decomposition constant $\geqslant 0$. If $K(x)$ is given by Eq. (8.1), we may write:

$$b'(x) - C \cdot b(x) = v \cdot \left[\frac{W'(x) \cdot \rho(x)}{100 - W(x)}\right]. \tag{8.6}$$

Integration from $x = 1$ to b_z yields

$$b(x) = b_1 \cdot e^{C(x-1)} + v \cdot \int_1^x e^{C(x-t)} \cdot \frac{W'(t)}{100 - W(t)} \cdot \rho(t) \cdot dt. \tag{8.7}$$

This is a general equation of state which is assumed to be valid for x = 1 to be b_z.

If C is assumed to be zero, which is probably valid in many cases, then the equation becomes:

$$b(x) = b_1 + v \cdot \int_1^x \frac{W'(t)}{100 - W(t)} \cdot \rho(t) \cdot dt. \quad (8.8)$$

This equation can be transformed to an explicit form for any given sediment core after empirical determinations of the water content and the loss on ignition (or organic content). This will now be shown by means of empirical data from Lake Hjälmaren, Sweden (see Håkanson 1981b). The results will be used later on to discuss lake sediment age.

We will utilize a set of empirical data from 20 sediment cores from areas of accumulation in Lake Hjälmaren. Table 8.1 gives mean values for the water content, W(x), the organic content [IG(x) in % and $IG^0(x)$ in % ws] and the bulk density, ρ(x), for different vertical levels, x. The relationship between the organic content in % ws and ds is given by:

$$IG^0(x) = IG(x) \cdot \frac{100 - W(x)}{100}. \quad (8.9)$$

Table 8.1. Mean values on organic content [IG(x) in % ds and IG^0 (x) in % ws], water content, W(x), and bulk density, ρ(x), for different vertical levels, x, in Lake Hjälmaren, Sweden (Håkanson 1981b)

Level (cm)	x (cm)	IG(x) (% ds)	W(x) (% ws)	IG^0(x) (% ws)	ρ(x) ($g\ cm^{-3}$)
0– 1	0.5	12.3	90.4	1.18	1.049
2– 3	2.5	12.1	87.2	1.55	1.082
4– 5	4.5	11.6	85.7	1.66	1.095
6– 7	6.5	11.0	84.0	1.76	1.103
8– 9	8.5	11.1	82.4	1.95	1.109
10–11	10.5	10.5	80.0	2.10	1.113
12–13	12.5	9.9	79.4	2.04	1.117
14–15	14.5	9.6	78.2	2.09	1.121
16–17	16.5	9.0	79.7	1.83	1.124
18–19	18.5	9.6	78.8	2.04	1.126
20–21	20.5	9.1	77.5	2.05	1.129

The density of the inorganic material (ρm) in Lake Hjälmaren is, on average, 2.53 $g\ cm^{-3}$. The bulk density, ρ(x), can be determined according to Eq. (4.7) by using these data on W(x), IG^0(x) and ρm, i.e.:

$$\rho(x) = \frac{100 \cdot 2.53}{100 + [W(x) + IG^0(x)] \cdot (2.53 - 1)}. \quad (8.10)$$

The best fit regression line for the vertical distribution of the water content in Lake Hjälmaren is:

$$W(x) = 90.4 - 2.99 \cdot \ln(2x) \quad (8.11)$$
$$r = 0.93.$$

Thus, the sediment constant K_s is equal to -2.99 and the correlation coefficient between the model data and the empirical data is very good ($r = 0.93$). The relationship between the organic content, IG(x) and the sediment depth, x, can now be determined as:

$$IG(x) = 0.318 \cdot [90.4 - 2.99 \cdot \ln(2x) - 51.2] \quad (8.12)$$
$$r = 0.89.$$

To be able to use the equation of state (8.8), we must dispose of an expression relating the bulk density, $\rho(x)$, to the water content, W(x), and the sediment depth, x. Such an expression can be derived from Eqs. (8.9), (8.10), (8.11) and (8.12). We obtain:

$$\rho(x) = \frac{K_2}{K_3 - [W_{0-1} + K_1 \cdot \ln(x) + K_4]^2} . \quad (8.13)$$

where, for Lake Hjälmaren:

$K_1 = K_s = -2.99;$
$K_2 = 52{,}000;$
$K_3 = 69{,}390;$
$K_4 = -233.93.$

It should be noted that these are calculations based on empirical data on water content and organic content from different vertical levels. From the four given constants (K_1, K_2, K_3 and K_4) and from data on water content of surficial sediments, $W_{0-1} = 90.4$, and bulk density, $\rho_{0-1} = 1.049$, the requested explicit expression for the equation of state can be determined accordingly:

$$b(x) = b_1 + v \cdot K_2 \cdot \left[\alpha_1 \cdot \ln\left|\frac{A - K_1 \cdot \ln(x)}{A}\right| + \alpha_2 \cdot \ln\left|\frac{K_1 \cdot \ln(x) + B - C}{B - C}\right| + \right.$$
$$\left. + \alpha_3 \cdot \ln\left|\frac{K_1 \cdot \ln(x) + B + C}{B + C}\right|\right], \quad (8.14)$$

where $$\alpha_1 = \frac{1}{(A + B + C)(A + B - C)} ; \quad (8.15)$$

$$\alpha_2 = \frac{1}{2 \cdot C \cdot (A + B - C)} ; \quad (8.16)$$

$$\alpha_3 = \frac{1}{2 \cdot C \cdot (A + B + C)} ; \quad (8.17)$$

and
$$A = 100 - K_1 \cdot \ln(2) - W_{0-1} ; \quad (8.18)$$
$$B = W_{0-1} + K_4 ; \quad (8.19)$$
$$C = \sqrt{K_3} . \quad (8.20)$$

With the data for Lake Hjälmaren, this gives:

$$b(x) = b_1 + v \cdot [0.249 \cdot \ln(1 + 0.00735 \cdot \ln x) + 0.759 \cdot \ln(1 - 0.0249 \cdot \ln x)$$
$$- 1.0036 \cdot \ln(1 + 0.285 \cdot \ln x)]. \quad (8.21)$$

This explicit expression shows that the net biotransport, b(x), depends on b_1 (i.e., the net biotransport at 1 cm sediment depth), on v (i.e., the rate of sedimentation) and on the sediment depth (x). For x = 1, we obtain:

$$b(1) = b_1. \tag{8.22}$$

For x = 2, the equation yields:

$$b(2) = b_1 - 0.185 \cdot v. \tag{8.23}$$

Thus, to determine the net biotransport, b(x), for an arbitrary sediment depth (x), we must know b_1 and v. The rate of sedimentation on the bottom, v, can be determined by means of bottom sediment traps or radioisotope techniques (Pb-210 or Cs-137). The b_1-value is causally linked to the bioturbation limit, b_z.

This limit can, for example, be estimated during sediment sampling with the sampling equipment developed by Milbrink (1973) (see Fig. 8.16), as the lowest sediment level where tubuficids are found. The idea is to determine the vertical distribution of, e.g., oligochaetes, and then establish, by extrapolation, the sediment depth underneath which no living worms exist. The sampler illustrated in Fig. 8.16 does not

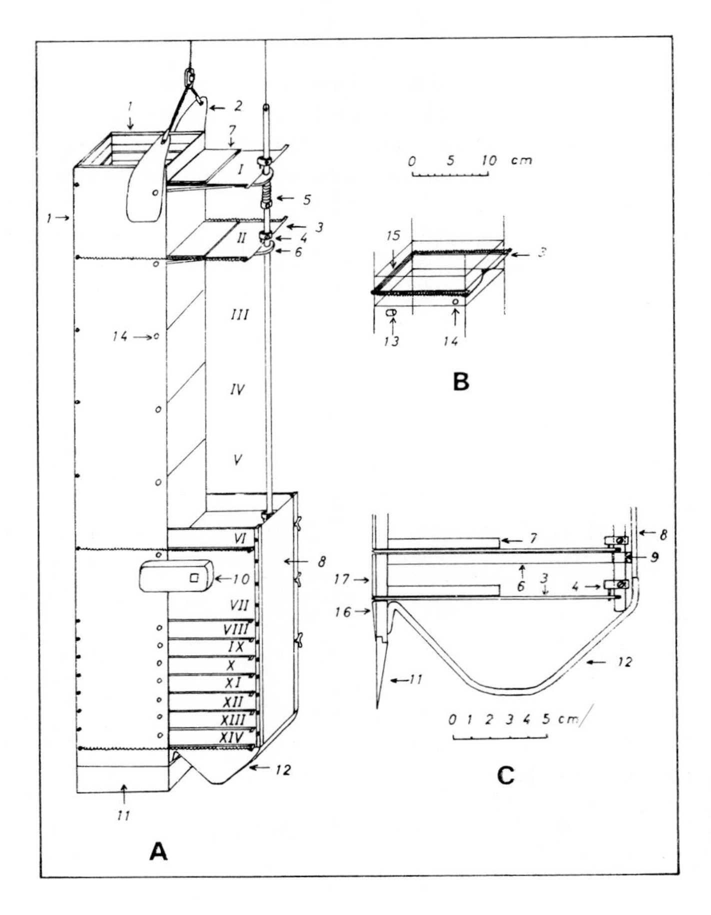

Fig. 8.16. A sub-sampling equipment for in situ stratification of loose sediments (Milbrink 1973)

meet the given requirement for outer clearance, but it provides rapid in situ enclosement, which is important since chironomid larvae, crustaceans and some oligochaetes may be mobile. An alternative to this technique is to equip the sampler (e.g., the modified Ekman sampler) with one side adapted for partitioning slides applied immediately after retrieval.

In Lake Hjälmaren, *Tubifex* has been observed down to 25 cm, which will be used as an estimate of the bioturbation limit in this lake. The rate of sedimentation is, on average, in the range of 0.3–0.5 cm yr^{-1}, and the value 0.4 will be used as a mean to illustrate the use of the model (see Håkanson 1981b).

From these premises, Fig. 8.17 illustrates the relationship between b(x), b_d(x), b_u(x) in Lake Hjälmaren for v = 0.4 cm yr^{-1}, C = 0 and b_z = 25 cm (corresponding to a b_1-value of 0.27, see computer programmes in Appendix 2). From this figure, we can see that the upward biotransport b_u(x) is comparatively large for small x-values and decreases logarithmically with sediment depth to concur with the curve for the net biotransport, b(x), at about 8 cm, when the downward component, b_d(x), is practically negligible.

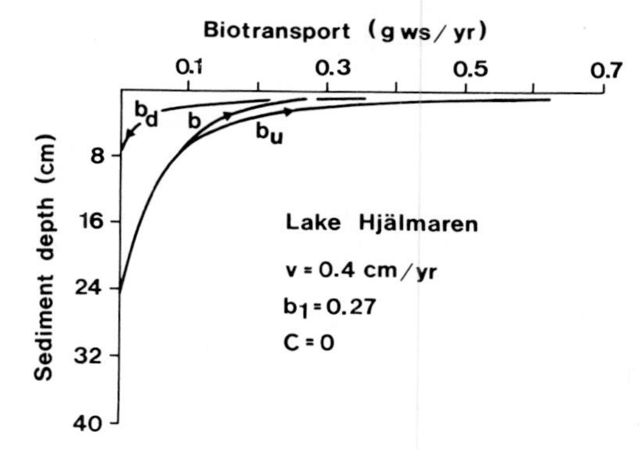

Fig. 8.17. Curves illustrating the relationship between net biotransport *(b)*, upward biotransport *(b_u)* and downward *(b_d)* biotransport in Lake Hjälmaren, Sweden (Håkanson 1981b)

It should be stressed that this model represents a simplification of very complex processes and that it may be difficult to establish the bioturbation limit (b_z) and the corresponding b_1-value. Hence, these discussions are primarily conducted to illustrate important principles and to highlight problems and limitations concerning the possibilities to determine sediment mixing and sediment age. With these things in mind, we will proceed to illustrate some fundamental principles concerning the impact of varying rate of sedimentation, varying bioturbation and varying substrate decomposition on the biotransport of materials in lakes. Using the data from Lake Hjälmaren, Fig. 8.18 illustrates what happens when the rate of sedimentation varies and when the bioturbation and the substrate decomposition are kept constant (b_1 = 0.27 and C = 0). As can be seen from this figure, the curves illustrating the net biotransport will depend on the rate of sedimentation. A high v-value, i.e., a high amount of substrate available for the bottom fauna, will imply a low bioturbation limit, and vice versa. That is, if the v-values are small, the bottom fauna must burrow deep to maintain a given steady supply of food.

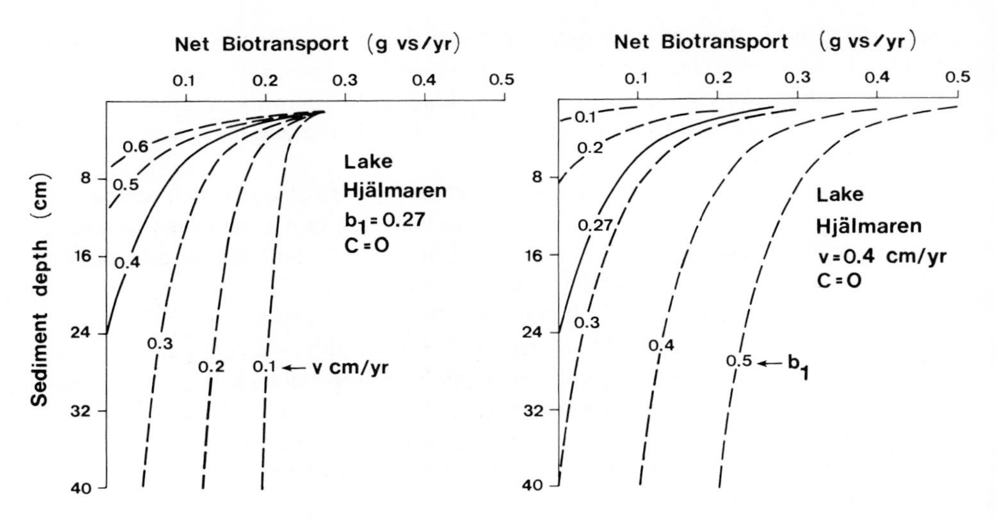

Fig. 8.18. Nomograms illustrating the use of the bioturbation model. In this case the relationship between net biotransport and sediment depth when the rate of deposition *(v)* and the bioturbulent activity *(b_1)* are varied while the substrate decomposition is held constant (C = 0) (Håkanson 1981b)

Figure 8.18B demonstrates, for a constant rate of sedimentation (v = 0.4) and substrate decomposition (C = 0), that there exists a close linkage between the intensity of the bioturbation (as given by b_1) and the depth of the active layer.

8.2.3.2 An Empirical Model

Since it is very difficult to derive practical, general dynamic models of bioturbulent mixing in lake sediments, it is natural to look the other way around and try to establish practically working empirical or semi-empirical models based on input data that can be established with inexpensive standard methods.

The main objective with this section is to discuss one such simple empirical formula based on readily accessible parameters which describe the major causal relationships in a more or less indirect way. Thus, in this case the objective is not to describe the processes of bioturbation but rather the effects, in terms of sediment age, of the processes of bioturbation. In anoxic sediments, with negligible bioturbation (and steady-state assumption), the age of the sediments at level x is simply:

$$t_x = \frac{x}{v}, \qquad (8.24)$$

where t_x = the mean age of the sediments in years;
x = the sediment depth in cm;
v = the rate of sedimentation in cm yr^{-1}.

In sediments with bioturbulent activity, this simple formula would yield erroneous results, as illustrated in Table 8.2 with data from Lake Hjälmaren. The corrected

Table 8.2. The age of the sediments at various vertical levels within areas of accumulation in Lake Hjälmaren

Sediment depth (cm)	Median age (t_x^m)	Diff. ($t_{x+1}^m - t_x^m$)	Uncorrected age (t_x^{-1})	Diff. ($t_x^m - t_x^{-1}$)	Correction factor (t_x^m/t_x^{-1})
0– 1	3.2	0.9	1.25	1.95	2.56
1– 2	4.1	2.3	3.75	0.35	1.09
2– 3	6.4	2.5	6.25	0.15	1.02
3– 4	8.9	2.5	8.75	0.15	1.02
4– 5	11.4	2.6	11.25	0.15	1.01
5– 6	14.0	2.5	13.75	0.25	1.02
6– 7	16.5	2.7	16.25	0.25	1.02
7– 8	19.2	2.6	18.75	0.45	1.02
8– 9	21.8	2.7	21.25	0.55	1.03
9–10	24.5	2.5	23.75	0.75	1.03
10–11	27.0	2.7	26.25	0.75	1.03
11–12	29.7	2.6	28.75	0.95	1.03
12–13	32.3	2.7	31.25	1.05	1.03
13–14	35.0	2.8	33.75	1.25	1.04
14–15	37.8	2.5	36.25	1.55	1.04
15–16	40.3	2.7	38.75	1.55	1.04
16–17	43.0	2.9	41.25	1.75	1.04
17–18	45.9	2.4	43.75	2.15	1.04
18–19	48.3	2.7	46.25	2.05	1.05
19–20	51.0	2.7	48.75	2.25	1.04
20–21	53.7	2.6	51.25	2.45	1.05
21–22	56.3	2.7	53.75	2.55	1.05
22–23	59.0	2.7	56.25	2.75	1.05
23–24	61.7	2.6	58.75	2.95	1.05
24–25	64.3	2.6	61.25	3.05	1.05
25–26	66.9	2.7	63.75	3.15	1.05
26–27	69.6	2.7	66.25	3.35	1.05
27–28	72.3	2.7	68.75	3.55	1.05

median age, t_x^m, in this table has been determined from the equation of state (see also Table 8.3). We may note that the ratio between the corrected (for bioturbation and compaction) median age and the uncorrected mean age (t_x^m/t_x^{-1}) is 2.56 for the surficial sediments (0–1 cm) where the bioturbation is most intense, and close to 1 at all sediments depth below 2 cm. This is important information, since it suggests two things: That the major processes determining the age of the sediments, i.e., upward biotransport versus downward biotransport plus compaction, would level each other out at a certain interval in the sediment column; close to the sediment-water interface, where the sediments are comparatively loose and the bioturbation intense, a correction factor (t_x^m/t_x^{-1}) should be applied to obtain a good determination of the mean age; deeper down in the sediments, where the compaction increases in importance compared to b_u and b_d, another correction factor might be required. However, in this context, we will only focus attention on the surficial sediments of ecological rather than geological interest.

Table 8.3. Rates of sedimentation (v in cm yr^{-1}), mean number of oligochaetes (N ind. m^{-2} during the summer season), mean water content of surficial sediments within areas of accumulation (W_{0-1}), uncorrected mean age of the sediments (t_x^{-1}), corrected median age (t_x^m) and correction factor ($K_t = t_x^m/t_x^{-1}$) for lakes Ekoln, Hjälmaren and Vänern

Lake	v (cm yr^{-1})	N (ind. m^{-2})	W_{0-1} (% ws)	Sed. depth (cm)	Uncorrected mean age (t_x^{-1})	Corrected median age (t_x^m)	Correction factor ($K_t=t_x^m/t_x^{-1}$)
Ekoln	0.8	15,000	80.9	0– 1	0.625	0.91	1.46
				5– 6	6.875	6.85	1.00
				10–11	13.125	11.99	0.91
				15–16	19.375	18.31	0.95
Hjälmaren	0.4	2,000	90.4	0– 1	1.25	3.2	2.56
				5– 6	13.75	14.0	1.02
				10–11	26.25	27.0	1.03
				15–16	38.75	40.3	1.04
Vänern	0.2	750	85.5	0– 1	2.5	6.45	2.58
				5– 6	27.5	28.48	1.04
				10–11	52.5	54.96	1.05
				15–16	77.5	79.96	1.03

Based on empirical data (Table 8.3) from lakes Hjälmaren, Vänern and Ekoln, the following formula, describing the correction factor for the mean age of surficial sediments (K_t^{0-1}) in relation to a v-factor for the rate of sedimentation, a N-factor for the potential bioturbation and a W-factor for the physical sediment character, has been derived (see Håkanson 1982e):

$$K_t^{0-1} = 1 + \frac{(1.78 - v)^3 \cdot 0.77 \cdot \log N}{(100 - W_{0-1})^{0.64}}, \tag{8.25}$$

where v = the rate of sedimentation (in cm yr^{-1});
N = the number of oligochaetes during the bioproductive period (summer);
W_{0-1} = the water content of surficial sediments.

v, N and W_{0-1} can all be determined with standard methods. The main reasons for using the number of oligochaetes (N) are: these animals have a high direct potential for biotransport (see Lindeström 1979), and the number of oligochaetes may also presumably be used as an indirect measure of bioturbation caused by, e.g., bottom-dwelling fishes like bream, who feed on this worm. In the future, it is probable that a more general measure of potential biotransport could be derived by combining some more important species, and not only use the oligochaetes.

The water content (W_{0-1}) is used as a simple and accurate measure of the physical character of the sediments. The general idea is: the looser the sediments, the higher potential for bioturbulent mixing. Thus, to conclude:

– K_t should decrease from a maximum at the top centimetre to 1 at about 3 cm sediment depth. This holds for Lake Hjälmaren (see Table 8.2) as well as for the two other lakes investigated, Vänern and Ekoln. Table 8.3 gives data on mean rate of sedimentation (v), number of oligochaetes (N) and surficial water content

(W_{0-1}). Since these lakes differ in many limnological and sedimentological respects, it is probable that the conclusion about the correction factor (K_t) would also hold for, at least, similar lakes from Nordic environments, but that remains to be demonstrated.

- K_t should increase with N; the more bottom organisms, the more bioturbation, the larger the correction factor.
- K_t should decrease with v; when the rate of sedimentation is very large and/or the deposits are very coarse, one can presuppose a small potential effect of bioturbation, i.e., a low K_t-value.
- K_t should increase with W_{0-1}.

When K_t is determined, the corrected mean age of surficial sediments (0–1 cm) can be determined from the relationship:

$$t_{0-1} = \frac{0.5}{v} \cdot K_t^{0-1} \qquad (8.26)$$

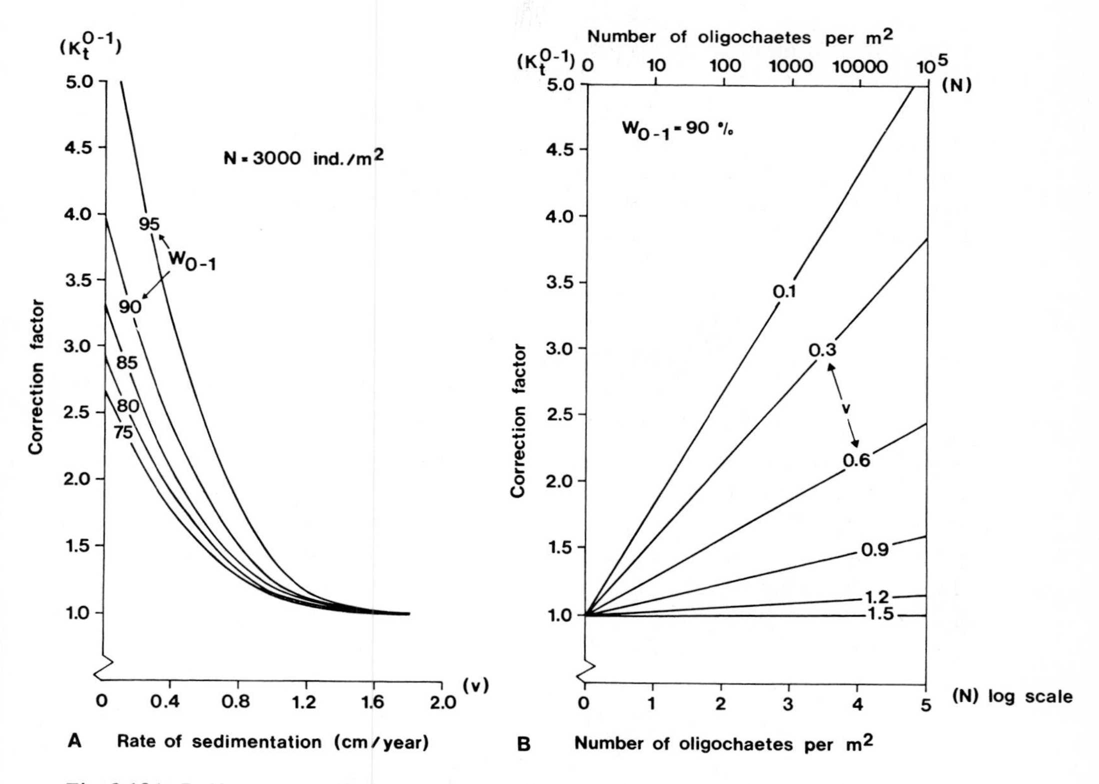

Fig. 8.19A, B. Nomograms illustrating the use of the simple model to determine the correction factor for the determination of the mean age of surficial sediments. **A** The number of oligochaetes is held constant (N = 3000 ind. m^{-2}); **B** the water content of the sediments (W_{0-1} = 90%) is held constant

The correction factor for the interval 1–2 cm, K_t^{1-2}, may be estimated by the following rule of thumb:

$$K_t^{1-2} = 1 + 0.1 \cdot K_t^{0-1}. \tag{8.27}$$

The correction factor for the interval 2–3 cm, K_t^{2-3}, may, in the same manner, be estimated to be:

$$K_t^{2-3} = 1 + 0.01 \cdot K_t^{0-1}. \tag{8.28}$$

No correction factor would be needed below 3 cm (down to about 25 cm).

The empirical relationship (8.25) indicates that the bioturbation of the surficial sediments (0–1 cm) cannot significantly influence the mean age if the rate of sedimentation exceeds 1.78 cm yr^{-1}, see Fig. 8.19. This nomogram demonstrates how the curves for W_{0-1} = 95, 90, 85, 80 and 75 asymptotically reach 1 as v increases towards 1.78. In this particular example, the number of oligochaetes was held at N = 3000 ind. m^{-2}. Figure 8.19B shows how the formula describes the correction factor, K_t^{0-1}, when the water content is constant (90%), but the number of oligochaetes and the rate of sedimentation vary.

From these results it is possible to construct a map showing the areal distribution of the corrected mean age of surficial sediments (0–1 cm). One such map is given in Fig. 8.20 for Lake Ekoln.

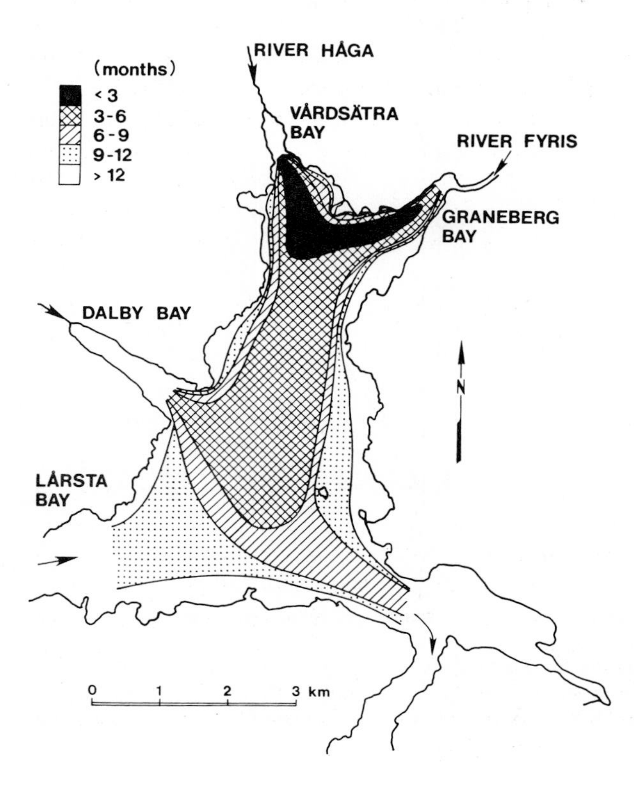

Fig. 8.20. Areal distribution of corrected mean age of surficial sediments (0–1 cm) in Lake Ekoln

8.3 Sediment Age and Age Determination

It is evident that the problem concerning biotransport of material is closely linked to the problem of sediment age and age distribution. We shall therefore discuss principles and qualitative aspects of the bioturbation/time stratification problem and give some empirical examples, illustrating basic causal relationships. We will utilize the equation of state [Eq. (8.7)] as applied to data from Lake Hjälmaren [Eq. (8.21)] to determine the vertical age distribution or the time stratification of deposits from given layers (in this case 1 cm thick) in sediment cores. From the given model and in-data on water content, W(x), and organic content (loss on ignition), IG(x), the age distribution for Lake Hjälmaren sediments can be calculated by means of programme b (time stratification) in Appendix 2 (see Fig. 8.21). It appears in this lake that the median year at the 0–1 cm level is 3.2 years (sediment sampling done in 1977), that 97.3% of the material at this level is younger than 21 years (1956), etc. The limit at 21 years corresponds to the year which does not represent more than 0.5% of the material at the given level. In the same way, the value 0.5% has been used as the limit to the "left" at, e.g., the 3–4 cm level, where the deposits from 1977 for the first time constitute less than 0.5% of the material. The curves for the various sediment intervals illustrate how the median age increases successively with the sediment depth and the increased flatness of the age frequencies. Below the bioturbation limit, at 24–25 cm, the shape of the curves is only influenced by compaction. It should be noted that the median age at, e.g., the 24–25 cm level, only represents about 3% of the material. Thus, knowledge of only the median or the mean age, and not of the age distribution, may be of limited value for sediments affected by bioturbation.

The equation of state may also be used when the bioturbation is negligible, e.g., during anaerobic conditions. Then b(1) is simply zero and the result is an equation which only accounts for compaction. Such test series have been run on the data from Lake Ekoln and Lake Vänern (see Fig. 8.22). Of course, when bioturbation is negligible, the age of the laminated sediments can be determined in a very straightforward and accurate manner by "counting varves". These are important "vertical" aspects of the age problem. The "areal" aspects, linked to the fact that the rate of sedimentation varies within a lake, were illustrated in Fig. 8.20.

8.3.1 Methods of Age Determination

The objective in this section is to discuss some radioisotope techniques to date recent sedimentary deposits. Since the focus is on recent sediments, i.e., deposits from principally the industrial era, we will not discuss paleolimnological techniques (see Reeves 1968), such as the radiocarbon method (C-14) or paleomagnetic dating (see Goudie 1981, for review in these matters). Instead, we will try to give brief outlines on two specific methods in radionuclide limnochronology, namely on lead-210 cesium-137. For a more detailed account on the use of radioactive nuclides to date lake sediments, see Krishnaswami and Lal (1978) or Dominik et al. (1981). Based on origin, the radioactivites may be classified into three categories:

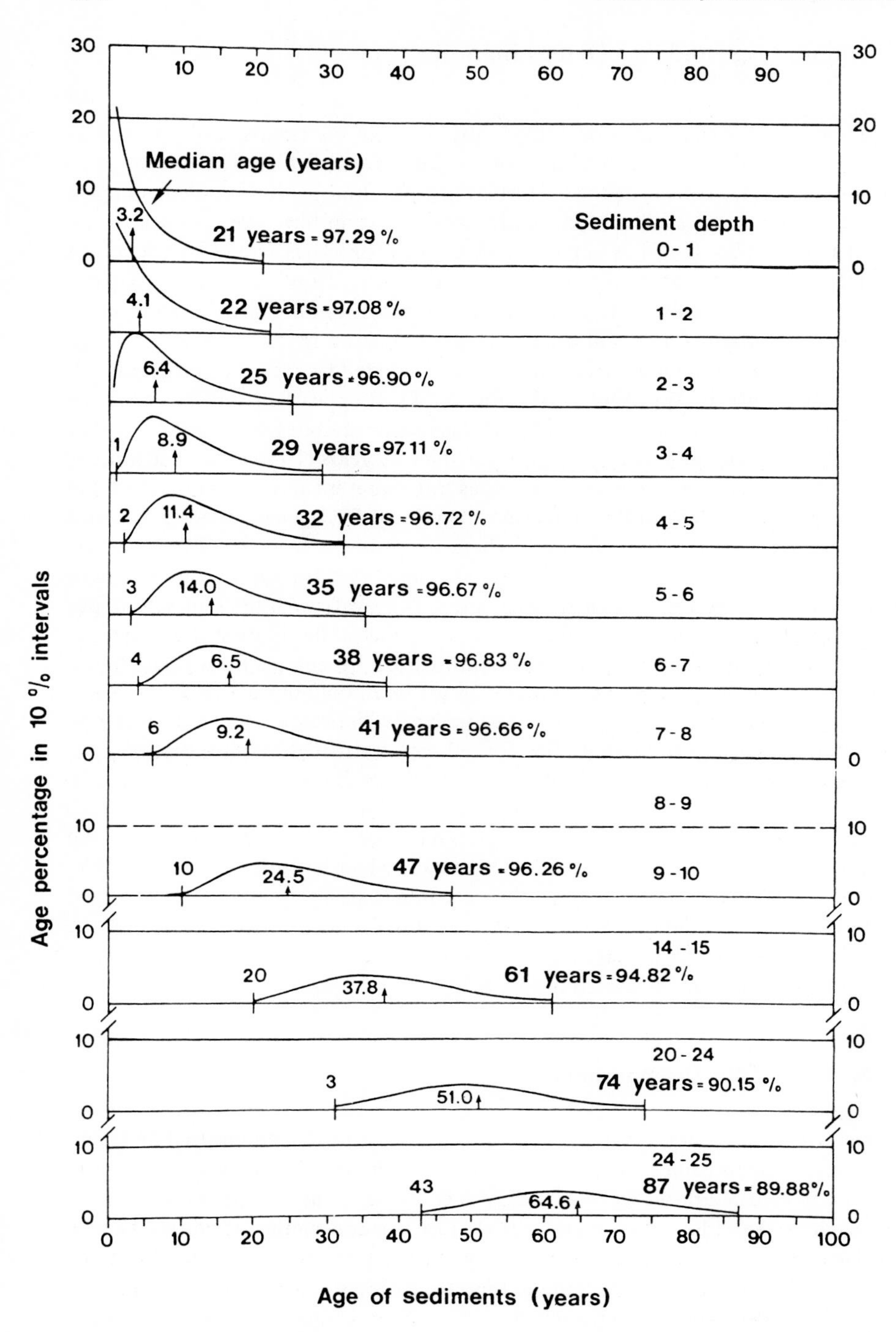

Fig. 8.21. Age frequencies at various vertical sediment intervals in Lake Hjälmaren

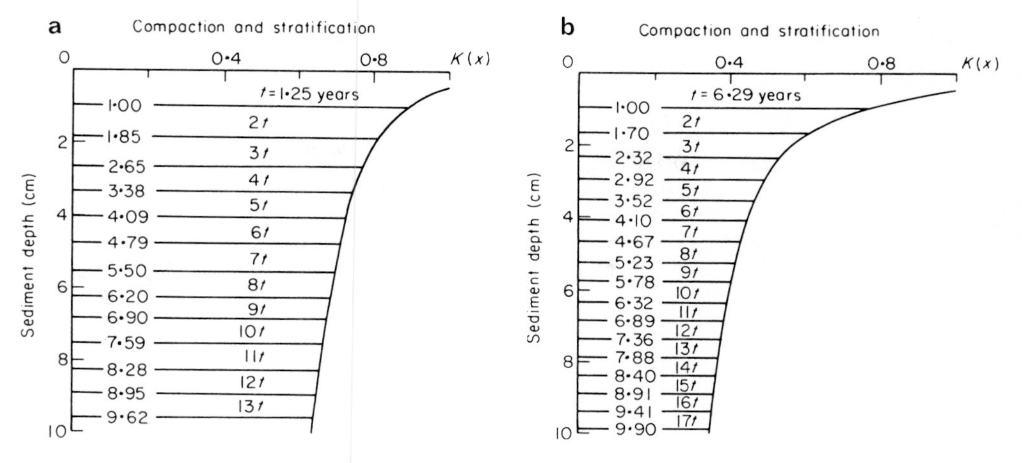

Fig. 8.22a, b. Average vertical sediment compaction and calculated time stratification when the effect of bioturbation is neglected in **a** Lake Ekoln, rate of sedimentation 0.8 cm yr^{-1}, and **b** Lake Vänern, rate of sedimentation 0.2 cm yr^{-1} (Håkanson and Källström 1978)

– *Primordial* nuclides, which were present during the formation of the earth and are still present due to their very long half-lives ($> 10^8$ years), e.g., K-40, Pb-87, Th-232, U-235 and U-238. Some of the daughter nuclides produced from these primordial nuclides have half-lives short enough to be of interest in our present context. Figure 8.23 illustrates the natural mode of decay for Uranium-238 to Lead-206 via Lead-210, the latter of great interest in dating recent deposits, since it has a half-life of 22.3 years.

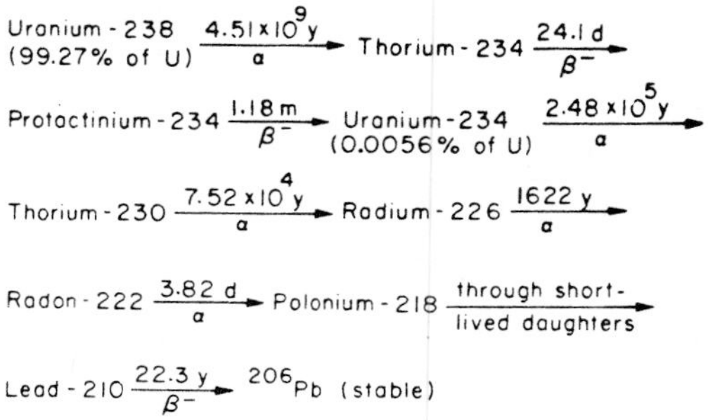

Uranium-238 (99.27% of U) $\xrightarrow[\alpha]{4.51 \times 10^9 y}$ Thorium-234 $\xrightarrow[\beta^-]{24.1 d}$

Protactinium-234 $\xrightarrow[\beta^-]{1.18 m}$ Uranium-234 (0.0056% of U) $\xrightarrow[\alpha]{2.48 \times 10^5 y}$

Thorium-230 $\xrightarrow[\alpha]{7.52 \times 10^4 y}$ Radium-226 $\xrightarrow[\alpha]{1622 y}$

Radon-222 $\xrightarrow[\alpha]{3.82 d}$ Polonium-218 $\xrightarrow[\text{lived daughters}]{\text{through short-}}$

Lead-210 $\xrightarrow[\beta^-]{22.3 y}$ ^{206}Pb (stable)

Fig. 8.23. The natural U-series of radioactive nuclides, their half-lives and mode of decay (Krishnaswami and Lal 1978)

– *Cosmic ray-produced* nuclides are created continuously in and on the earth from components of the cosmic radiation. These nuclides have a range in half-life spanning from a few minutes to millions of years. Table 8.4 lists isotopes with half-lives longer than one day.

– *Man-made* nuclides have appeared during the last decades primarily due to testing of nuclear weapons. The most important of these isotopes are given in Table 8.5.

Table 8.4. Isotopes with half-lives longer than one day produced by cosmic rays in the atmosphere (Krishnaswami and Lal 1978)

Isotope	Half-life	Main radiation	Main target nuclide(s)
^{3}He	Stable	–	N, O
^{10}Be	1.5×10^{6} years	β^- – 550 KeV	N, O
^{26}Al	7.2×10^{5} years	β^+ – 1.17 MeV	Ar
^{36}Cl	3.0×10^{5} years	β^- – 714 KeV	Ar
^{81}Kr	2.1×10^{5} years	K – X-ray	Kr
^{14}C	5730 years	β^- – 156 KeV	N, O
^{32}Si	~ 300 years	β^- – 100 KeV	Ar
^{39}Ar	270 years	β^- – 565 KeV	Ar
^{3}H	12.3 years	β^- – 18 KeV	N, O
^{22}Na	2.6 years	β^+ – 540 KeV γ – 1.3 MeV	Ar
^{35}S	87.5 days	β^- – 167 KeV	Ar
^{7}Be	53.4 days	γ – 480 KeV	N, O
^{37}Ar	35 days	K – X-ray	Ar
^{33}P	25.3 days	β^- – 250 KeV	Ar
^{32}P	14.3 days	β^- – 1.7 MeV	Ar

Table 8.5. Principal artificial radionuclides (Krishnaswami and Lal 1978)

(a) Fission products

Nuclide	Half-life	Fission yield from ^{235}U (%)
^{89}Sr	50.4 days	4.8
^{90}Sr	28.5 years	5.8
^{95}Zr	63.3 days	6.3
^{106}Ru	1 year	0.4
^{137}Cs	30 years	6.0
^{141}Ce	32.5 days	6.0
^{144}Ce	285 days	5.7
^{147}Pm	2.5 years	2.4

(b) Induced Activities

Nuclide	Half-life	Nuclide	Half-life
^{3}H	12.3 years	^{55}Fe	2.7 years
^{14}C	5730 years	^{57}Co	270 days
^{35}S	87.5 days	^{58}Co	72 days
^{51}Cr	27.8 days	^{60}Co	5.3 years
^{54}Mn	312 days	^{65}Zn	244 days
		^{239}Pu	2.43×10^{4} years
		^{241}Pu	14.9 years

Lakes receive these nuclides either from land (rivers, ground water, weathering) or from the atmosphere (wet precipitation and dry fallout). Fig. 8.24 illustrates major pathways of lead-210 to lake sediments. Radioactive dating is generally based upon either the decay of a radioactive nuclide or the build-up of a daughter nuclide.

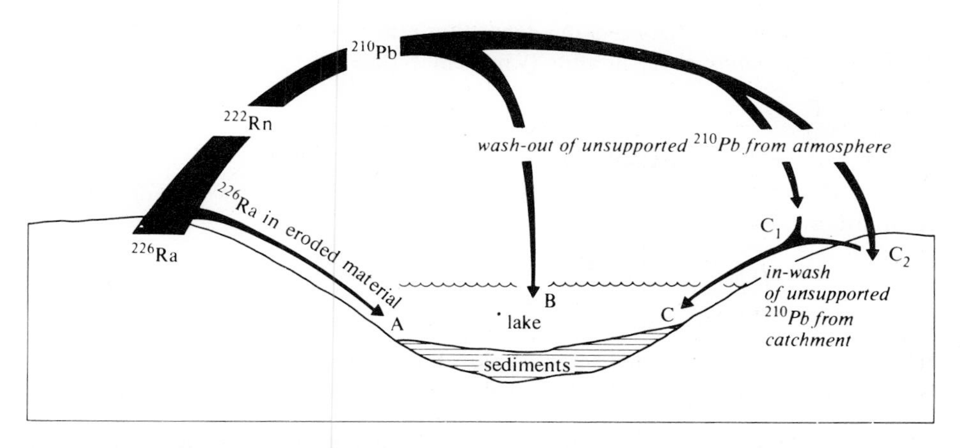

Fig. 8.24. Pathways of Pb-210 input to lake sediments. (Modified from Wise 1980, presented in Goudie 1981). *A* ^{210}Pb, the radioactivity of which is 'supported' by its parent isotope ^{226}Ra. *B* 'Unsupported' ^{210}Pb falling directly on the lake. *C* 'Unsupported' ^{210}Pb falling on the catchment; C_1 reaches the lake immediately via the drainage net; C_2 is adsorbed onto clay and organic matter in soils and may reach the lake after a time lag. *A + B + C* Total ^{210}Pb; *B + C* 'unsupported' ^{210}Pb

8.3.1.1 Lead-210

Lead-210 reaches lake sediments in two forms:

– an "unsupported" component from the atmosphere;
– a "supported" component derived directly from the soil/substrate of the drainage area.

To determine dates by lead-210 analysis, it is necessary to establish the total lead-210 as well as the "supported" lead. The "supported" lead, not having been in atmospheric contact, emanates from radium-226 in the U-series (Fig. 8.23). The amount of "supported" lead-210 may vary due to varying sedimentological conditions. It is important to measure the radium-266 content of the sample to get an adequate figure of the "supported" lead-210. Total lead-210 can be determined by direct gamma-ray assay or indirectly by measuring daughter (bismuth-210) or grand-daughter (polonium-210) isotopes. Subtraction of "supported" from the total yields "unsupported" lead-210. The procedure to actually date the sediments using lead-210 depends on the assumptions made concerning the way in which lead-210 is incorporated in the deposits (see Robbins 1978, Appleby and Oldfield 1979). Commonly, it is agreed that lead-210 does not migrate in the sediment column, and it is also assumed that the flux to the sediments is constant. Then the activity (A) of the nuclide in the sediments at depth x is given by:

$$A(x) = A_0 \cdot e^{-\lambda \cdot t}, \tag{8.29}$$

where A_0 = the activity (in disintegrations per minute per gram of sediments = dpm g^{-1}) in a freshly deposited sediment at time t = 0;

λ = the radioactive decay constant. (To avoid confusion, it may be noted that λ, traditionally, also stands for wave length.)

A_0 is the quotient between the deposition rate of the nuclide (F in dpm cm^{-2} $time^{-1}$) to the rate of sedimentation (v in g cm^{-2} $time^{-1}$), i.e.:

$$A_0 = \frac{F}{v}. \tag{8.30}$$

The age (t in years) of sediments at depth (x) can be written as (under steady state assumptions):

$$t = \frac{1}{\lambda} \cdot \ln \left(\frac{Q_0}{Q(x)} \right), \tag{8.31}$$

where Q_0 = the integrated activity below the sediment-water interface (x = 0);
$Q(x)$ = the integrated activity below sediment depth x.

Figure 8.25 shows age/depth relationships in a dated sediment core from the central area of Lake Constance (Dominik et al. 1981). The term *excess* lead-210 in this figure stands for "unsupported" lead-210. In this particular example, the rate of sedimentation was determined to be 0.062 g cm^{-2} yr^{-1} on the average, or 2.4 mm yr^{-1} as freshly deposited sediments, or 1.4 mm yr^{-1} on the average.

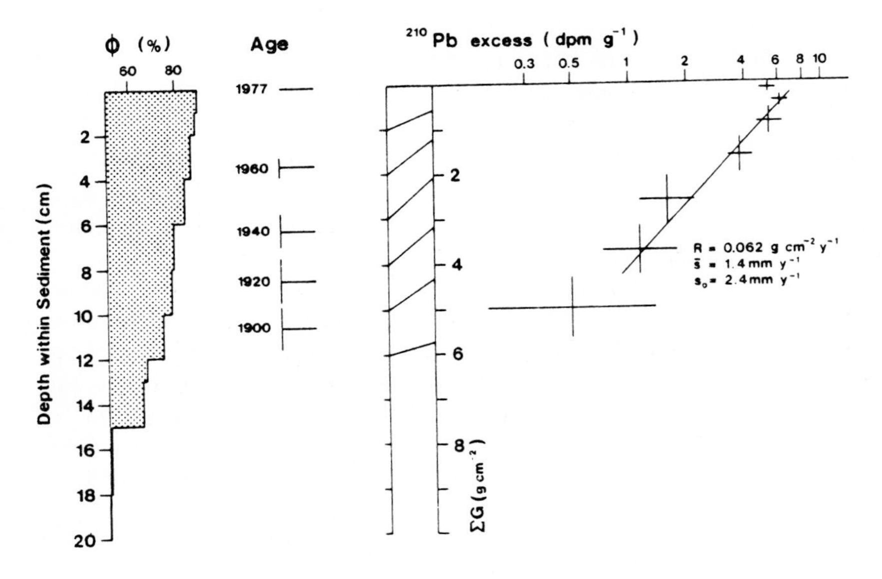

Fig. 8.25. Lead-210 distribution, sediment age and sediment characteristics from the central part of Lake Constance (Bodensee). Pb-210 excess is given against the cumulative weight of dry sediments deposited on one cm^2 (Σ G). *R* rate of sedimentation in g/cm^2 yr^{-1}; $\bar{s}$ average rate of sedimentation (compaction neglected) in mm yr^{-1}; s_o freshly deposited sediments per year in mm ϕ porosity. *Vertical bars* denote errors of calculation from Pb-210 data. (Modified from Dominik et al. 1981)

8.3.1.2 Cesium-137

Cesium-137 is an artificial radioisotope emanating from nuclear explosions. Figure 8.26 shows the atmospheric fall-out of cesium-137 from the 1950's and onwards. The fall-out began in 1954 and the peak activity took place around 1963. This curve is used to calibrate cesium-137 concentrations in lake sediments. The half-life of Cs-137 is 30 years (see Table 8.5). As with other dating techniques, bioturbation may imply interpretational problems. Unlike lead-210, cesium-137 can migrate in the sediment column, and downward diffusion combined with delayed cesium-137 input from soil erosion may be a source of error concerning evaluations of data.

The example given in Fig. 8.26 shows that the 1963 peak is clearly indentified in the sediments of Lake Windermere. Ideal conditions for Cs-137 dating is rapid and predominantly authochthonous sedimentation and minimum bioturbation.

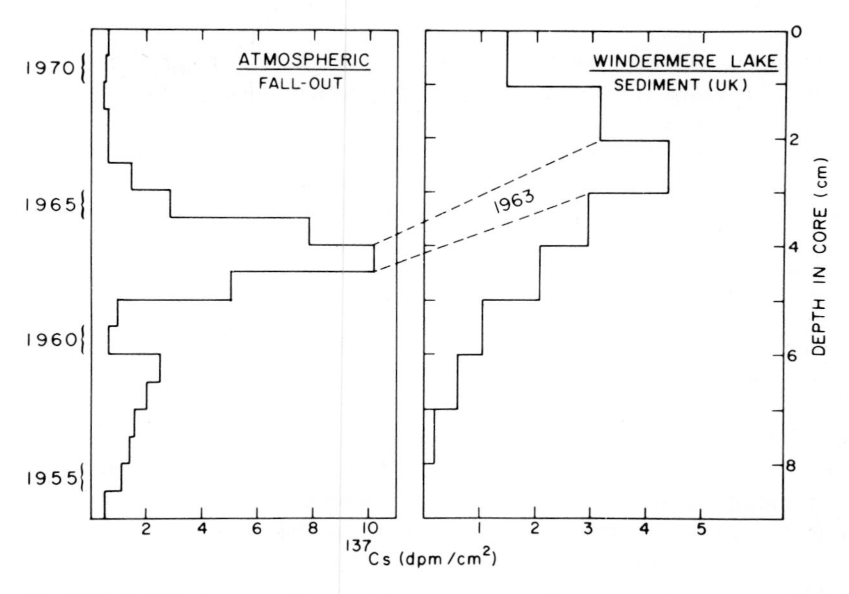

Fig. 8.26. Cs-137 depth profile in sediments from Lake Windermere and its atmospheric fall-out at the same location. (Data from Pennington et al. 1973, diagram from Krishnaswami and Lal 1978)

9 Release of Substances from Lake Sediments – the Example of Phosphorus

9.1 Background and Presuppositions

So far, this book has described sedimentation processes and conditions within the sediments. In this chapter, we shall discuss some of the factors and processes which can influence the transport of elements from the sediments back to the lake water. From an ecological standpoint, such fluxes can be very important when the recirculated elements are essential nutrients or toxic substances, e.g., heavy metals.

In many cases nutrients and other elements are returned to lake water in gaseous form either by diffusion of dissolved gas or as gas bubbles. Gas formation is essentially the result of microbial processes in the sediments and is discussed in Chapter 5.4. Here, we shall concentrate on how an element without a gaseous phase is recirculated to the lake water by the influence of physical, chemical and biological factors. Phosphorus will be used as the model element in this context for several reasons:

- the prerequisites necessary for phosphorus release from sediments are in many features representative for recirculation processes for most elements without a gas phase;
- the release of phosphorus is a result of a complicated pattern of interacting and counteracting processes which deserve a detailed treatment to be fully understood;
- release of phosphorus from lake sediments is an important environmental problem since this flux often significantly contributes to the overall phosphorus loading of lake water and thereby increases the trophic level.

Thus, we consider that the complicated network of factors and processes behind the release of elements to lake water may be pedagogically illustrated and understood from a rather complete discussion of a single and in this context important element, namely phosphorus.

For a closer penetration of the phosphorus-sediment interactions, the literature surveys of Syers et al. (1973) and Boström et al. (1982) can be recommended. The conclusions and illustrations presented in the latter paper constitute the backbone of the following presentation. Related problems concerning heavy metals are described by Förstner and Wittmann (1979).

The knowledge that phosphorus under certain conditions is released from the sediments is of old date. In the years around 1940, it was very convincingly demonstrated how lake sediments released phosphorus to the superimposed water when the sediment surface became reduced (Einsele 1936, 1937, 1938, Ohle 1937, Mortimer 1941,

1942). The phosphorus release was explained by the reduction of iron (III) to iron (II) whereby phosphorus bound to iron (III) in salts or adsorbed to iron complexes was returned to solution. These results had an enormous impact, and to some degree it seems justified to say that these early findings paralysed the research in this field, at least concerning soft water lakes. For nearly 40 years one of the limnological-sedimentological dogmas stated that phosphorus was bound in the sediments when these were superimposed by oxygenated water and was released from the sediments when the bottom water became anoxic. However, in recent years, there has been an increasing awareness that phosphorus is released from sediments also under apparently well-oxygenated conditions. This has, for example, been observed in a number of shallow lakes which have received sewage water. Highly reduced loadings of phosphorus seldom gave the expected results since the decrease in external input was compensated by a net flux from the sediments. To gain knowledge and obtain correctives concerning release of phosphorus from lake sediments is therefore an important task in contexts related to lake management.

It should be emphasized that in all oligotrophic lakes and also in many eutrophic waters, the net flux of phosphorus on a yearly basis is towards the sediments. However, under certain conditions, particularly in nutrient-rich lakes, the release from the sediments may exceed the input by sedimentation; a phenomenon commonly termed *internal loading.*

The magnitude of the potential phosphorus source in the sediments is illustrated in Fig. 9.1. It is evident that only limited parts of the phosphorus pool in the sediments has to be released to the lake water in order to affect the concentration of phosphorus in the water column considerably.

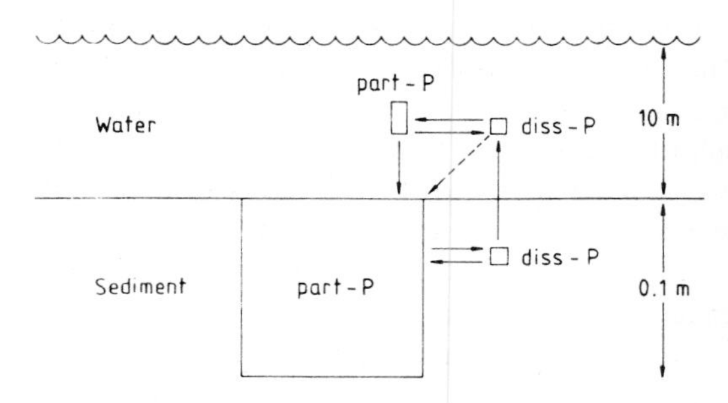

Fig. 9.1. Schematic illustration of the distribution of particulate and dissolved phosphorus in a lake-sediment system. A hypothetical example assuming a mean lake depth of 10 m, an "active" sediment layer of 0.1 m, that the concentration of total-P is 50 $\mu g\ l^{-1}$ in the lake water and 2 mg g^{-1} ds in the sediments, and a water content of the sediments of 90% (Boström et al. 1982)

The transport of phosphorus from sediments (except from resuspension) occurs from the comparatively small pool of phosphorus dissolved in the sediment-pore water. The size of this pool is in turn regulated by the equilibria with particulate phosphorus. Consequently, before a substantial release can occur, two fundamentally different processes must function more or less simultaneously. Phosphorus must be *mobilized* from particulate to dissolved form by the action of physical, chemical and biological reactions and the dissolved phosphorus must be *transported* to the lake mainly through physical processes (Fig. 9.2).

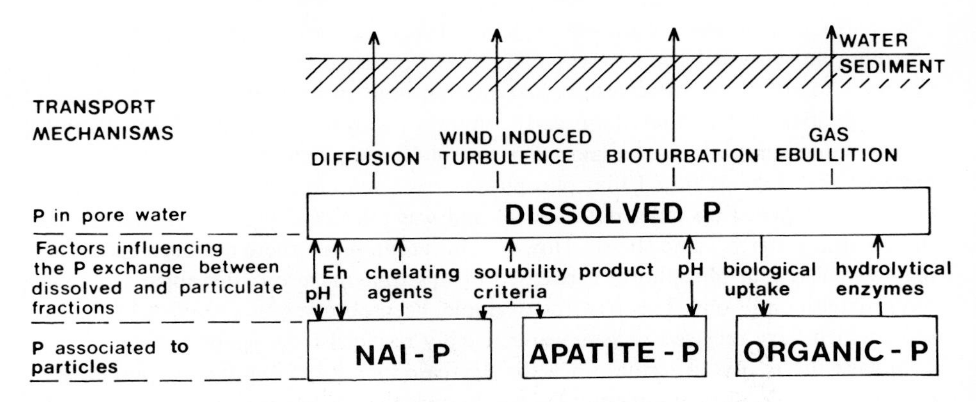

Fig. 9.2. Illustration of the dominating processes regulating the release of phosphorus from lake sediments

9.2 Factors of Importance for Mobilization of Phosphorus

9.2.1 Fractional Distribution of Particulate Phosphorus

The more phosphorus that is supplied to sediments, the lower the sediment's capacity to store and retain the phosphorus. With an excessive loading, such as in lakes polluted by sewage water, the sediments will eventually become saturated. If the sediments in all lakes had similar capacity to bind phosphorus, the tendency to release phosphorus should be proportional to the concentration of phosphorus in the sediments. In reality, however, it is more complicated since the binding capacity depends on the chemical associations of phosphorus in the sediments.

The forms of phosphorus in lake sediments were presented in Chapter 4. In this section, we shall only briefly repeat some major features necessary for the understanding of the subsequent text.

Analyses of discrete phosphorus minerals, type of complexes, or organic phosphorus compounds, etc., in sediments have not been made in great detail. Existing knowledge of sediment phosphorus composition results from chemical extraction schemes (Fig. 4.20), which only account for a limited number of fractions. According to Williams et al. (1976), existing extraction schemes are valid only to yield NAI-P (non-apatite inorganic phosphorus, which mainly consists of phosphorus associated with iron and aluminium complexes), apatite-P and organic-P.

Some authors prefer to characterize the fractions by the extraction media used to dissolve them. Hieltjes and Lijklema (1980) use the following terms:

- NaOH-extractable phosphorus (approximately equal to NAI-P),
- HCl-extractable phosphorus (approximately equal to apatite-P),
- residual-P (mainly organic-P),
- NH_4Cl-P (very loosely sorbed phosphorus).

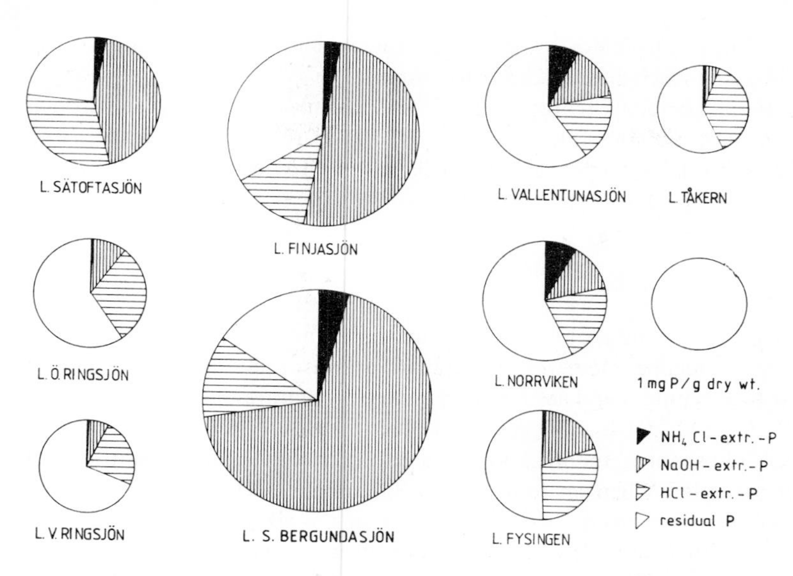

Fig. 9.3. Total concentration and fractional composition of phosphorus in the sediments of nine Swedish lakes. The total concentrations are proportional to the *size of the circles* (Boström 1982)

One example of different total phosphorus concentrations and fractional distribution from some eutrophic Swedish lake sediments is given in Fig. 9.3. The different fractions are highly variable in their reactions vis-à-vis different environmental changes. Generally, the iron or aluminium phosphorus are the most unstable, while "organic" or residual phosphorus is more stable.

9.2.2 Redox Conditions

As already emphasized, the classical model describing the phosphorus exchange between sediments and lake water is based on the interaction between iron and phosphorus during aerobic and anaerobic conditions.

In oxidized situations, phosphorus is sorbed to iron (III) mainly in the form of iron (III) hydroxide gels. Such aggregates may be formed in the oxidized surface sediments or precipitated from overlying waters. When the sediments become reduced, iron (III) is transformed to iron (II), whereby both iron and sorbed phosphate is returned to solution. Dissolved phosphate is then transported back to the water column from the sediment pore water or the sediment surface.

These early findings are still valid and it is beyond doubt that this redox-dependent process is one of the most important reasons why phosphorus can be released from lake sediments. It is most common in productive lakes during stagnation periods (summer and winter) when oxygen is depleted in the bottom water.

The reduction of iron (III) takes place at redox potentials around 200 mV, which correspond to an O_2 concentration of approximately 0.1 mg l^{-1}. It usually entails exhaustion of oxygen, but the process can be hindered by high concentrations of nitrate or mangenese, which are reduced before iron.

Iron reduction is generally considered to be a chemical process where iron is reduced by, e.g., organic compounds or S^{2-}. However, recent research (see, e.g., Sørensen 1982) suggests that iron can be reduced also by bacteria, which utilize iron (III) as electron acceptor during anaerobic respiration.

9.2.3 pH

Redox-regulated phosphorus mobilization occurs independent of pH in the neutral and slightly alkaline regime. However, the strength by which phosphorus is bound to iron, as well as to aluminium and calcium compounds, is in different ways dependent on pH.

pH-mediated phosphorus mobilization is a rather common and quantitatively important phenomenon. When an iron (III) or aluminium (III) hydroxide gel with adsorbed phosphorus is subjected to a pH increase in the alkaline regime, adsorbed phosphate is substituted by hydroxide ions and returned to solution (Fig. 9.4). This type of phosphorus mobilization is most likely to occur in sediments of shallow productive lakes, where the pH in the lake water often increases by two or three units due to the photosynthetic activity of planktonic algae, and where the whole water mass is in more or less continuous contact with the sediment surface. The commonly observed increase in pH during summer in such lakes is often paralleled by raised levels of phosphorus (Fig. 9.5). The covariation of these two variables, as illustrated in Fig. 9.5, is evidently no proof that one of them regulates the level of the other; in this case such observations can be seen as valuable complements to well-controlled laboratory experiments, like those given in Fig. 9.4.

Calcium-bound phosphate is influenced by pH changes in quite the opposite manner, as compared to iron and aluminium. Thus, the formation of hydroxy-apatite (Ca-P mineral) upon settling $CaCO_3$ particles is favoured by high pH (Stumm and Leckie 1971). This means that a pH increase in calcium-rich waters may have the effect that phosphate is precipitated out of the water column and fixed in the sediments.

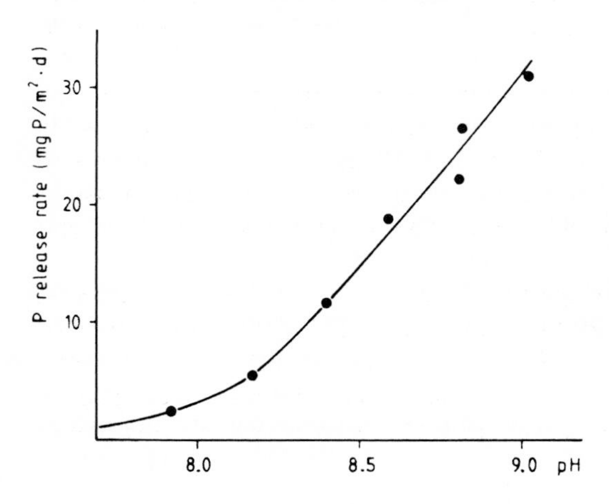

Fig. 9.4. The relationship between release rate of phosphorus and pH for sediments from Lough Neagh (Rippey 1977)

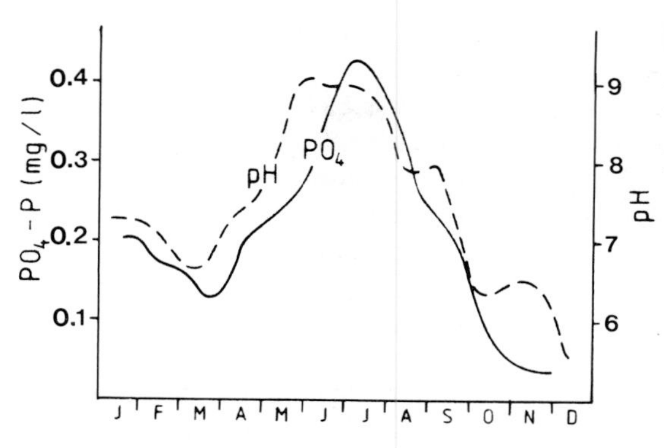

Fig. 9.5. Seasonal variation in pH and phosphate concentration in Lake Glaningen, Sweden. (From Boström et al. 1982, as modified from Ryding and Forsberg 1977)

Consequently, calcium-bound phosphorus in the sediments may be redissolved by decreasing pH, which often follows the CO_2 production associated with microbial degradation of organic matter.

9.2.4 Microbial Mineralization

In this context, mineralization stands for production of dissolved inorganic phosphate by means directly related to bacterial metabolism. It should be stressed that bacterial activity affects the mobilization of phosphorus also in other indirect ways, as mentioned earlier and which will be discussed later in this chapter.

In mineralization processes, organic phosphate esters are hydrolyzed by enzymes called phosphatases which are often located on the bacterial cell surfaces or occurring as free dissolved enzymes. Hydrolysis rates are often optimal at slightly alkaline conditions, hence the term alkaline phosphatases. Very few studies have been conducted on these enzymes in lake sediments and the role of phosphatases for the total mobilization of phosphorus is unknown. However, a number of investigations on phosphatases have been made in lake water and although the ecological relevance of the phosphatases remains to be proved, it seems evident that they are very efficient in hydrolyzing natural organic substrates (see, e.g., Jansson 1977). On the other hand, the "organic phosphorus" fraction of lake sediments appears to be very resistant. The analytical routines used for the characterization of organic phosphorus in lake sediments are, however, crude and rarely include compounds suitable as phosphatase substrates. It is therefore not unlikely that mineralization of organic phosphorus by phosphatases is important for the mobilization of phosphorus supplied from dead or dying organisms or from the sediment bacteria themselves. The turnover rate of this process may be so rapid that detectable amounts of these easily degradable compounds are never accumulated.

Another way by which bacteria might be able to mobilize phosphate associated to particulate matter is by excreting chelating agents. Contrary to phosphatases, these act on inorganic compounds. Harrison et al. (1972) found that chelating organic acids were formed by bacteria isolated from lake sediments and that these replaced phosphorus as ligands to metal ions analogous to EDTA. The order of P-solubilization on the phosphates tested were:

$$CaHPO_4 > Ca_3(PO_4)_2 > Fe(PO_4) > Mg_3(PO_4)_2 > Al_3(PO_4)_2 .$$

9.2.5 Equilibrium Reactions

The stability of inorganic precipitates of phosphorus has been found to depend on factors like redox potential and pH. When their formation is a result of equilibrium reactions, these inorganic compounds are sensitive to concentration changes in the sediment pore water. Thus, if the phosphate concentration decreases in the pore water, it is likely that this decrease is compensated by dissolution of particulate phosphates.

It is not yet possible to give predictions on how, or at which concentrations, these reactions occur; not even for discrete mineral forms, such as hydroxy-apatite, with well-defined solubility products. The reason for this is that the sediments are too complex an environment to allow the application of laboratory results, like solubility product criteria, which are obtained in more controlled and "pure" systems.

9.3 Transport Mechanisms

Under this heading, we shall discuss some means by which phosphorus can be transported to the lake water after being dissolved in the pore water.

9.3.1 Phosphorus in Sediment Pore Water

As illustrated in Fig. 9.1, the phosphorus in pore water generally constitutes a very small portion (less than a few percent) of the total sediment phosphorus. However, this dissolved phosphorus is very important, since it is highly mobile and in direct contact with lake water. Knowledge of pore water concentrations is therefore essential when studying and predicting the exchange of phosphorus (and other elements) across the sediment-water interface.

Whatever technique is used to separate pore water from sediment particles (Chap. 3.6), it is essential to handle and interpret pore water data with great care. Enell (1979) studied the influence of several parameters on the sampling and analysis of pore water phosphorus and concluded the following items as crucial for the result:

- time between sampling and processing or analysis;
- temperature during storage or processing;
- redox conditions during sampling, storage and processing.

Since standard procedures do not exist, it is evident that difficulties arise when results from different investigations should be interpreted. Thus, the pore water phosphorus concentrations from different lakes given in Table 9.1 should be regarded only as information on the "order of magnitude". The data in Table 9.1 does, however, reflect two very important features. Firstly, the concentration of phosphorus in pore water is generally considerably higher (5–20 times) than corresponding values in lake water. This is important since it implies that a pronounced concentration gradient exists at the sediment-water interface, which will favour diffusion of phosphorus from sediments to lake water.

Table 9.1. Phosphorus concentrations in sediment pore water from different lakes (Boström et al. 1982). MRP, molybdate reactive phosphorus

	MRP (mg/l)	Tot-P (mg/l)	Trophic state	Reference
L. Stugsjön (0–5 cm)	0.02	0.09	Oligotrophic	Jansson (unpubl.)
L. Hymenjaure (0–5 cm)	0.02	0.10	Oligotrophic	Jansson (unpubl.)
8 lakes in ELA, Canada (0–20 cm)		0.320 (0.012–1.64)	Oligotrophic	Brunskill et al. (1971)
L. Fiolen "surface"	0.02		Oligotrophic	Ripl and Lindmark (1979)
L. Skärshultssjön "surface"	0.02		Oligotrophic	Ripl and Lindmark (1979)
L. Hinnasjön "surface"	0.02		Oligotrophic	Ripl and Lindmark (1979)
Castle Lake (0–5 cm)		0.02	Mesotrophic	Neame (1977)
L.S. Bergundasjön "surface"	3.16		Eutrophic	Ripl and Lindmark (1979)
L. Vombsjön "surface"	1.19		Eutrophic	Ripl and Lindmark (1979)
L. Lillesjön "surface"	12.7		Eutrophic	Ripl and Lindmark (1979)
L. Oxundasjön (0–4 cm)	2.20	2.25	Eutrophic	Hillerdal (1975)
L. Kinneret (0–22 cm)	0.12		Eutrophic	Serruya et al. (1974)
L. Schöhsee (0–8 cm)	–	0.14	Eutrophic	Ohle (1964)
L. Schlüensee (0–8 cm)	–	0.27	Eutrophic	Ohle (1964)
L. Gr. Plönersee (0–8 cm)	–	1.20	Eutrophic	Ohle (1964)
L. Plußsee (0–8 cm)	–	2.10	Eutrophic	Ohle (1964)

Secondly, it is clear that the difference in concentration is generally great between oligotrophic and eutrophic lake sediments. In contrast to the total phosphorus content of lake sediments, the pore water concentration thus reflects the trophic state of the lake. The values in Table 9.1 are all derived from one sampling occasion with the exception of Lake Kinneret, Israel, where the value is a mean from several samples during a year. Such data can be misleading since the pore water phosphorus concentration, at least in productive lakes, can be highly variable due to changing environmental conditions. This has been demonstrated in an investigation in Lake Mendota (Fig. 9.6). Pore water phosphorus concentrations were followed during the years 1974 and 1975 in different sediment layers at two locations, one shallow and one well below the thermocline.

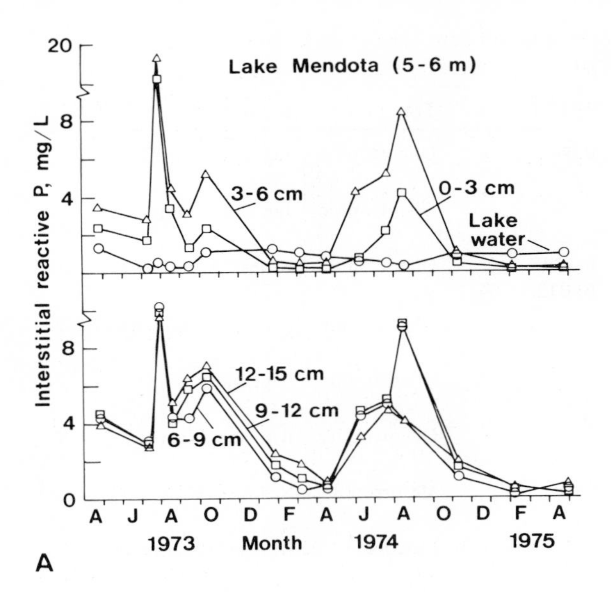

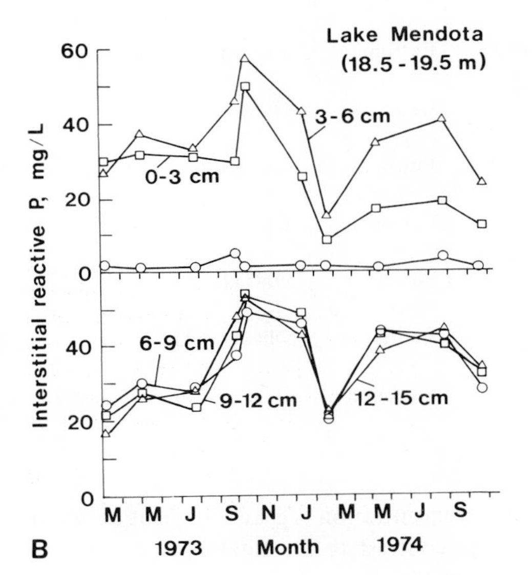

Fig. 9.6A, B. Seasonal variation in the concentration of phosphorus in pore water of Lake Mendota sediments. **A** Data from 5 to 6 m water depth. **B** Data from 18 to 19 m water depth (Holdren et al. 1977)

There were considerable variations during the year, particularly in the shallow sediments. The generally higher values in the deep sediments were interpreted as reflecting the restricted mixing and consequent higher tendency for iron (III) reduction in these areas.

Finally, it should be mentioned that very little is known about the chemical composition of pore water phosphorus. Most investigations have focused on analysis of either or both of the parameters MRP (molybdate reactive phosphorus) or total dissolved phosphorus. MRP is usually interpreted to reflect the concentration of ortophosphate. It is likely, however, that MRP analysis overestimates ortophosphate concentrations by including, e.g., dissolved organic species. A more critical approach to the analytical problems may therefore produce more adequate results, since knowledge about the real distribution of phosphorus components is highly essential when discussing different mobilization or transport processes.

9.3.2 Diffusion

Diffusion is the major transport mechanism from sediments superimposed by stagnant bottom waters, e.g., hypolimnetic sediments. The diffusion flux is primarily determined by the concentration gradient between the sediment pore water and the lake water (Fig. 9.7). The greater the difference in concentration, the higher the rate of diffusion. Diffusion is the transport process intimately linked to the classic anaerobic phosphate release, since reduced sediments usually are formed under stagnant water masses.

The diffusion coefficient for phosphate in sediment-water systems has been calculated to be approximately 10^{-6} $cm^2 \cdot s^{-1}$ (Stumm and Leckie 1971, Kamp-Nielsen 1974).

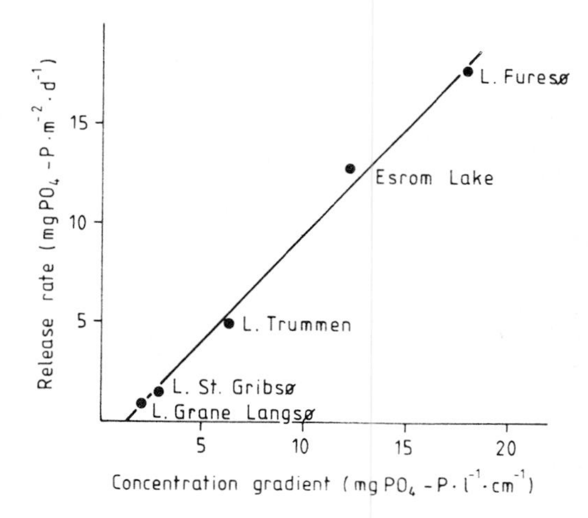

Fig. 9.7. The relationship between the anaerobic release of phosphorus and the concentration gradient across the sediment-water interface. The concentrations are calculated as the difference between the phosphate concentrations in the water and the ammonium chloride-extractable concentration of phosphate in the 0–1 cm sediment layer. (Boström et al. 1982, as modified from Kamp-Nielsen 1974)

9.3.3 Turbulent Mixing/Bottom Dynamics

If the transport of phosphorus from sediments superimposed by non-turbulent water is well described as a diffusion process, the conditions are entirely different for sediments in contact with circulating water. It is not yet possible to determine the exact influence of turbulence at the sediment-water interface, but there is no doubt that the transport of dissolved species in the sediments is generally enhanced by turbulent mixing.

Experiments (e.g., Holdren and Armstrong 1980) have demonstrated that stirring in laboratory sediment-water systems significantly increases the release of phosphorus as compared to undisturbed experiments.

Field observations in shallow lakes have indicated that wind-induced turbulence accelerates internal phosphorus loading (Andersen 1974, Ryding and Forsberg 1977, Ahlgren 1980). The effect of turbulent mixing of the surface sediments is several-fold:

- The diffusion rate will increase since phosphorus leaving the sediments is rapidly dispersed in the water column, whereby a steep concentration gradient is maintained.
- Dissolved phosphorus in the sediments is mechanically transported to the overlying waters, thus increasing the overall transport from the sediments.
- Increased output of dissolved phosphorus enhances the phosphorus mobilization from equilibrium-regulated reactions.
- Resuspension of sediment particles may facilitate an increase in particulate-bound phosphate in the lake water. Part of this phosphorus can be utilized for biological production, but resuspension may also cause an exhaustion of available phosphorus by adsorption and precipitation of dissolved inorganic phosphorus.

The impact of turbulent mixing for the overall release rates of phosphorus from lake sediments has to be strongly emphasized. Direct and indirect effects of turbulent mixing may be the key regulating factor of the flow of phosphorus (and other elements as well) from sediments in shallow lakes.

9.3.4 Bioturbation

Bioturbation is treated in detail in Chapter 8. Here we shall only briefly mention the impact of stirring animals on release processes. Several tests in the laboratory have shown that presence of burrowing animals enhances the release of our model substance, phosphorus (e.g., Granéli 1979, Holdren and Armstrong 1980). The effect of bioturbation seems to depend mainly on physical mixing of the sediment surface layer and the consequent impact on chemical parameters like pH and redox potential. The result of bioturbation may either be increased phosphorus outflow due to the mixing or decreased release because of, e.g., increased redox potential in reduced sediments.

9.3.5 Gas Convection

The anaerobic microbial metabolism in sediments leads to the formation of several gaseous products (see Chap. 5.4), like N_2, CO_2 and CH_4. When gas production and pressure conditions allow bubble formation, the gases rise upwards rather rapidly. Release of gases from surficial sediments predominates in productive shallow areas. The ascending gas bubbles cause vertical currents by which sediment particles and dissolved substances in the pore water will be transported upwards to the water column. Methane has been attributed a particularly important role in this context and the phenomenon is sometimes called methane convection. Gas ebullition can be highly significant for the recycling of elements from sediments and a dominating transport system in highly eutrophic lake sediments (Ohle 1958, 1978).

9.4 A General View of Phosphorus Release

Hitherto, we have in some detail presented important mechanisms by which a type element, phosphorus, can be returned to the lake water from the sediments. At first sight, it may seem as if all these singular processes are not interrelated and only have limited impact on the release of phosphorus from lake sediments. Therefore, in this section, we shall try to "knit together" the individual pieces, whereby (hopefully) the complex pattern of phosphorus release will be more easily understood.

In the first place, it should be realized that before phosphorus is released in substantial amounts, the sediments have to be "saturated" with phosphorus. Oligotrophic sediments seldom release phosphorus, but generally have a high capacity to retain excess amounts of the element. If the supply to the sediment increases, for example by introduction of sewage water, it will take some time (years or decades) before the saturation level of the sediment is reached.

This has been demonstrated in whole lake fertilization experiments with phosphorus in Canada (Schindler et al. 1977) and Sweden (Jansson 1978). A major part of the added phosphorus was bound in the sediments throughout several years of fertilization without any substantial release taking place. On the other hand, if a lake has been heavily loaded with phosphorus for several years and the external supply is cut off, then the sediments can be "oversaturated" with phosphorus and a net flux to lake water may occur. This means that the sediments and the lake water can be regarded as a two-phase system where the direction of the phosphorus flux is directed in the direction yielding an equilibrium between the two phases.

The different roles of oligotrophic and very eutrophic lake sediments in this respect are illustrated in Fig. 9.8. Lake Hymenjaure, an oligotrophic lake, was fertilized for several years with phosphorus, which increased the external input approximately 15 times. The sediments served as an efficient phosphorus trap and retained approximately 80% of the annual loading. Eutrophic Lake Ryssbysjön, on the other hand, was polluted with sewage water for several decades. A sewage treatment plant was built in 1973 and in the following years the external loading was reduced to about the same level as in the fertilized Lake Hymenjaure. Instead of retaining phosphorus, the sediments in Lake Ryssbysjön supplied phosphorus to the lake water and the net

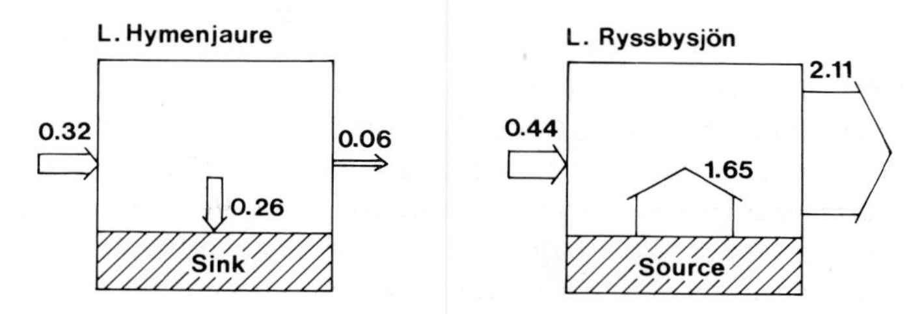

Fig. 9.8. Net budgets of phosphorus in two shallow lakes showing the sediments as a sink (Lake Hymenjaure) and as a source (Lake Ryssbosjön). Values from these two Swedish lakes are given in g per m^3 per year. (Modified from Boström et al. 1982, based on data from Ryding and Forsberg 1977 and Jansson 1978)

release exceeded the external input approximately four times. Both lakes in Fig. 9.8 have similar morphometric characteristics.

It should be noticed that the saturation level of sediments differs from one lake to another depending on the chemical composition of the deposits. Sediments rich in compounds with high phosphorus sorption capacity tolerate longer periods of high loading before the flux turns towards the lake water. Thus, the prehistory, the trophic state of the lake and the sorption capacity of the sediments, are essential concepts for the tendency to release phosphorus. But the morphometry may also influence the release of phosphorus.

The important difference in this context is between deep and shallow lakes. Deep lakes are stratified, with accumulation areas beneath stagnant bottom water during large parts of the year. In shallow lakes, on the other hand, all parts of the sediments are superimposed by circulating water at least during spring, summer and autumn. This means that of the mechanisms and factors favouring the release of phosphorus (Fig. 9.2), the reduction of iron (III) complexes and subsequent transport of phosphorus by diffusion or gas ebullition will predominate in deep lakes. In shallow lakes, on the other hand, all processes mobilizing phosphorus from the sediments can theoretically contribute to the outflux from the sediments:

- As discussed in Chapter 5.4, it is not unlikely that reduced zones occur in the surface sediments also when the sediments are in contact with well-oxygenated water.
- Mobilization triggered by a photosynthetic pH increase is likely to occur during summer in shallow lakes where the entire sediment areas may be in contact with the trophogenic water layer.
- The enzymatic and chelating mobilization (if important) by bacterial activity should be of potentially greater magnitude in shallow lakes, since the temperature during summer is high in the surficial sediments.
- Turbulent mixing increases the transport of phosphorus from the sediments in shallow lakes.

Thus, at comparable phosphorus saturation levels, the sediments in shallow lakes should release significantly more phosphorus than the sediments of deep lakes. Furthermore, the effect of the released phosphorus should be more drastic in a shallow lake since released nutrients are continuously supplied to productive water layers instead of being trapped below the thermocline.

Finally, we will briefly comment on the driving forces behind phosphorus release from lake sediments. It is evident that the mobilization of phosphorus in lake sediments (Table 9.2) is governed by several chemical factors, like pH, redox potential, chelating agents and biochemical enzymatic reactions. It is also clear that most changes of these parameters depend on biological activity; the redox potential is decreased by bacterial decomposition of organic matter; enzymes and chelating agents can be produced by bacteria; and pH variations are caused by photosynthetic processes. Some of the transport processes (e.g., bioturbation and gas ebullition) is also the result of biological activity. Therefore, the more productive the lake, the higher the overall biological metabolism and the tendency of the sediments to release phosphorus. In this context, temperature must also be recognized as a crucial factor regulating the intensity of biological reactions.

Table 9.2. Summary of the causal relationships determining the release of phosphorus from lake sediments. (Modified from Boström et al. 1982)

Environmental parameters	Effects on phosphorus mobilization
Redox potential	Iron-bound phosphorus is released at potentials below 200 mV when iron (III) is reduced to iron (II)
pH	An increase in pH decreases the phosphorus-binding capacity of iron and aluminium compounds, primarily due to ligand-exchange reactions where hydroxide ions replace phosphate. Calcite and apatite formation at higher pH-values increase the phosphorus-binding capacity of calcium
Temperature	An increase in temperature gives primarily indirect effects due to increased bacterial activity, which increases oxygen consumption and decreases the redox potential. The production of phosphate-mobilizing enzymes and chelating agents might increase accordingly
Equilibrium criteria	Affects adsorption-desorption and dissociations of precipitates
Chelating agents	Replace phosphate from salts with calcium, iron and aluminium. Chelating agents can be produced by bacteria and algae or occur as a pollutant
Processes	**Effects on phosphorus transport**
Diffusion	Phosphorus transported upwards to compensate the concentration gradient between the sediments and the lake water. Important when the sediments are superimposed by stagnant water
Turbulence	Linked to mechanical transport, resuspension and bottom dynamics. Enhances diffusion transport. Possibly the most efficient process of transport in shallow lakes
Bioturbation	Creates physical mixing and enhances transport in aerobic sediments
Gas convection	Mainly methane, which creates physical mixing and enhances transport. Important in highly reduced, organic sediments

Utilizing phosphorus as a type element, this chapter has highlighted the fact that in lake restoration programmes and in many other activities related to lake management, it is important to account for the role of the sediments.

10 Sediments in Aquatic Pollution Control Programmes

10.1 Introduction

The more industrialized and technically advanced society becomes, the more important it will be to have adequate systems or models for aquatic pollution control. Existing systems often have their roots in the 1960's, since there is a general time-lag between discovery and application. These traditional systems are frequently based upon either water samples (point samples of 1 litre or so, taken more or less at random in a river or a lake) and subsequent laboratory analysis of standard parameters like pH, O_2, major ions, nutrients, BOD, turbidity, or on broad and rather superficial biological investigations of bottom animals, phyto- and zooplankton, bottom dwelling and "pelagic" fishes, etc., throughout as many levels of the lake or river ecosystem as economically possible without too much consideration paid to sample representativity and natural fluctuations.

The purpose of aquatic pollution control is to supervise the status of the receiving waters, the recipient, and see that no inappropriate contamination takes place. The ultimate aim should be to diagnose possible negative ecological effects. But this is a very difficult task: Ecological effects? – On what species? On what organ? What effect would an increase of the substance NN in water, sediments or biota imply in itself and relative to other contaminants? How is it possible to prove that NN from AA is the problem and not MM from BB? These types of questions are generally impossible to answer in a strictly scientific way, since present knowledge on pathways and effects of toxic substances in aquatic systems is too limited. Moreover, this often implies that the ecologists and the protecting authorities will be in an inferior position relative to the opposite party. Thus, it is obvious that more basic and applied research and better methods for aquatic pollution control are needed. The purpose with this section is to demonstrate how lake sediments can be used for aquatic pollution control; to discuss benefits and limitations, causal relationships and to give some practical examples.

10.2 Why Use Sediments?

It has already been demonstrated by many maps, figures and tables in previous sections of this book that lake sediments can be regarded as a bank of environmental information, as an environmental "tachometer". Sediment samples can reveal which areas are polluted and unpolluted, spread patterns, historical development, which substances contaminate and how much. These are very good reasons for the attractiveness of the sediments in aquatic pollution control programmes. It should, however, be emphasized that the sediment concept includes biological species, like microscopic organisms, which cannot conventionally be separated from inorganic components or dead organic matter in the sediments. These microorganisms may affect the sediment metabolism of many pollutants, particularly organic substances, and hence also the concentrations and distributions of pollutants in sediments. However, disregarding the problems to determine the partitioning of contaminants into living and dead materials, we will proceed here with bulk analysis of sediment samples and presuppose that, in general, most pollutants are attached to abiotic fractions of the sediments.

Sediment samples can provide time-integrated, highly informative data of high local representativity. Chemical and biological parameters in lake water generally yield low local representativity for the sample site due to great temporal and spatial variations of water masses in lakes, and also low concentrations of contaminants, which can cause analytical problems. Analyses of biota are often difficult to interpret due to the fact that a proper interpretation requires full, or at least adequate, control of governing factors. Population dynamics, ecological interrelationships, biological uptake and distribution of pollutants are parts of a much more complicated system than the factors governing the information value of sediment samples. Therefore, biological samples linked to, e.g., fish, bottom fauna or plankton, often yield low local representativity and involve great patchiness. Furthermore, the concentrations are often lower in biota than in sediments. This is exemplified for zinc in Fig. 10.1 and for cadmium in Fig. 10.2. These two figures give concentrations in the following average sequence for the following media: water (generally lowest) – fish muscle – fish liver – algae, macrophytes, insects, bivalves, etc. – sediments (highest). The two figures also show the enormous spread in terms of concentrations causing short-term and long-term effects on test organisms. That spread is, for example, in the order of 10^4 for Cd concentrations depending on the choice of the test organism. High sediment concentrations mean a facilitated laboratory analysis and more reliable data, as compared to water samples and many biological samples.

Sediment samples are comparatively easy to collect, analyze and interpret as compared to most alternatives, which often require more specialized personnel. This is not a trivial aspect in water pollution control programmes. Subsequently, we will discuss principles, methods and standardized schemes on lake sediments in aquatic pollution assessment programmes.

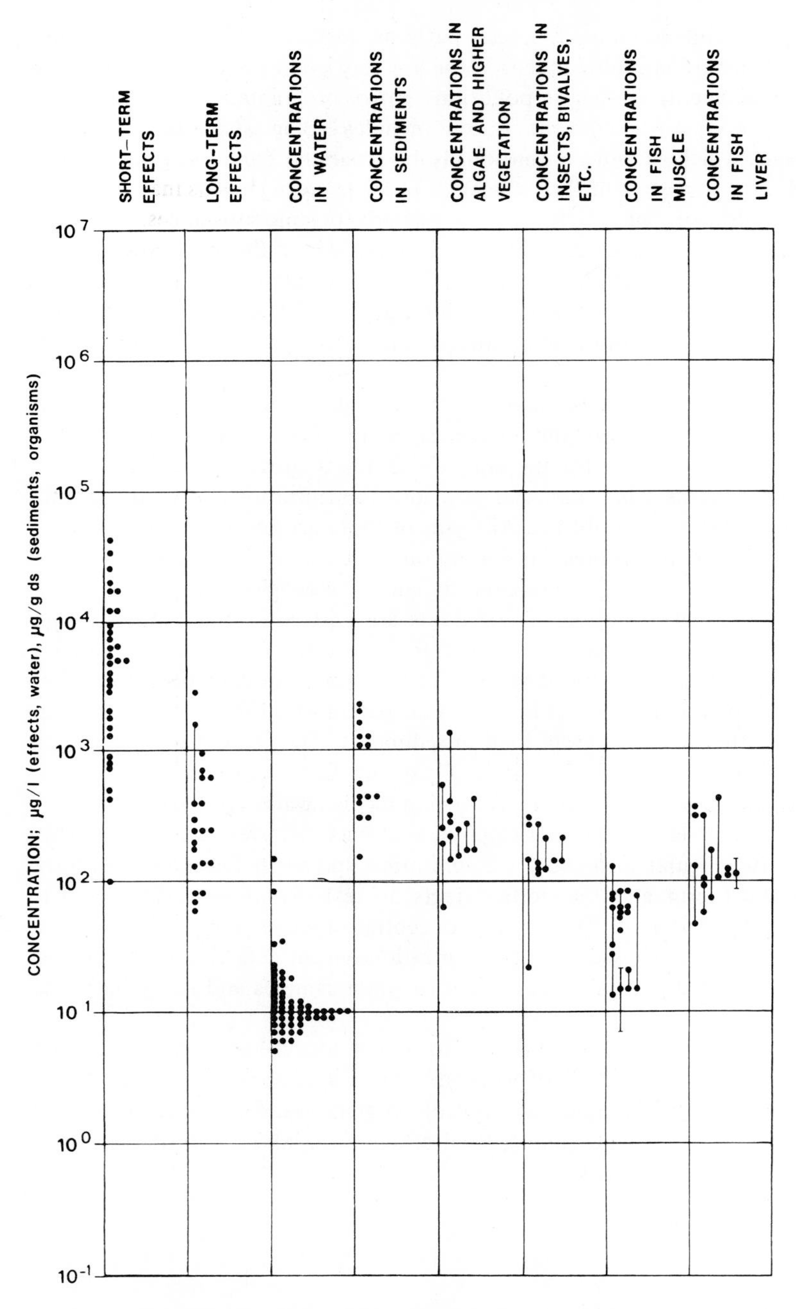

Fig. 10.1. Zinc. Concentrations yielding short-term and long-term toxic effects, and concentrations in river water, lake sediments, attached algae, macrophytes (higher vegetation), invertebrates, fish muscle and fish liver (dry substance of organisms put at 20% of wet weight) (SNV 1980a)

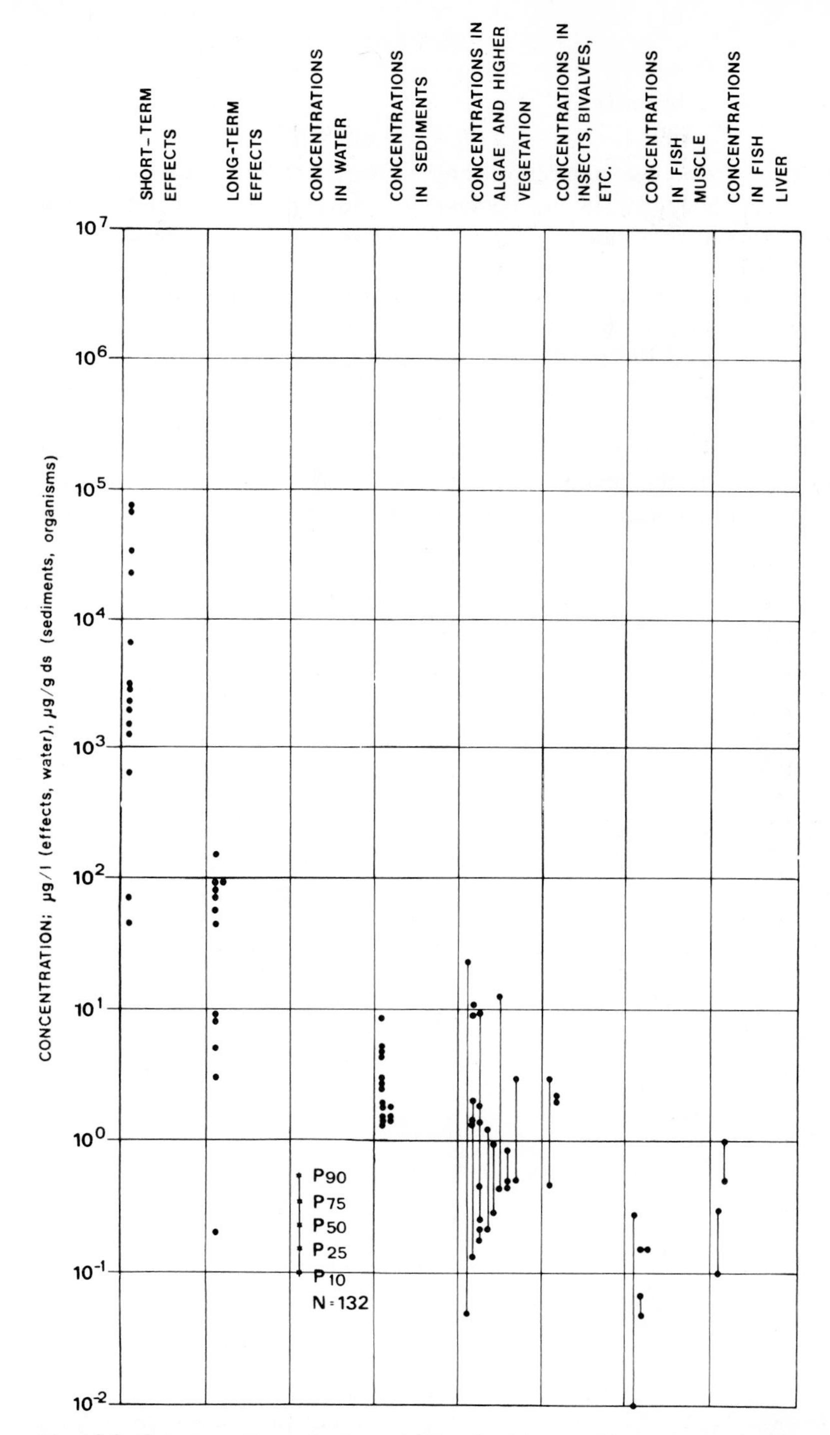

Fig. 10.2. Cadmium. Concentrations yielding short-term and long-term toxic effects, and concentrations in river water, lake sediments, attached algae, macrophytes (higher vegetation), invertebrates, fish muscle and fish liver (dry substance of organisms put at 20% of wet weight) (SNV 1980b)

10.3 How to Use Sediments

There must always be a close connection between the purpose of the investigation and the utilized methodology. In this context, it is not possible or appropriate to discuss all types of contaminants and/or principles in great detail (see Baker 1980). Instead, we will focus on heavy metals (see also Chap. 4.2.3.6), and try to present only the most fundamental concepts on "the road from dose to response". The reason why we consider the metals and not any other group of substances, like chlorinated hydrocarbons (PCB, DDT, etc.), is that the available information on metals is comparatively good, which minimizes the possibilities for error and maximizes the possibilities to make justifiable simplifications (a very good review on metals has been given by Förstner and Wittmann 1979; about 20,000 papers are available on mercury in the environment, see Taylor 1975), and that we focus on principle aspects, which often are analogous for different types of contaminating substances.

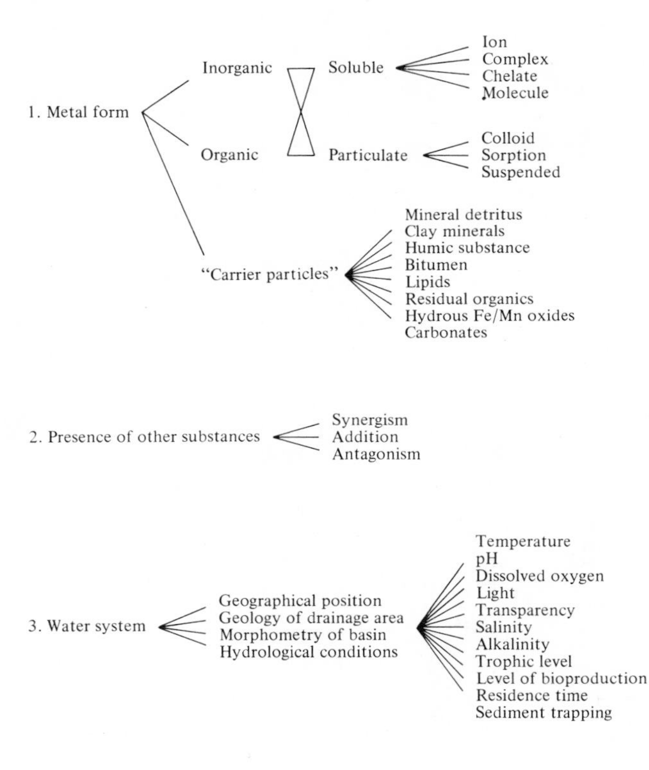

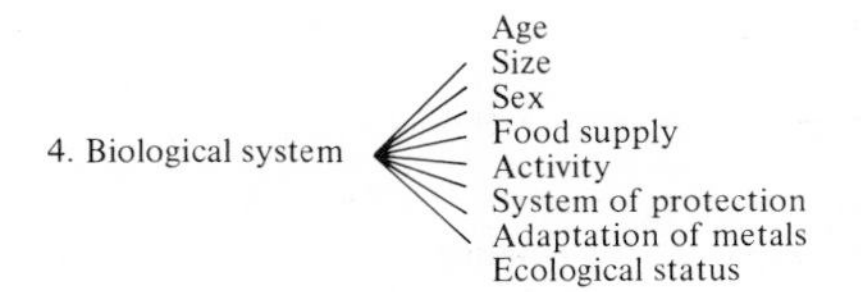

Fig. 10.3. Factors affecting the toxicity of metals in aquatic systems. (After Beijer et al. 1977)

10.3.1 Principles of Metal Distribution in Aquatic Systems

Very many factors influence the spread and toxicity of metals in aquatic systems. This is evident from Fig. 10.3. Most metals appear in low concentrations in water where they are found not primarily in solution but attached to various types of "carrier particles", i.e., suspended organic and inorganic particles/aggregates of different origin and chemical character. The way the metals are bound to these "carrier particles" will determine not only the distribution in water and sediments but also the potential ecological effects. There may be considerable differences between total concentration and biologically available amount (see Förstner and Prosi 1979).

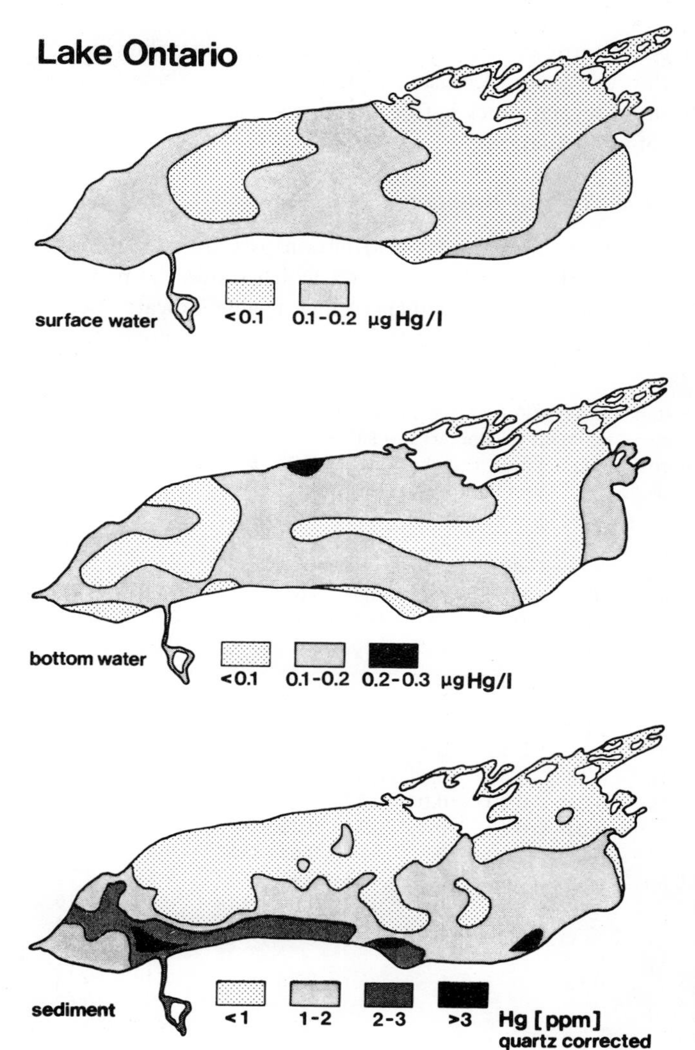

Fig. 10.4. The mercury concentration in surface water *(upper)*, in bottom water *(middle)* and in surficial sediments *(lower)* in Lake Ontario (Förstner and Wittmann 1979, from Thomas 1972, Chau and Saitoh 1973)

Figure 10.4 illustrates that there may be a very obscure linkage between the metal concentration in surface water, bottom water and surficial sediments. This figure from Lake Ontario clearly demonstrates the hazards that may befall interpretations of data from water sampling systems in aquatic pollution control. The sediment samples, on the other hand, yield very good information on spread patterns from the sources of pollution. Chau and Saitoh (1973) conclude:

"– High mercury concentrations were found in the vicinity of some industrial activities associated with the major tributary inputs on the American, southern side of the lake (primarily Niagara River).
– There is no apparent relationship between mercury in water and in sediments.
– 30–90% of the total mercury in lake waters was non-extractable by dithizo-chloroform and, thus, held tightly by suspended matter."

The mercury and its "carrier particles" will be distributed by the prevailing hydrological flow. If settled within the zones of erosion or transportation, these "carrier particles" may be redistributed several times before finally coming to rest below the "critical limit", in the area of accumulation.

10.3.1.1 The Type of Metal and Type of Pollution

Metals from the same source often show similar distribution patterns in sediments. This may be revealed by means of a correlation matrix which shows the pair-wise correlation, in terms of the correlation coefficient (r). A high r-value indicates a common source and distribution pattern (see Table 10.1) Even metals with quite different

Table 10.1. Mean values ($\bar{x}$), standard deviations (s_X), mean values from accumulation areas ($\bar{x}$-A) and correlation matrix for various sediment parameters from 12 surficial samples (0–1 cm) from areas affected by diffuse internal contamination in Lake Vättern. DDT, PCB and metals given in $\mu g\ g^{-1}$ ds, N and P in $mg\ g^{-1}$ ds. D is the water depth in m (Håkanson 1977d)

	D	W	IG	Hg	Cu	Ni	Pb	Zn	N	P	DDT	PCB
$\bar{x}$	60.5	79.6	10.4	0.17	50	44	150	483	3.6	1.3	0.081	0.170
s_X	18.6	9.5	1.9	0.08	16	16	41	152	1.4	0.2	0.036	0.092
$\bar{x}$-A	64.6	83.3	10.5	0.16	54	43	155	510	4.0	1.4	0.083	0.165
Correlation matrix												
D	–	0.26	0.14	–0.31	0.16	–0.10	0.21	0.25	0.47	0.39	0.21	–0.43
W		–	0.45	0.16	0.68	–0.40	0.11	0.17	0.78	0.57	0.43	0.31
IG			–	0.79	0.04	–0.80	–0.07	–0.27	0.51	0.23	0.29	0.44
Hg			0.79	–	–0.04	–0.60	–0.36	–0.39	0.22	0.19	0.34	0.77
Cu		0.68			–	–0.10	0	–0.07	0.68	0.58	0.35	0.03
Ni			–0.80	–0.60		–	0.20	0.56	–0.46	–0.11	–0.45	–0.32
Pb							–	0.54	0.07	0.14	–0.30	–0.46
Zn						0.56	0.54	–	0.03	0.38	–0.28	–0.22
N		0.78	0.51		0.68				–	0.78	0.48	0.16
P		0.57			0.58				0.78	–	0.40	0.18
DDT											–	0.62
PCB				0.77							0.62	–

chemical properties "in the laboratory" may appear with similar distribution patterns in nature, e.g., due to the fact that they are linked to the same type of "carrier particles" with similar sedimentological properties. Table 10.1 illustrates the correlation matrix for those areas in Lake Vättern, Sweden, affected by diffuse contamination, i.e., areas close to industrialized/urban areas without any single, obvious source of pollution. In this particular example, there is a strong positive correlation between the PCB-content and the Hg-content (total Hg) because these two substances emanate from the same tributary (urban source) and show a rather high affinity for organic matter (the "carrier particle"), in spite of the fact that Hg and PCB are very different from a chemical point of view. There is, for example, a strong negative correlation between the Hg-content and the Ni-content, because Ni, unlike Hg, does not appear as a contaminating metal but as a non-contaminating or a conservative one; it is not discharged to the lake from noticeable sources within the drainage area. But that is most certainly the case in Lake Ekoln, Sweden, where a strong positive correlation exists (r = 0.81) between Hg and Ni, since, in this case, they emanate from the same source (Table 10.2).

Table 10.2. Correlation matrix for various sediment parameters in surficial sediments (0–1 cm) of Lake Ekoln, Sweden, affected by diffuse internal contamination. Z is distance from pollution source (i.e., from the mouth of River Fyris), D is water depth (Håkanson 1977d)

	D	Z	W	IG	Hg	Cu	Ni	Pb	Zn	Cr	N	P
D	–	–0.24	0.22	0.18	–0.01	0.02	–0.20	–0.23	0.31	0.62	0.02	0.43
Z		–	–0.56	–0.72	–0.71	–0.62	–0.45	–0.75	–0.80	0.06	–0.74	–0.26
W		–0.56	–	0.74	0.16	0.08	0.28	0.42	0.24	–0.03	0.27	0.11
IG		–0.72	0.74	–	0.66	0.55	0.68	0.42	0.57	–0.28	0.71	–0.17
Hg		–0.71		0.66	–	0.89	0.81	0.48	0.90	–0.23	0.91	0.04
Cu		–0.62		0.55	0.89	–	0.67	0.45	0.86	–0.15	0.68	–0.10
Ni				0.68	0.81	0.67	–	0.28	0.63	–0.35	0.81	–0.09
Pb		–0.75						–	0.47	–0.29	0.42	0.09
Zn		–0.80		0.57	0.90	0.86	0.63		–	0.09	0.80	0.32
Cr	0.62									–	–0.25	0.76
N		–0.74		0.71	0.91	0.68	0.81		0.80		–	0.14
P										0.76		–

10.3.1.2 The "Carrier Particles"

Several techniques exist to determine and define the various fractions or phases that constitute the "carrier particles" (see, e.g., Chester and Hughes 1967, Tessier et al. 1979, Salomons and Förstner 1980). Table 10.3 summarizes several common chemical techniques used for metals in sediments. The following main phases are often used (see, e.g., Tessier et al. 1980):

- organic phases; humic and fulvic acids, solid organic material,
- carbonate phases,
- Mn and Fe phases,
- detrital/non-detrital; authigeneous/lithogeneous phases,
- adsorption and cation exchange phases.

Table 10.3. Common methods for extractions of metals associated with different chemical phases in sediments (Salomons and Förstner 1980)

Adsorption and cation exchange
Extractions with: BaCl, MgCl, NH_4OAc

Detrital/non-detrital; authigeneous/lithogeneous fractions
Extractions with: EDTA, 0.1 M HCl, 0.3 M HCl, 0.5 M HCl, 0.1 M HNO_3

Manganese and iron phases; reducible, easily and moderately reducible phases
Extractions with (in approximate order or release of iron): Acidified hydroxylamine, ammonium-oxalate, hydroxylamine-acetic acid, dithionite/citrate

Carbonate phases
Extractions with: CO_2 treatment, acidic cation exchange, NaOAc/HOAc (pH 5)

Organic phases; humic and fulvic acids, solid organic material
Extractions with: H_2O_2, H_2O_2-NH_4OAc, H_2O_2-HNO_3, organic solvents, 0.5 M NaOH, 0.1 M NaOH/H_2SO_4, Na hypochlorite-dithionite/citrate

Salomons and Förstner (1980) have suggested a simple standard extraction method to meet demands for routine use and accuracy. The procedure is as follows:

A. An extraction with acidified hydroxylamine hydrochloride (0.1 M $NH_2OH \cdot HCl$ + 0.01 M HNO_3, pH 2), which includes the extraction of exchangeable cations and carbonates. Metals extracted by this procedure are lightly bound and readily available for biological uptake. This group may be called the *easily exchangeable metals.*
B. An extraction with acidified hydrogen peroxide (30% H_2O_2, 90°C) followed by an extraction of ammonium acetate (1 M NH_4OAc, pH 2) to remove any reabsorbed metal ions. Metal extracted by this procedure are bound to the *organic fraction* of the sediment sample.
C. A total analysis ($HF/HClO_4$ digestion) of the residue remaining after the first two extractions. This may be called the *inert fraction.* Metals found in this group may not be expected to constitute an major ecotoxicological threat.

From Fig. 10.5, which illustrates the speciation of various trace elements in samples from 18 different river sediments in the world, it is evident that different metals have affinity for different fractions. This figure also includes a special fraction for the easily exchangeable cations (extraction by 1 M NH_4Ac, pH 7), which otherwise is included under A, the hydroxylamine fraction. Figure 10.6 summarizes more thoroughly the forces and factors regulating the metal associations in sediments.

Much more can be said on this very interesting and important topic on metal associations (see Hart 1982), but in this context it may be sufficient to conlude that:

– There is a general increase of metals in the inert fraction from polluted to less polluted rivers.
– The inert fraction constitutes an important sink for metals.

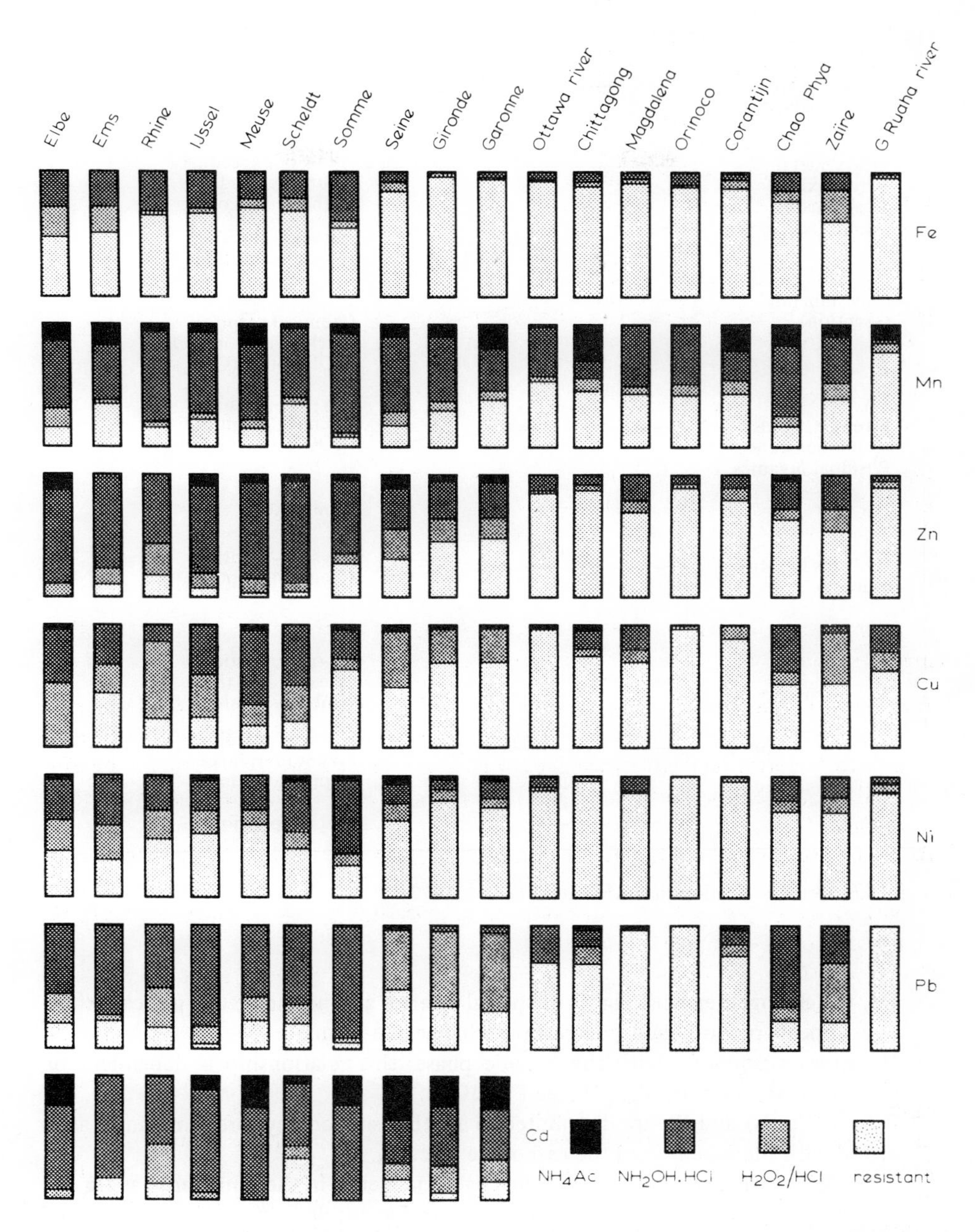

Fig. 10.5. The speciation of 7 heavy metals in 18 different river sediments. The rivers are arranged according to approximate geographical position from north to south. Most tropical rivers contain too low amounts of cadmium to obtain a reliable speciation. For the Rio Magdalena and the Orinoco River, insufficient material was available to determine the NH_4Ac-fraction. Thus the acidified hydroxylamine step contains also these metals (Salomons and Förstner 1980)

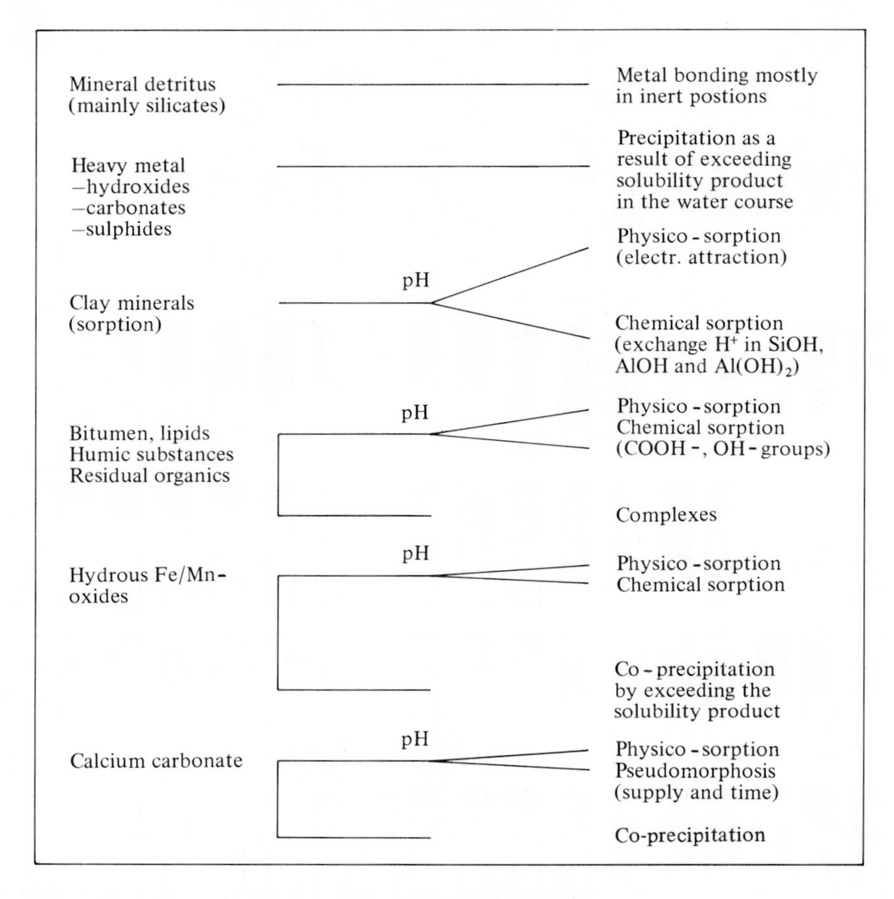

Fig. 10.6. Metal associations in sediments (Förstner 1977)

- Cd, which together with Hg, is of special interest among the metals in ecotoxicology, appears predominantly in easily exchangeable forms.
- Hg appears primarily with the organic phase; this relationship is dependent on environmental factors, like pH (see Fig. 10.7).
- Cu, Cr, Co, Ni and Pb are distributed in a rather similar way, predominantly associated with the organic and the carbonate fraction.
- Fe and Mn, which cannot be regarded as toxic metals in the same sense as Hg and Cd, primarily appear in the inert and easily exchangeable forms and not associated with the organic phase.

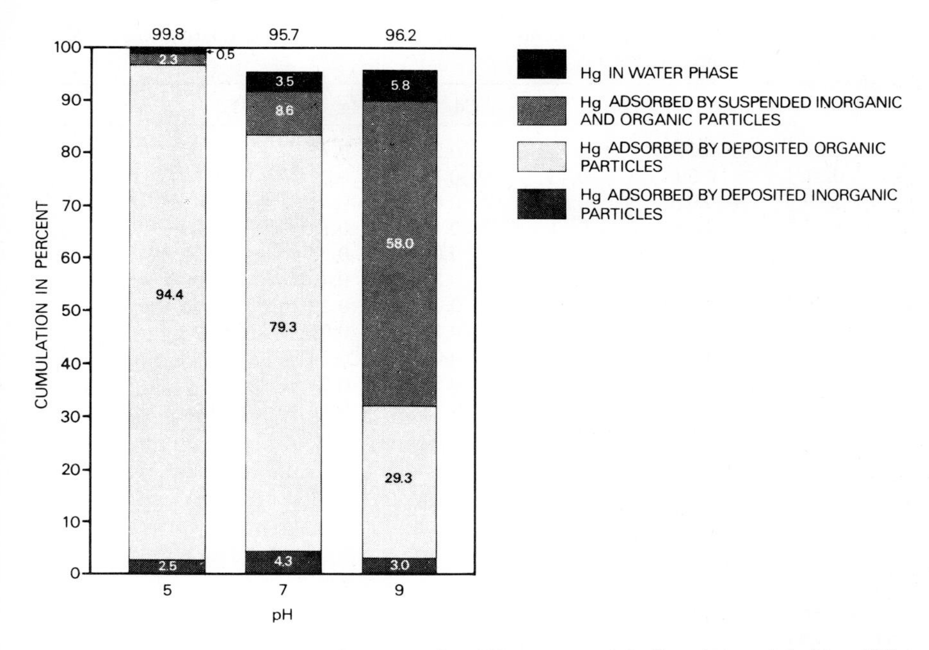

Fig. 10.7. The distribution of mercury for different materials ("carrier particles") at different pH in a laboratory experiment (Håkanson 1972)

10.3.1.3 The Environmental Characteristics

Several environmental factors, like the pH of the water, alkalinity, salinity, trophic level, water retention time, influence the spread and fate of the metals (see Håkanson 1983b). Table 10.4 illustrates a correlation matrix between some selected metals in surficial sediments (Pb, Cu, Zn, Cd, Hg, and Fe) and some selected environmental factors (water content, loss on ignition, size of drainage area, percentage of lakes within the drainage area, water residence time, pH, transparency, colour, NO_3-N, tot-P, trophic level and phytoplankton biomass). The data emanate from 49 forest lakes in south-central Sweden (Johansson 1980). All these lakes have low productivity and alkalinity, but vary rather widely in terms of pH (from 4.9 to 7.5), colour (from 5 to 150 mg Pt l^{-1}) and retention time (from 0.05 to 9.0 years). The main point of this table is that most of these environmental factors may have an influence on the metal distribution in water and sediments, and that different environmental factors may influence different metals. Too little research has yet been done in this very important sector, and Table 10.4 demonstrates that it may be a fruitful future area of investigation.

It has even been demonstrated that a negative correlation may exist between the dose factor, as this can be given by the sediment contamination (C_f), and the response factor, i.e., the quotient between the natural background level and the actual concentration in the fish organ, which is pike liver (PIL) in the example given in Table 10.5 from the River Kolbäcksån water system. Here we find a positive correlation

Table 10.4. Correlation matrix: metal concentrations in surficial sediments relative to various environmental factors (Johansson 1980)

	Pb	Cu	Zn	Cd	Hg	Fe
W	0.24	0.39 [b]	−0.25	−0.30	0.41 [b]	−0.04
IG	−0.21	0.09	−0.43 [b]	−0.11	0.37 [b]	0.21
D. area	0.09	−0.01	0.19	0.07	−0.02	0.47 [a]
Lake %	0.46 [a]	0.55 [a]	0.11	0.10	0.15	−0.07
Time	0.65 [a]	0.71 [a]	0.04	0.03	0.25	0.05
pH	−0.11	−0.26	0.53 [a]	0.21	−0.40 [b]	−0.09
Transp.	0.76 [a]	0.64 [a]	0.08	0.00	0.27	0.15
Colour	−0.36 [b]	−0.29	−0.03	0.10	−0.07	−0.22
NO_3	0.60 [a]	0.61 [a]	0.00	0.00	0.24	0.07
Tot-P	0.05	0.05	0.35 [b]	0.37 [b]	−0.01	−0.11
TI	−0.43 [b]	−0.30	0.08	0.05	−0.13	−0.06
Pb	−0.32	−0.13	−0.22	−0.25	−0.06	0.13

W = water content of sediments, IG = loss on ignition, D. area = drainage area, Lake % = lake area in percent of the drainage area, Time = water retention (residence) time, Transp. = Secci-Disk transparency, TI = trophic index and Pb = phytoplankton biomass

a = correlation certain at the 99.9% (r > 0.45) level
b = correlation certain at the 99% (r > 0.40) level
correlation certain at the 98% (r > 0.26) level
correlation certain at the 95% (r > 0.30) level
correlation certain at the 90% (r > 0.25) level
correlation certain at the 80% (r > 0.19) level
(n = 47)

Table 10.5. Comparison between sediment contamination factor (C_f) and fish contamination factor (in this case pike liver, PIL) for six metals in seven lakes from the River Kolbäcksån water system (Håkanson 1983a)

	Hg		Cd		Pb		Cu		Cr		Zn		r
	C_f	PIL	C_f	PIL	C_f	PIL	C_f	PIL	C_f	PIL	C_f	PIL	(n=6)
Bysjön	3.9	10.2	2.5	4.6	1.8	3.5	0.8	7.6	0.8	4.5 [a]	1.3	3.2	0.57
Väsman	28.4	9.8	6.5	14.7	18.3	2.2	4.8	1.4	1.6	1.8 [a]	6.1	3.1	0.29
Ö. Hillen	20.0	11.8	6.1	22.5	5.0	3.9	4.2	3.0	1.9	2.3 [a]	7.1	1.9	0.33
N. Barken	10.7	6.4	10.7	5.3	16.7	2.8	5.2	2.2	2.0	3.4 [a]	13.8	1.8	−0.01
St. Aspen	5.4	7.6 [a]	6.1	2.0 [a]	18.0	1.9 [a]	5.1	1.7 [a]	90.0	5.4	9.7	0.7 [a]	0.35
Östersjön	6.9	4.4	2.3	1.4	3.0	1.1	6.0	3.3	30.2	34.3	6.0	3.9	0.997
Freden	4.0	7.8 [a]	3.4	1.5 [a]	1.8	0.9 [a]	3.4	2.3 [a]	16.4	13.0 [a]	5.4	3.6	0.89
r (n = 7)	0.48		0.29		0.03		−0.81		0.19		−0.65		0.24 (n=42)

[a] Data based on less than 3 analyses

between C_f and PIL for mercury (r = 0.48), which is to be expected (see Håkanson 1980b), but a strong negative correlation between C_f and PIL for copper (r = −0.81) and zinc (r = −0.65). This could not, however, be interpreted to imply that the dose is unimportant in general terms for Cu and Zn, only that other factors, lake sensitivity factors, may be more important than the dose, and that these environmental factors must be accounted for in matters related to the question of how much of a given toxic substance a given lake can endure? That is, evidently, a crucial question in contexts related to aquatic pollution control and ecotoxicology.

10.3.1.4 Natural Background Levels

To be able to quantify the degree of anthropogeneous contamination and compare different metals which appear in different ranges of concentration in lake sediments, it is essential to establish a natural reference, a preindustrial background level. Table 10.6 gives reference data of various metals in different types of geological and biological media. Fig. 10.8 illustrates schematically how a given element can be distributed in a sediment core.

Table 10.6. The abundance of various elements (in ppm) in igneous rocks, soils, fresh water, land plants and land animals (Bowen 1966)

	Igneous rocks	Soils	Fresh water	Land plants	Land animals
Ag	0.07	0.1	0.00013	0.06	0.006
Al	82,000	71,000	0.24	500	4–100
As	1.8	6.0	0.004	0.2	⩽ 0.2
Cd	0.2	0.06	<0.08	0.6	⩽ 0.5
Co	25	8.0	0.0009	0.5	0.03
Cr	100	100	0.00018	0.23	0.075
Cu	55	20	0.01	14	2.4
Fe	56,300	38,000	0.67	140	160
Hg	0.08	0.03–0.8	0.00008	0.015	0.046
Mn	950	850	0.012	630	0.2
Mo	1.5	2.0	0.00035	0.9	< 0.2
Ni	75	40	0.01	3.0	0.8
Pb	12.5	10	0.005	2.7	2.0
Sn	2.0	10	0.00004	< 0.3	< 0.15
V	135	100	0.001	1.6	0.15
Zn	70	50	0.01	100	160

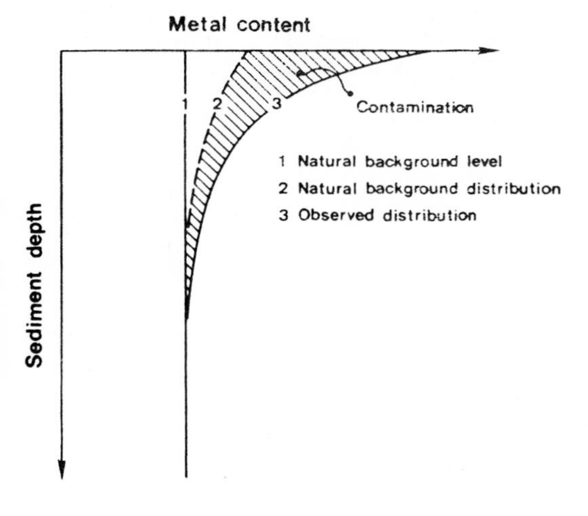

Fig. 10.8. Illustration of natural background level, natural background distribution, observed distribution and sediment contamination in a sediment core

The natural background level is, according to this definition, the lowest value, the asymptotic value, obtained in a sediment core. The natural background distribution, which depends on compaction, diffusion and bioturbation in the sediments and release to the water (see, e.g., Lee 1970, Cline and Upchurch 1973, Tessenow 1975), may be extremely difficult to establish correctly. Different metals are known to have differential vertical mobility, e.g., lead is regarded as rather immobile whereas iron and manganese are known to be mobile and highly dependent on chemical character in the sediments (pH, Eh, O_2) (see, e.g., Förstner and Müller 1974, Edgington and Robbins 1976). There are no general models available, to our knowlege, that can predict the natural background distribution in such a simple way that it may be utilized in more standardized programmes in aquatic pollution control. This is an interesting field of research and while waiting for applicable knowledge, we will disregard the position of curve (2) in Fig. 10.8 and accept a cruder definition of the concept sediment contamination. According to this rough definition, the total area (the integral) betweeen the observed, empirically established curve (3) and the natural background level (line 1) constitutes the sediment contamination. From a principal point of view, it would be more correct to define the contamination as the shaded area in Fig. 10.8, i.e., as the area between curves (3) and (2). This simplification would not, however, mean any major drawback in aquatic pollution control programmes, since the difference between line (1) and curve (2) probably would be insignificant relative to the difference between line (1) and curve (3) when the sediment contamination is substantial, and also because simplicity is essential in standardized aquatic control programmes. Natural background levels can be established according to several methods (see, e.g., Förstner and Salomons 1981), but in this context we will only very briefly mention two alternatives:

1. The "average geochemical background", or "average shale" (see Turekian and Wedepohl 1961, Müller 1979), which is often denoted B_n and represents the standard level of the metals in fossil argillaceous sediments (see Table 10.7).

Table 10.7. Concentrations (in ppm) of selected heavy metals in preindustrial or natural environments

	Fossil lake sediments (Ries-Lake) [a]	Recent lake sediments from remote areas [a]	European and American lakes [b]	Shale standard [c]
Fe	18,200	43,400	26,700	46,700
Mn	406	760	860	850
Sr	252	151	–	300
Zn	105	118	111	95
Cr	59	62	49	90
Ni	51	66	50	68
Cu	25	45	29	45
Pb	16	34	35	20
Co	15	16	19	19
Hg	0.5	0.35	0.12	0.4
Cd	0.2	0.40	0.58	0.3

[a] Förstner and Salomons (1981)
[b] Håkanson (1980a)
[c] Turekian and Wedepohl (1961)

2. The site, lake or local specific "natural background level", K_n. The concentration in preindustrial sediments can vary between different lakes and between sediment types in one and the same lake. Table 10.8 exemplifies this with data from the four largest Swedish lakes, Mälaren, Vänern, Vättern and Hjälmaren, and for three different preindustrial sediment types from Lake Vänern. Both alternatives to define background levels, the standardized one and the one yielding local representativity, imply advantages and disadvantages from the perspective of aquatic pollution control. The first alternative is rough but simple, since no analysis from the recipient would be required. The second alternative presupposes core sampling and laboratory analysis, but yields more accurate data.

Table 10.8 A. Natural background levels (determined as the mean, $\bar{x}$, plus one standard deviation, s_x) from the four largest Swedish lakes Mälaren, Vänern, Vättern and Hjälmaren for nine selected elements. **B** Natural background levels ($\bar{x} + s_x$) for nine elements (note V instead of Cr) in three different preindustrial sediment types in Lake Vänern, and data on water content (W) and loss on ignition (IG). Concentrations in $\mu g\ g^{-1}$ ds. (Data from Håkanson 1977d, 1981b)

A.	Mälaren	Vänern	Vättern	Hjälmaren
Hg	0.095	0.030	0.040	0.09
Cd	< 1	0.5	< 1	0.9
Pb	135	40	80	40
Cu	65	25	25	45
Zn	145	100	200	200
Cr	130	< 50	< 50	50
Ni	85	35	40	35
N	3000	1500	1500	5300
P	1500	2000	1000	1000
B. Type:	**1)**	**2)**	**3)**	
Hg	0.042	0.024	0.042	
Cd	1	1	1	
Pb	52	33	63	
Cu	27	25	20	
Zn	114	91	114	
Ni	38	32	22	
V	179	149	124	
N	1600	1100	1800	
P	1500	2000	3900	
W	63.3	56.1	65.1	
IG	4.1	3.3	8.6	

10.3.2 The Contamination Factor

The contamination factor or the enrichment factor expresses the relationship between the concentration of a pollutant in the recent, surficial sediment layer and the natural background concentration, i.e.:

$$C_f^i = \frac{C_{0-x}^i}{K_n^i}, \tag{10.1}$$

where C_f^i = the contamination factor for the element (i);

C_{0-x}^i = the empirically determined concentration of the substance in question (i) from surficial sediments (0–x cm);

K_n^i (or B_n) = the empirically determined (or standard natural background level for the given element (i).

Before demonstrating the practical use of the contamination factor, we will very briefly summarize certain requirements that must be met on sampling and sample preparation to achieve optimal information:

A. The sampling equipment should provide minimum disturbance, which is especially tricky when the surficial sediments are very loose (see Chap. 3.1).
B. The sample formula (see Chap. 3.2.1 and 3.2.2) indicates how many samples that may be required in aquatic pollution control contexts in lakes.
C. In most cases, the top centimeter layer would be preferable, i.e., 0–1 cm [x = 1 in Eq. (10.1)], since thicker layers provide less resolution and thinner layers require special instruments (see Chap. 3.2.4).
D. The water content (W_{0-1}) and the loss on ignition (organic content, IG_{0-1}) should be determined directly on the fresh samples. The water content can be used as an indicator of the prevailing bottom dynamics (erosion, transportation, accumulation, see Chaps. 4.1.1 or 7.3). These two co-parameters are needed to express the load of contaminants (per volume unit, see Chap. 3.5).
E. The "informative" fraction of the sediments is obtained after simple wet sieving of the sediments through a 63 μm mesh. Fine, loose sediments from areas of accumulation will pass this mesh more or less unaltered, whereas coarser/mixed deposits from areas of erosion and transportation would be separated into two fractions. The finer fraction, the "informative" fraction, is analyzed for contaminants. If load calculations are to be made, then analysis of fresh (unsieved) sediments would also be required.
F. The contamination factor can be expressed for any given site or sediment core and as a mean for a basin or a lake. Subsequently, we will use empirical data for entire lakes.

10.3.3 Case Study – River Kolbäcksån

The aim of this section is to give a practical example of a sedimentological programme for aquatic pollution control, to demonstrate the great potentials of sedimentological studies in such matters, and to point out some limitations of the methodology.

The River Kolbäcksån water system in central Sweden receives waste water from several metal industries and mines along its course (Fig. 10.9). The limnological and contaminational character of this water system have been thoroughly described in several Swedish reports. Table 10.9 gives some basic data on lake area, volume, mean depth, water retention (or residence) time, mean pH, level of bioproduction (BPN) and mean rate of deposition from seven selected lakes (Bysjön, Väsman, Övre Hillen, Norra Barken, Stora Aspen, Östersjön and Freden) throughout the stretch of the water system. The utilized set of data on metals (and nutrients and oil) in sediments

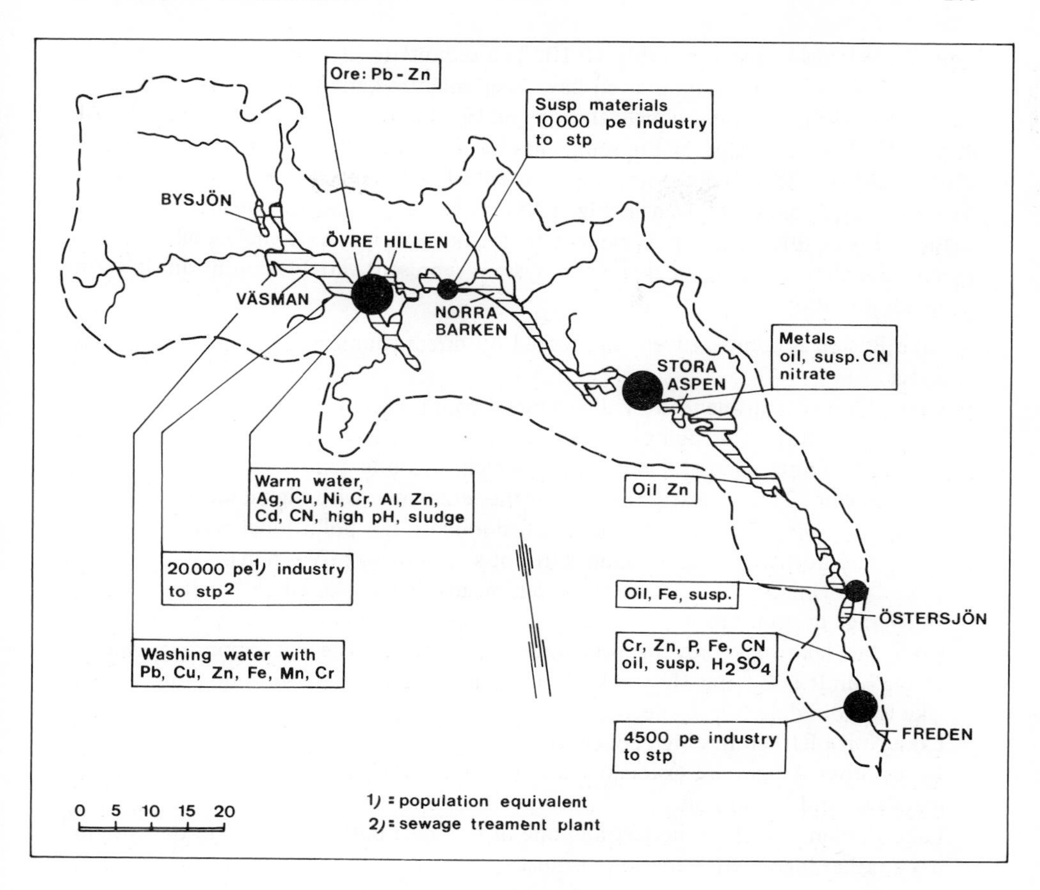

Fig. 10.9. Location map showing investigated lakes and sites of major water-polluting industries

Table 10.9. Limnological data on the seven selected lakes from the River Kolbäcksån water system (Håkanson 1977d, 1981b)

Lake	Area (a, km^2)	Volume (V, km^3)	Mean depth ($\bar{D}$, m)	Retention time (T, months)	Mean pH	Level of bio-production (BPN)	Mean rate of deposition (v, cm yr^{-1})
Bysjön	5.1	0.032	6.3	–	6.86	0.37	0.081
Väsman	38.6	0.41	10.6	13	6.84	0.38	0.15
Övre Hillen	5.0	0.053	3.7	1.7	7.02	0.49	0.31
Norra Barken	19.5	0.20	10.1	4.5	7.04	0.49	0.63
Stora Aspen	5.9	0.041	7.0	3.2	7.12	0.49	0.54
Östersjön	1.3	0.004	3.0	0.06	7.09	0.58	0.61
Freden	3.3	0.018	5.5	0.3	7.11	0.52	0.70

from these lakes is given in Table 10.10. The sediments (0–1, 4–5, 9–10, and 19–20 cm) from areas of accumulation have been analyzed for water content (W), loss on ignition (IG), and total concentrations of Hg, Cd, Pb, Cu, Zn, Cr, Ni, Fe, Mn, Al, P, and N. The mean age of the sediments has been determined (see Håkanson and Uhrberg 1981). Mean concentrations in surficial sediments (0–1 cm) in all of the 24 investigated lakes are given in Fig. 10.10 as a background to the subsequent discussion. From this figure, it is evident that Lake Saxen is extremely polluted with metals and that the lakes in this water system represent variable conditions in terms of metal pollution.

- Lake Bysjön is comparatively unaffected by direct municipal and industrial waste water.
- Lake Väsman is influenced by direct metal contamination from River Gonäsån in the south and from dredging operations carried out between 1976 and 1979 in the extremely contaminated Lake Saxen, which is connected to Lake Väsman via a short stream. This is clearly revealed in the sediments, with high concentrations of specific metals, like Pb and Cu, which dominate the pollutional profile in Lake Saxen and northern Lake Väsman (but not southern Lake Väsman).
- Lake Övre Hillen is "black-listed", which means that a 1-kg pike from this lake has an average concentration of 1.0 mg Hg per kg wet weight (muscle), and that fish from the lake may only be sold under restrictions. The corresponding "black-listing" limit is 0.5 mg Hg per kg in USA and Canada. The contamination in this lake is of old date (see Table 10.10.).
- Lake Norra Barken has high concentrations of most metals and oil.
- Lakes Stora Aspen and Östersjön have very high levels of pollution, especially the metals Cr and Ni, and oil.
- Lake Freden, which is the last and one of the most eutrophic lakes in the system, has a rather high contamination of most metals.

It should be noted that the sample formula was developed after the planning and execution of the field work in the River Kolbäcksån water system, but the correspondence between "suggested" and collected number of samples is still acceptable (see Table 10.11). Table 10.12 gives a statistical summary in terms of mean values, standard deviations, minimum and maximum values on water content, organic content and six metals from surficial sediments (0–1 cm) from the seven given lakes. Subsequently, we will discuss only these six metals (Hg, Cd, Pb, Cn, Cr, and Zn), which ought to be adequate from the present pedagogical perspective. In this context, we will use natural background levels for the whole water system, as determined from preindustrial sediment data from nine sediment cores from four of the lakes (Bysjön, Väsman, Stora Aspen and Östersjön, see Table 10.13). For this particular set of data (n = 14), our requested K_n-values have been determined as the mean value ($\bar{x}$) plus the 95% confidence interval ($t_{0.975} \cdot s_x/\sqrt{n-1}$). Then these values have been rounded off accordingly (in $\mu g\ g^{-1}$ ds or ppm):

$$Hg\ (0.090) < Cd\ (0.83) < Cu\ (21.5) < Cr\ (23.0) < Pb\ (47.0) < Zn\ (210)$$

This procedure accounts for existing geological fluctuations in a statistically relevant manner; i.e., the fact that different preindustrial sediment types, even within one and the same water system, may have somewhat different natural metal concentrations.

Table 10.10. Sediment data from seven lakes in the River Kolbäcksån water system (Håkanson 1983a)

Lake	n	Level	W	IG	Hg	Cd	Pb	Cu	Zn	Cu	Ni	Fe	Mn	Al	P	N	Oil	Mean age	
		(cm)	(% ws)	(% ds)	($\mu g\ g^{-1}$ ds)							(mg g^{-1} ds)					($\mu g\ g^{-1}$ ds)	(age)	(year)
Bysjön	4	0– 1	88.4	13.1	0.35	2.1	84	18	280	17	17	64	7.9	17	1.4	5.1	–	15.4	1962
	1	4– 5	81.7	8.1	0.17	0.9	46	121	295	15	8	62	4.9	–	1.1	3.0	–	56	1923
	1	9–10	84.4	10.3	0.08	0.6	30	15	155	13	8	85	5.8	–	1.2	4.9	–	117	1862
	1	19–20	83.1	9.9	0.06	0.5	27	24	155	14	8	75	4.9	–	1.1	3.0	–	241	1738
Väsman	11	0– 1	90.8	14.7	2.56	5.36	864	104	1270	38	19	82	15.1	23	2.7	4.9	784	7.5	1970
	4	4– 5	87.4	14.9	0.70	2.48	141	47	758	25	12	79	7.1	–	2.7	4.7	–	30	1949
	4	9–10	86.6	15.3	0.11	0.90	38	19	230	20	9	108	7.3	–	1.9	3.9	–	63	1916
	4	19–20	86.4	15.4	0.09	0.93	35	17	188	19	10	111	6.2	–	1.9	4.3	–	130	1849
Övre Hillen	4	0– 1	90.8	12.6	1.80	5.1	235	90	1470	44	18	55	2.7	40	3.9	6.9	1210	3.5	1975
	1	4– 5	66.5	4.7	1.63	2.3	562	155	1060	13	21	81	3.0	–	–	1.4	–	15	1964
	1	9–10	69.4	4.9	3.59	2.1	83	120	915	17	30	83	4.6	–	1.3	1.8	–	31	1948
	1	14–15	78.8	8.2	2.47	1.8	82	86	835	22	27	93	6.1	–	0.9	2.4	–	47	1932
Norra Barken	9	0– 1	88.8	11.5	0.96	8.9	785	112	2860	46	54	60	6.7	18	2.4	6.3	1715	1.2	1978
	3	4– 5	84.1	9.7	1.25	13.2	680	104	4080	42	49	80	4.8	–	1.3	5.2	–	7	1972
	3	9–10	77.7	8.4	0.71	7.5	566	87	1543	36	27	59	2.7	–	1.2	3.8	–	15	1964
	3	19–20	77.4	9.9	0.44	1.8	183	63	773	29	23	66	2.4	–	1.1	4.2	–	31	1948
Stora Aspen	7	0– 1	83.1	9.6	0.49	5.1	848	109	2014	2079	476	85	1.9	28	1.4	4.4	3160	–	–
	2	4– 5	69.0	5.6	0.07	0.6	41	16	260	22	22	37	1.1	–	0.7	2.5	–	–	–
	2	9–10	69.1	6.5	0.06	0.6	67	21	275	26	19	37	1.2	–	0.8	2.6	–	–	–
	2	19–20	71.5	7.3	0.06	0.5	39	16	230	27	18	33	1.2	–	0.9	2.5	–	–	–
Östersjön	13	0– 1	83.9	12.0	0.62	1.9	139	128	1256	698	604	79	1.4	15	1.8	7.9	2400	1.0	1978
	6	3– 7	77.0	10.8	0.63	1.7	119	163	841	583	473	71	1.0	17	2.1	4.8	4000	8	1971
	6	9–13	75.1	11.9	0.61	1.1	82	133	543	169	136	53	0.7	17	1.5	4.3	2000	18	1961
	6	15–20	71.9	9.9	0.40	0.6	54	100	283	100	30	41	0.5	23	0.9	4.7	–	29	1950
Freden	4	0– 1	83.8	10.9	0.36	2.8	87	73	1128	378	190	59	1.5	32	2.0	5.4	1020	1	1978
	1	4– 5	79.1	13.1	0.38	4.1	100	112	1220	287	200	60	1.5	–	1.9	5.2	–	6	1973
	1	9–10	79.2	12.4	0.28	4.8	99	120	1270	354	270	57	1.8	–	2.3	5.3	–	14	1965
	1	23–25	76.5	11.5	0.32	7.3	126	160	1320	322	175	64	1.6	–	3.0	5.1	–	34	1945

n = number of analyses, w = water content and IG = loss on ignition

Lakes along the main watercourse

	Hg	Cd	Pb	Cu	Zn	Cr	Ni	Fe	Mn	Al	P	N	Oil
Björken	0.34	1.4	63	15	297	16	< 20	63	2.6	19	1.2	4.9	–
○ Bysjön	0.35	2.1	84	18	280	17	17	64	7.9	17	1.4	5.1	–
○ Väsman	2.35	5.1	794	97	1210	37	19	87	15.5	22	2.6	4.6	784
○ Ö. Hillen	1.80	5.1	235	90	1470	44	18	55	2.7	40	3.9	6.9	1210
N. Hillen	1.54	2.7	196	56	1090	39	⩽ 20	60	1.5	17	2.1	5.2	183
Leran	0.65	3.0	268	51	870	27	40	32	1.4	13	1.3	3.5	413
○ N. Barken	0.96	8.9	785	112	2860	46	54	60	6.7	18	2.4	6.3	1720
S. Barken	0.47	5.4	260	45	1500	31	⩽ 20	41	1.9	19	1.5	4.3	99
○ St. Aspen	0.49	5.1	848	109	2010	2080	476	85	1.9	28	1.4	4.4	3160
L. Aspen	0.39	4.3	840	80	1930	1380	457	68	2.9	33	1.5	5.0	–
Åmänningen	0.28	2.5	378	51	1610	443	169	65	7.4	37	2.1	6.6	282
Virsbosjön	0.30	1.8	170	47	1310	172	75	44	1.2	34	1.1	5.2	306
St. Nadden	0.14	1.9	135	44	1100	152	85	44	2.4	35	0.9	7.4	–
L. Nadden	0.33	2.3	180	53	1360	158	110	45	1.7	39	0.9	6.3	–
Gnien	0.31	1.9	100	44	940	141	70	31	1.0	23	1.0	6.2	–
Magsjön	0.29	1.8	128	37	1010	156	63	35	1.9	34	1.1	5.2	247
○ Östersjön	0.62	1.9	139	128	1260	698	604	79	1.4	15	1.8	7.9	2400
○ Freden	0.36	2.8	87	73	1130	378	190	59	1.5	32	2.0	5.4	1020

Lakes outside the main watercourse

	Hg	Cd	Pb	Cu	Zn	Cr	Ni	Fe	Mn	Al	P	N	Oil
Saxen	0.95	105	16100	4030	35430	3340	7	78	1.3	–	2.7	5.1	–
Haggen	0.20	2.7	136	27	308	16	43	60	6.5	19	1.9	5.2	150
Noren	0.46	1.0	98	74	325	34	< 20	48	2.1	34	1.1	6.7	–
Trätten	0.60	1.2	58	62	273	22	< 20	37	1.0	32	2.3	9.7	–
Snyten	0.19	0.7	65	32	203	11	< 20	43	7.5	39	2.0	4.8	–
Västersjön	0.20	1.1	65	38	450	52	35	32	0.4	36	1.4	14.4	50

Fig. 10.10. Mean concentrations in surficial sediments (0–1 cm) from areas of accumulation of Hg, Cd, Pb, Cu, Zn, Cr, Ni and oil (in $\mu g\ g^{-1}$ ds, ds = dry sediments) and Fe, Mn, Al, P and N (in $mg\ g^{-1}$ ds) in 24 lakes belonging to the River Kolbäcksån water system. Lakes *marked with a ring* are in focus in this context (Håkanson 1983a)

Table 10.11. Comparison between required number of samples according to the sample formula and the actual number of sediment samples taken in the seven actual lakes

	Required number according to sample formula	Actual number of samples
Bysjön	5	3
Väsman	8	9
Övre Hillen	5	3
Norra Barken	7	6
Stora Aspen	5	5
Östersjön	4	11
Freden	4	3

Table 10.12. A statistical representation of sediment data from seven lakes form the River Kolbäcksån water system

	Surficial sediments (0–1 cm)				
	n	$\bar{x}$	s_x	x_{min}	x_{max}
Water content (% ws)	53	86.9	4.8	76.6	94.2
Organic content (% ds)	53	12.0	2.6	6.6	17.7
Hg (ng g^{-1} ds)	53	1100	1460	70	7910
Cd (ng g^{-1} ds)	53	4550	3570	400	20300
Pb ($\mu g\ g^{-1}$ ds)	53	490	610	20	3570
Cu ($\mu g\ g^{-1}$ ds)	53	100	97	11	360
Cr ($\mu g\ g^{-1}$ ds)	53	490	980	11	4000
Zn ($\mu g\ g^{-1}$ ds)	53	1550	1050	135	5600

From these premises, we can determine the requested contamination factors (C_f^i) for the seven lakes and the six metals in focus (Table 10.14). There is a substantial increase in comparability and information value from concentrations to contamination factors, which enables direct comparisons to be made between metals relative to natural background levels. The contaminational profiles of the given lakes are clear from Table 10.14. The contamination factor (C_f^i) accounts for the contamination of single elements. If $C_{0-1}^i \geqslant K_n^i$, the substances are defined as contaminating or enriched; if $C_{0-1}^i < K_n^i$, then the element should not be characterized as contaminating but as non-contaminating or conservative. The following terminology may be used to obtain a uniform way of describing the contamination factor:

$$C_f^i < 1 \Rightarrow \text{low contamination factor}$$

$$1 \leqslant C_f^i < 3 \Rightarrow \text{moderate contamination factor}$$

$$3 \leqslant C_f^i < 6 \Rightarrow \text{considerable contamination factor}$$

$$C_f^i \geqslant 6 \Rightarrow \text{very high contamination factor.}$$

Table 10.13. Determination of the average natural background levels (K_n) for various elements (in $\mu g\ g^{-1}$ ds) in preindustrial lake sediments from the River Kolbäcksån water system

Lake:	Bysjön		Väsman						
Station:	5		8		9		11	14	
Level (cm):	9–10	19–20	9–10	19–20	9–10	19–20	19–20	9–10	19–20
Mean age:	1862	1738		9–10 ≈ 1916;		19–20 ≈ 1849			
Hg	0.08	0.06	0.12	0.10	0.10	0.06	0.10	0.07	0.10
Cd	0.6	0.5	0.6	0.4	0.6	0.7	1.8	0.5	0.8
Pb	30	27	32	7	10	7	25	41	100
Cu	15	24	24	16	27	25	14	11	14
Zn	155	155	215	180	175	150	215	150	205
Cr	13	14	20	20	24	20	20	13	16
Ni	8	8	10	8	11	8	13	6	9
Fe	85	75	164	179	104	115	52	68	96
Mn	5.8	4.9	11.3	10.2	3.0	3.0	3.4	5.6	8.3
P	1.2	1.1	1.9	2.5	1.6	1.6	1.7	2.4	1.7
N	4.9	3.0	4.9	4.1	2.8	3.6	3.8	3.2	5.7

[a] $t_{0.975} = 2.16$, $\nu = 13$

Table 10.14. Sediment contamination factors for selected metals in the seven lakes in the River Kolbäcksån water system

	Sediment contamination factor (C_f)					
	Hg	Cd	Pb	Cu	Cr	Zn
Bysjön	3.9	2.5	1.8	0.8	0.8	1.3
Väsman	28.4	6.5	18.3	4.8	1.6	6.1
Övre Hillen	20.0	6.1	5.0	4.2	1.9	7.1
Norra Barken	10.7	10.7	16.7	5.2	2.0	13.8
Stora Aspen	5.4	6.1	18.0	5.1	90.0	9.7
Östersjön	6.9	2.3	3.0	6.0	30.2	6.0
Freden	4.0	3.4	1.8	3.4	16.4	5.4

10.3.4 The Degree of Contamination

The degree of contamination (C_d) can be defined as the sum of all contamination factors larger than 1 for a given lake/recipient, i.e.:

$$C_d = \sum_{i=1}^{n} C_f^i = \sum_{i=1}^{n} \bar{C}_{0-1}^i / K_n^i . \qquad (10.2)$$

With this approach, we have a contamination factor accounting for single substances and a degree of contamination (C_d), accounting for the total sediment pollution within a lake or a defined recipient. The following terminology may be used to describe the C_d-values:

$C_d < n \Rightarrow$ low degree of contamination
$n \leqslant C_d < 2n \Rightarrow$ moderate degree of contamination
$2n \leqslant C_d < 4n \Rightarrow$ considerable degree of contamination
$C_d \geqslant 4n \Rightarrow$ very high degree of contamination.

Stora Aspen			Östersjön		n	$\bar{x}$	s_x	Natural background level, K_n
46		47	76	82				
9–10	19–20	19–20	19–20	17–18				$\bar{x} + t_{0.975}(s_x/\sqrt{n-1})$ [a]
0.06	0.06	0.06	0.05	0.05	14	0.076	0.022	0.089 ≈ 0.090
0.4	0.5	0.4	0.5	0.3	14	0.61	0.36	0.826 ≈ 0.83
50	59	18	23	25	14	32.4	24	46.8 ≈ 47.0
20	15	17	19	19	14	18.6	4.7	21.4 ≈ 21.5
235	260	200	165	145	14	186	35	207 ≈ 210
23	24	29	25	22	14	20.2	4.6	23.0 ≈ 23.0
17	15	20	17	14	14	11.7	4.1	14.2
35	32	34	26	28	14	78.1	48	107
1.1	1.1	1.2	0.7	0.6	14	4.3	3.5	6.4
0.7	1.0	0.8	0.5	0.2	14	1.35	0.66	1.7
2.6	2.6	2.4	3.0	2.9	14	3.54	0.98	4.1

Table 10.15. Ranking order of sediment contamination of the seven lakes from the River Kolbäcksån water system according to C_d-values, and sequence of contamination factors (C_f^i)

Lake	Degree of cont.	Contamination factor			
		Very high	Considerable	Moderate	Low
	C_d	$C_f^i \geqslant 6$	$3 \leqslant C_f^i < 6$	$1 \leqslant C_f^i < 6$	$C_f^i < 1$
Stora Aspen	134.1	Cr > Pb > Zn > Cd	Hg > Cu	–	–
Väsman	65.7	Hg > Pb > Cd > Zn	Cu	Cr	–
Norra Barken	59.1	Pb > Zn > Hg = Cd	Cu	Cr	–
Östersjön	54.4	Cr > Hg > Cu = Zn	Pb	Cd	–
Övre Hillen	44.3	Hg > Zn > Cd	Pb > Cu	Cr	–
Freden	34.4	Cr	Zn > Hg > Cd = Cu	Pb	–
Bysjön	11.1	–	Hg	Cd > Pb > Zn	Cu = Cr

To get a better view of the contaimination in the investigated lakes, the information in Table 10.14 is transported to Table 10.15, which illustrates the ranking order, the "contamination league", according to the C_d-values as well as the sequence of the contamination factors. Lake Stora Aspen leads the "pollution league" with a C_d-value of 134.1, which is very high. The contaminational profile in this lake is:

Cr > Pb > Zn = Cd > Hg > Cu.

The first four metals are seen to have very high contamination factors; Hg and Cu have considerable C_f^i-values. Lake Bysjön, which lies far up in the drainage area, is the only lake with a moderate degree of contamination (C_d = 11). The metal sequence in this lake is:

Hg > Cd > Pb > Zn > Cu = Cr.

Only Hg has a considerable contamination factor in Lake Bysjön.

To conclude, we may say that the contamination character can be described in a uniform, standardized and quantitative way by mean of the contamination factor and the degree of contamination. The terminology and limits between various classes can, of course, always be discussed and this set-up is only meant to serve as an example of one alternative. One basic requirement in aquatic pollution control is, however, to have a uniform framework, which facilitates comparisons between different areas and minimizes subjective speculations. The two main limitations to the use of sediment data in aquatic pollution control programmes concerns, firstly, the fact that sediments only provide indirect indications, whereas biological methods may yield direct ecological effects, and, secondly, the fact that sediment data give a resolution in time vis-à-vis a given discharge of contaminants "by the year", where sediment traps and biological techniques can yield a resolution "by the month" and water samples "by the hour".

11 Epilogue

On the outskirts of established scientific subjects, and at the cross-roads of traditional modes of thinking, good vibrations are generated and exciting new fields of research are formed. It is our belief that Environmental Sedimentology is a good, but perhaps not yet fully recognized, example of such an interesting multi-disciplinary subject. It has, for example, a central role in lake management, since nutrient fluxes and restoration programmes cannot be adequately handled without sedimentological considerations. Water quality control programmes offer another example, because sediments often provide time-integrated data with a high information value on distribution patterns of pollutants. The latter aspects emphasize the role of sedimentology in the discipline of ecotoxicology; most toxic substances in the aquatic environment are attached to suspended particles or aggregates which regulate not only the dispersal pattern of pollutants but also ecological effects.

Clearly, the discipline of Environmental Sedimentology is an important field of scientific endeavour from many "basic" and "applied" viewpoints. It is also exciting because much remains to be done and because several *major* problems have to be solved or systematically worked out, for example:

- The problem of representative sediment sampling in lakes, i.e., to reveal and rank the causal relationships regulating the information value of various sediment sampling systems, especially considering the problem of skew populations and small samples.
- The resuspension problem, i.e., to study the mechanisms regulating the difference between total and net deposition of materials in lakes.
- The classification problem, i.e., to make a thorough review of existing classification systems for lake sediments and then provide a reliable "new" system based on the "best" parts of the old ones and whatever it takes to establish a nomenclature and a classification that can gain general acceptance. This would be a major achievement, not only reducing much of the present confusion, but also facilitating future communications.
- The entrainment problem. The capacity of entrainment of cohesive sediments is regulated by several factors which may be separated into two distinct groups: a physical-chemical group, which may be represented by the water content of the sediments, the grain size and/or the bulk density, and a biological group, which should express the "glue" capacity of the sediments. But what is the "glue" capacity? How can it be measured? How can it be combined with, e.g., the water content, to reveal the capacity for entrainment? Is it possible to use specific trace

substances, such as uronic acid, to estimate the "glue" capacity? The problem of entrainment is, evidently, related to the resuspension problem.

- The bioturbation problem. The bottom fauna displays a considerable patchiness, i.e., variation in time and space in lakes. If this patchiness cannot be accounted for, then the real impact of the bottom fauna on the sediments cannot be numerically determined and the sediment age, the age distribution or the elemental mobility in and from the sediments cannot be evaluated in a strictly scientific manner – merely guessed. Very much remains to be done on the relationship between organisms and sediments.
- The *eco*toxicological problem, emphasizing ECO and the interrelationships between contaminants, "carrier particles" (such as humus, detritus, carbonates) and environmental factors (such as pH, alkalinity, residence time and morphometry) influencing the spread pattern and potential biological uptake and ecological effect.
- The problem on sediment-water exchange. It is not yet possible to rank and differentiate between the causal relationships determining the turnover and release of the most fundamental elements and compounds in different sediment types and lakes.
- The impact of biological metabolism on lake sediments. To what extent, in quantitative terms, do bacteria, plants and animals change the physical and chemical sediment milieu? What role do microorganisms have for the capacity of the sediments to bind and/or release nutrients and pollutants?

These are some important and central fields of research which will certainly attract the attention of scientists in this dynamic topic.

Appendix 1 Table for Student's t-Distribution

ν	$t_{0.995}$	$t_{0.99}$	$t_{0.975}$	$t_{0.95}$	$t_{0.90}$	$t_{0.80}$	$t_{0.75}$	$t_{0.70}$	$t_{0.60}$	$t_{0.55}$
1	63·66	31·82	12·71	6·31	3·08	1·376	1·000	0·727	0·325	0·158
2	9·92	6·96	4·30	2·92	1·89	1·061	0·816	0·617	0·289	0·142
3	5·84	4·54	3·18	2·35	1·64	0·978	0·765	0·584	0·277	0·137
4	4·60	3·75	2·78	2·13	1·53	0·941	0·741	0·569	0·271	0·134
5	4·03	3·36	2·57	2·02	1·48	0·920	0·727	0·559	0·267	0·132
6	3·71	3·14	2·45	1·94	1·44	0·906	0·718	0·553	0·265	0·131
7	3·50	3·00	2·36	1·90	1·42	0·896	0·711	0·549	0·263	0·130
8	3·36	2·90	2·31	1·86	1·40	0·889	0·706	0·546	0·262	0·130
9	3·25	2·82	2·26	1·83	1·38	0·883	0·703	0·543	0·261	0·129
10	3·17	2·76	2·23	1·81	1·37	0·879	0·700	0·542	0·260	0·129
11	3·11	2·72	2·20	1·80	1·36	0·876	0·697	0·540	0·260	0·129
12	3·06	2·68	2·18	1·78	1·36	0·873	0·695	0·539	0·259	0·128
13	3·01	2·65	2·16	1·77	1·35	0·870	0·694	0·538	0·259	0·128
14	2·98	2·62	2·14	1·76	1·34	0·868	0·692	0·537	0·258	0·128
15	2·95	2·60	2·13	1·75	1·34	0·866	0·691	0·536	0·258	0·128
16	2·92	2·58	2·12	1·75	1·34	0·865	0·690	0·535	0·258	0·128
17	2·90	2·57	2·11	1·74	1·33	0·863	0·689	0·534	0·257	0·128
18	2·88	2·55	2·10	1·73	1·33	0·862	0·688	0·534	0·257	0·127
19	2·86	2·54	2·09	1·73	1·33	0·861	0·688	0·533	0·257	0·127
20	2·84	2·53	2·09	1·72	1·32	0·860	0·687	0·533	0·257	0·127
21	2·83	2·52	2·08	1·72	1·32	0·859	0·686	0·532	0·257	0·127
22	2·82	2·51	2·07	1·72	1·32	0·858	0·686	0·532	0·256	0·127
23	2·81	2·50	2·07	1·71	1·32	0·858	0·865	0·532	0·256	0·127
24	2·80	2·49	2·06	1·71	1·32	0·857	0·685	0·531	0·256	0·127
25	2·79	2·48	2·06	1·71	1·32	0·856	0·684	0·531	0·256	0·127
26	2·78	2·48	2·06	1·71	1·32	0·856	0·684	0·531	0·256	0·127
27	2·77	2·47	2·05	1·70	1·31	0·855	0·684	0·531	0·256	0·127
28	2·76	2·47	2·05	1·70	1·31	0·855	0·683	0·530	0·256	0·127
29	2·76	2·46	2·04	1·70	1·31	0·854	0·683	0·530	0·256	0·127
30	2·75	2·46	2·04	1·70	1·31	0·854	0·683	0·530	0·256	0·127
40	2·70	2·42	2·02	1·68	1·30	0·851	0·681	0·529	0·255	0·126
60	2·66	2·39	2·00	1·67	1·30	0·848	0·679	0·527	0·254	0·126
120	2·62	2·36	1·98	1·66	1·29	0·845	0·677	0·526	0·254	0·126
∞	2·58	2·33	1·96	1·645	1·28	0·842	0·674	0·524	0·253	0·126

Appendix 2 Computer Programmes in BASIC for Determination of Biotransport, Time Stratification and Sediment Compaction

Steps marked by (*) are to be fed with empirical indata, e.g. K1=-2.99, K2=52000, K3=69390, K4=-233.93, W1=90.4 for Lake Hjälmaren in program a.

a) Biotransport

```
 10 OPEN "DK1:LHB" ON #1 FOR OUTPUT
 20 PRINT "Hjälmaren"
 30 PRINT #1,"Hjälmaren"
*40 K1=-2.99
*50 K2=52000
*60 K3=69390
*70 K4=-233.93
*80 W1=90.4
 90 PRINT "Rate of sed.=";
100 INPUT V
110 PRINT #1,"Rate of sed.=";V
120 PRINT "Biot. at surface=";
130 INPUT B1
140 PRINT #1,"Biot. at surface=";B1
150 PRINT "Substr. const.=";
160 INPUT C
170 PRINT #1,"Substr. const.=";C
180 PRINT "Level","Bioturbation","Rate"
190 PRINT #1,"Level","Bioturbation","Rate"
200 X=W1+K4
210 PRINT USING 355;1;B1;B1*(K3-X*X)/K2
220 PRINT USING #1,355;1;B1;B1*(K3-X*X)/K2
230 N=0
240 FOR Y=1 TO 40 STEP .1
250 N=N+1
260 W=K1*LOG(2*Y)+W1
270 W2=K1/Y
280 Z=W1+K4+K1*LOG(Y)
290 R=K2/(K3-Z*Z)
300 B=(1+.1*C)*B1+.1*V*W2*R/(100-W)
310 IF B<0 THEN 380
320 B1=B
330 IF N>10*INT(N/10) THEN 370
340 IF N<10*INT(N/10) THEN 370
350 PRINT USING 355;1+N*.1;B1;B1/R
355 ! ##              #.####          #.####
360 PRINT USING #1,355;1+N*.1;B1;B1/R
370 NEXT Y
380 CLOSE #1
390 END
```

```
HJälmaren
Rate of sed.= .4
Biot. at surface= .27
Substr. const.= 0


Level          Bioturbation  Rate

  1            0.2700        0.2533
  2            0.1967        0.1825
  3            0.1592        0.1465
  4            0.1344        0.1230
  5            0.1160        0.1058
  6            0.1016        0.0923
  7            0.0897        0.0812
  8            0.0796        0.0719
  9            0.0709        0.0639
 10            0.0633        0.0569
 11            0.0565        0.0507
 12            0.0503        0.0451
 13            0.0447        0.0400
 14            0.0396        0.0354
 15            0.0349        0.0311
 16            0.0305        0.0272
 17            0.0264        0.0235
 18            0.0226        0.0201
 19            0.0191        0.0169
 20            0.0157        0.0139
 21            0.0125        0.0111
 22            0.0095        0.0084
 23            0.0066        0.0058
 24            0.0039        0.0034
 25            0.0013        0.0011
```

b) Time stratification

```
  10 OPEN "DK1:LHT" ON #1 FOR OUTPUT
  20 PRINT "Biot. HJälmaren"
  30 PRINT #1,"Biot. HJälmaren"
  40 PRINT #1
 *50 K1=-2.99
 *60 K2=52000
 *70 K3=69390
 *80 K4=-233.93
 *90 W1=90.4
*100 M1=1.049
 110 R1=M1
 120 PRINT "Rate fo sed.=";
 130 INPUT V
 140 PRINT #1,"Rate of sed.=";V;"cm/yr"
 150 PRINT "Biot. at surface=";
 160 INPUT B1
 170 PRINT #1,"Biot. at surface=";B1;"gws/yr"
 180 PRINT "Substrat const.=";
 190 INPUT C
 200 PRINT #1,"Substrat const.=";C
 210 PRINT #1
 220 IF C=0 THEN 240
 230 C=1/C
```

```
240 DATA 1,2,3,4,5,6,7,8,9,10,11,12,13,14,15,16,17,18,19,20,21
250 DATA 22,23,24,25,26,27,28,29,30,31,32,33,34,35,36,37,38,39,40
260 PRINT "Level","Biot.","Biot.up","Biot.down","sws","Density"
270 PRINT #1,"Level","Biot.","Biot.up","Biot.down","sws","Density"
280 X1=1
290 FOR I=1 TO 40
300 READ X2
310 M1=M
320 FOR Y=X1 TO X2 STEP 0.01
330 W=K1*LOG(2*Y)+W1
340 W2=K1/Y
350 A=K4+W1+K1*LOG(Y)
360 R=K2/(K3-A*A)
370 IF I>1 THEN 400
380 M1=M
390 GOTO 410
400 M1=M1+R*.01
410 B=(1+.01*C)*B1+.01*V*W2*R/(100-W)
420 IF B<0 THEN 620
430 B1=B
440 NEXT Y
450 W3=K1*LOG(2*X2+1)+W1
460 W4=K1*LOG(2*X2-1)+W1
470 A=K4+W1+K1*LOG(X2+.5)
480 R3=K2/(K3-A*A)
490 A=K4+W1+K1*LOG(X2-.5)
500 R4=K2/(K3-A*A)
510 P1=(100-W1)*R1
520 P3=P1/((100-W3)*R3)
530 P=P1/((100-W)*R)
540 P4=P1/((100-W4)*R4)
550 U=B1*(P4-P)/(P-P3)
560 N=U-B1
570 PRINT USING 575;X2;B1;U;N;M1;R
575 ! ##          #.####          #.####          #.####          #.###          #.
580 PRINT USING #1,575;X2;B1;U;N;M1;R
590 X1=X2
600 NEXT I
610 GOTO 640
620 PRINT "Below Biot. Limit"
630 PRINT #1,"Below Biot. Limit"
640 PRINT
650 PRINT #1
660 RESTORE
670 GOTO 120
680 END
```

Biot. Hjälmaren

Rate of sed.= .4 cm/yr
Biot. at surface= .27 gws/yr
Substrat const.= 0

Level	Biot.	Biot.up	Biot.down	gws	Density
1	0.2689	0.6165	0.3476	0.000	1.066
2	0.1985	0.2750	0.0765	1.073	1.079
3	0.1615	0.2043	0.0428	1.094	1.087
4	0.1369	0.1627	0.0258	1.101	1.093
5	0.1188	0.1308	0.0120	1.095	1.097
6	0.1044	0.1123	0.0079	1.099	1.101
7	0.0926	0.0980	0.0053	1.103	1.104
8	0.0827	0.0863	0.0037	1.106	1.107
9	0.0739	0.0796	0.0057	1.120	1.110
10	0.0662	0.0708	0.0045	1.122	1.112
11	0.0594	0.0630	0.0037	1.124	1.114
12	0.0532	0.0562	0.0030	1.126	1.116
13	0.0476	0.0500	0.0025	1.128	1.118
14	0.0424	0.0445	0.0020	1.130	1.119
15	0.0377	0.0394	0.0017	1.131	1.121
16	0.0333	0.0347	0.0014	1.133	1.122
17	0.0292	0.0292	-.0001	1.123	1.123
18	0.0254	0.0253	-.0001	1.124	1.125
19	0.0219	0.0217	-.0001	1.125	1.126
20	0.0185	0.0184	-.0001	1.126	1.127
21	0.0153	0.0152	-.0001	1.128	1.128
22	0.0123	0.0122	-.0001	1.129	1.129
23	0.0094	0.0093	-.0001	1.130	1.130
24	0.0067	0.0066	-.0001	1.131	1.131
25	0.0041	0.0041	-.0001	1.132	1.132
26	0.0016	0.0016	-.0000	1.133	1.133

Below Biot. Limit

b) Time stratification continues

Output data from previous part are input data in the subsequent program.

Note, only data from 1,3,5,...cm have been included here to save space.

```
  10 OPEN "DK1:LHTC" ON #1 FOR OUTPUT
  20 DIM A(13,75)
  30 DIM U(13),N(13),M(13),D(13)
  40 DIM X(75),Y(75),Z(75)
  50 K1=13
  60 L1=75
 *70 R=1.049
 *80 V=.4
 *90 DATA .617,.348,1.066
*100 DATA .204,.043,1.087
*110 DATA .130,.012,1.097
*120 DATA .098,.005,1.104
*130 DATA .080,.006,1.110
*140 DATA .063,.004,1.114
*150 DATA .050,.003,1.118
*160 DATA .039,.002,1.121
*170 DATA .029,.000,1.123
*180 DATA .022,.000,1.126
*190 DATA .015,.000,1.128
*200 DATA .009,.000,1.130
*210 DATA .004,.000,1.132
*220 DATA .000,.000,1.133
 230 FOR K=0 TO K1
 240 READ U(K)
 250 READ N(K)
 260 READ M(K)
 270 D(K)=M(K)
 280 FOR L=0 TO L1
 290 IF L<>K THEN 320
 300 A(K,L)=M(K)
 310 GOTO 330
 320 A(K,L)=0
 330 NEXT L
 340 NEXT K
 350 FOR T=10 TO 100 STEP 10
 360 FOR I=T-9 TO T
 370 FOR K=0 TO K1
 380 X(0)=0
 390 FOR L=1 TO L1
 400 IF L=L1 THEN 430
 410 X(L)=A(K,L-1)
 420 GOTO 440
 430 X(L)=A(K,L-1)+A(K,L)
 440 NEXT L
 450 FOR L=0 TO L1
 460 A(K,L)=X(L)
 470 NEXT L
 480 NEXT K
 490 FOR K=K1 TO 0 STEP -1
 500 FOR L=0 TO L1
 510 IF K=0 THEN 540
 520 A(K,L)=(1-V)*A(K,L)+V*A(K-1,L)
 530 GOTO 580
 540 IF L=0 THEN 570
 550 A(K,L)=(1-V)*A(K,L)
 560 GOTO 580
 570 A(K,L)=R*V+(1-V)*A(K,L)
 580 NEXT L
```

```
590 NEXT K
600 FOR L=0 TO L1
610 S=A(0,L)*N(0)/D(0)-A(1,L)*U(0)/D(1)
620 X(L)=A(0,L)-S
630 S1=A(1,L)*N(1)/D(1)-A(2,L)*U(1)/D(2)
640 Y(L)=A(1,L)+S-S1
650 S=A(2,L)*N(2)/D(2)-A(3,L)*U(2)/D(3)
660 Z(L)=A(2,L)+S1-S
670 NEXT L
680 FOR K=3 TO K1
690 FOR L=0 TO L1
700 A(K-3,L)=X(L)
710 X(L)=Y(L)
720 Y(L)=Z(L)
730 S=A(K-1,L)*N(K-1)/D(K-1)-A(K,L)*U(K-1)/D(K)
740 S=S-A(K,L)*N(K)/D(K)
750 IF K=K1 THEN 770
760 S=S+A(K+1,L)*U(K)/D(K+1)
770 Z(L)=A(K,L)+S
780 NEXT L
790 NEXT K
800 FOR L=0 TO L1
810 A(K1-2,L)=X(L)
820 A(K1-1,L)=Y(L)
830 A(K1,L)=Z(L)
840 NEXT L
850 FOR K=0 TO K1
860 S=0
870 FOR L=0 TO L1
880 S=S+A(K,L)
890 NEXT L
900 FOR L=0 TO L1
910 IF K=K1 THEN 930
920 A(K+1,L)=A(K+1,L)+A(K,L)*(1-M(K)/S)
930 A(K,L)=A(K,L)*M(K)/S
940 NEXT L
950 NEXT K
960 NEXT I
970 PRINT "Time stratification in Hjälmaren after ";T;" years"
980 PRINT #1,"Time stratification in Hjälmaren after ";T;" years"
990 FOR K=0 TO K1
1000 PRINT "Level:";2*K+1;"cm"
1010 PRINT #1,"Level:";2*K+1;"cm"
1020 PRINT
1030 PRINT #1
1040 FOR L=0 TO L1
1050 P=INT(100*100*A(K,L)/M(K)+.5)
1060 PRINT USING 1080;P/100;
1070 PRINT USING #1,1080;P/100;
1080 ! ###.##
1090 IF L+1-10*INT((L+1)/10)=0 THEN 1110
1100 GOTO 1130
1110 PRINT
1120 PRINT #1
1130 NEXT L
1140 PRINT
1150 PRINT
1160 PRINT #1
1170 PRINT #1
1180 NEXT K
1190 FOR K=1 TO 5
1200 PRINT
1210 PRINT #1
1220 NEXT K
1230 NEXT T
1240 CLOSE #1
1250 END
```

```
Example for v=0.4, b1=0.27 and C=0 for Lake Hjälmaren.

Only data from 1,9 and 19 cm have been included to save space.

Time stratification in Hjälmaren after  10  years

Level: 1 cm

  21.32  15.18  11.66   9.20   7.38   5.99   4.90   4.03   3.33   2.76
   6.87   5.12   1.83   0.38   0.05   0.00   0.00   0.00   0.00   0.00
   0.00   0.00   0.00   0.00   0.00   0.00   0.00   0.00   0.00   0.00
   0.00   0.00   0.00   0.00   0.00   0.00   0.00   0.00   0.00   0.00
   0.00   0.00   0.00   0.00   0.00   0.00   0.00   0.00   0.00   0.00
   0.00   0.00   0.00   0.00   0.00   0.00   0.00   0.00   0.00   0.00
   0.00   0.00   0.00   0.00   0.00   0.00   0.00   0.00   0.00   0.00
   0.00   0.00   0.00   0.00   0.00   0.00

Level: 9 cm

   0.01   0.15   0.86   2.38   4.09   5.53   6.52   7.04   7.19   7.05
  17.83  16.96  13.12   7.45   2.92   0.76   0.12   0.01   0.00   0.00
   0.00   0.00   0.00   0.00   0.00   0.00   0.00   0.00   0.00   0.00
   0.00   0.00   0.00   0.00   0.00   0.00   0.00   0.00   0.00   0.00
   0.00   0.00   0.00   0.00   0.00   0.00   0.00   0.00   0.00   0.00
   0.00   0.00   0.00   0.00   0.00   0.00   0.00   0.00   0.00   0.00
   0.00   0.00   0.00   0.00   0.00   0.00   0.00   0.00   0.00   0.00
   0.00   0.00   0.00   0.00   0.00   0.00

Level: 19 cm

   0.00   0.00   0.00   0.00   0.00   0.00   0.01   0.03   0.09   0.22
   0.62   1.46   4.30   9.79  17.18  22.57  21.57  14.30   6.15   1.52
   0.18   0.01   0.00   0.00   0.00   0.00   0.00   0.00   0.00   0.00
   0.00   0.00   0.00   0.00   0.00   0.00   0.00   0.00   0.00   0.00
   0.00   0.00   0.00   0.00   0.00   0.00   0.00   0.00   0.00   0.00
   0.00   0.00   0.00   0.00   0.00   0.00   0.00   0.00   0.00   0.00
   0.00   0.00   0.00   0.00   0.00   0.00   0.00   0.00   0.00   0.00
   0.00   0.00   0.00   0.00   0.00   0.00
```

c) Compaction

```
 10 OPEN "DK1:LHC" ON #1 FOR OUTPUT
 20 PRINT "Compaction Hjälmaren"
 30 PRINT #1,"Compaction Hjälmaren"
*40 K1=-2.99
*50 K2=52000
*60 K3=69390
*70 K4=-233.93
*80 W1=90.4
 90 V=.4
100 M=1.049
110 I=0
120 I1=0
130 FOR X=.5 TO 1.STEP .01
140 W=K1*LOG(2*X)+W1
150 A=K4+W1+K1*LOG(X)
160 R=K2/(K3-A*A)
170 IF X>.5 THEN 200
180 W2=W
```

```
190 R2=R
200 I=I+2*W
210 I1=I1+2*R
220 NEXT X
230 I=(I-W-W2)*5.00000E-03
240 I1=(I1-R-R2)*5.00000E-03
250 W=(W1-I)*2
260 R2=(M-I1)*2
270 PRINT "Water content 0-0.5 cm=";W2;"Density 0-0.5 cm=";R2
280 PRINT #1,"Water content 0-0.5 cm=";W2;"Density 0-0.5 cm=";R2
290 PRINT
300 PRINT "Level","Red Level","sds","Acc sds","sws","Acc sws","Comp"
310 PRINT #1
320 PRINT #1,"Level","Red Level","sds","Acc sds","sws","Acc sws","Comp"
330 J=0
340 I5=(100-W1)*R2*V/100
350 J1=0
360 FOR D=V TO 40 STEP V
370 I=0
380 I1=0
390 FOR X=D-V TO D STEP .01
400 IF X>.5 THEN 440
410 W=W2
420 R=R2
430 GOTO 470
440 W=K1*LOG(2*X)+W1
450 A=K4+W1+K1*LOG(X)
460 R=K2/(K3-A*A)
470 A=R*(1-W/100)
480 IF X>D-V THEN 510
490 A1=A
500 R1=R
510 I=I+2*A
520 I1=I1+2*R
530 NEXT X
540 I=(I-A-A1)*5.00000E-03
550 I1=(I1-R-R1)*5.00000E-03
560 IF D>V THEN 580
570 T=I
580 D1=D-V+(D*T-J*V)/I
590 J=J+I
600 J1=J1+I1
610 I6=I5/I
620 PRINT USING 625;D;D1;I;J;I1;J1;I6
625 ! ##.#     ##.###      #.###      #.###      #.###     ##.###      #.###
630 PRINT USING #1,625;D;D1;I;J;I1;J1;I6
640 NEXT D
650 PRINT
660 PRINT #1
670 CLOSE #1
680 END
```

Example. Note, only data to 12 cm have been included to save space.

Compaction Hjälmaren
Water content 0-0.5 cm= 90.4 Density 0-0.5 cm= 1.059

Level	Red Level	sds	Acc sds	sws	Acc sws	Comp
0.4	0.400	0.040	0.040	0.413	0.413	1.026
0.8	0.779	0.042	0.082	0.413	0.826	0.971
1.2	1.110	0.048	0.130	0.416	1.241	0.841
1.6	1.417	0.053	0.183	0.418	1.659	0.769
2.0	1.710	0.056	0.239	0.420	2.079	0.722
2.4	1.992	0.061	0.300	0.432	2.512	0.671
2.8	2.259	0.063	0.363	0.434	2.945	0.646
3.2	2.520	0.065	0.428	0.435	3.380	0.625
3.6	2.775	0.067	0.495	0.436	3.816	0.609
4.0	3.026	0.068	0.563	0.437	4.253	0.595
4.4	3.255	0.068	0.631	0.427	4.679	0.597
4.8	3.505	0.069	0.700	0.427	5.107	0.586
5.2	3.751	0.071	0.771	0.428	5.535	0.577
5.6	3.995	0.072	0.843	0.429	5.963	0.568
6.0	4.235	0.073	0.915	0.429	6.392	0.560
6.4	4.474	0.074	0.989	0.430	6.822	0.553
6.8	4.710	0.074	1.063	0.430	7.252	0.546
7.2	4.944	0.075	1.139	0.431	7.683	0.540
7.6	5.176	0.076	1.215	0.431	8.114	0.534
8.0	5.406	0.077	1.292	0.432	8.546	0.529
8.4	5.694	0.080	1.371	0.443	8.989	0.511
8.8	5.915	0.080	1.452	0.443	9.432	0.507
9.2	6.135	0.081	1.533	0.444	9.876	0.502
9.6	6.354	0.082	1.614	0.444	10.320	0.498
10.0	6.571	0.082	1.696	0.445	10.765	0.495
10.4	6.787	0.083	1.779	0.445	11.210	0.491
10.8	7.002	0.083	1.863	0.445	11.655	0.488
11.2	7.216	0.084	1.947	0.446	12.101	0.484
11.6	7.429	0.084	2.031	0.446	12.547	0.481
12.0	7.641	0.085	2.116	0.446	12.993	0.478

References

Ahlgren I (1980) A dilution model applied to a system of shallow eutrophic lakes after diversion of sewage effluents. Arch Hydrobiol 89:17–32

Allen JRL (1971) Physical processes of sedimentation. Allen and Unwin, London, 248 p

Aller RC (1977) The influence of macrobenthos on chemical diagenesis of marine sediments. Unpubl. Ph D Diss, Yale Univ, 600 p

Aller RC (1980) Diagenetic processes near the sediment-water interface of Long Island Sound. I. Decomposition and nutrient element geochemistry (S, N, P). In: Saltzman B (ed) Estuarine physics and chemistry: studies in long Island Sound. Academic Press, London New York, pp 237–350

Andersen JM (1974) Nitrogen and phosphorus budgets and the role of sediments in six shallow Danish lakes. Arch Hydrobiol 74:528–550

Andersen FØ, Lastein E (1981) Sedimentation and resuspension in shallow eutrophic Lake Arreskov, Denmark. Verh Int Ver Limnol 21:425–430

Ansell AD (1974) Sedimentation of organic detritus in Lochs Etive and Creran, Argyll, Scotland. Mar Biol 27:263–273

Appleby PG, Oldfield F (1979) Letter on the history of lead pollution in Lake Michigan. Environ Sci Technol 13:478–480

Aston SR, Bruty D, Chester R, Padgham R (1973) Mercury in lake sediments. A possible indicator of technological growth. Nature 241:450–451

Axelsson V (1967) The Laitaure delta. A study of deltaic morphology and processes. Geogr Ann 49:1–127

Axeisson V (1980) Transport och avlagring av sediment vid flodmynningar. Univ Uppsala, UNGI Rap 52, pp 391–399

Axelsson V, Håkanson L (1971) Sambandet mellan kvicksilverförekomst och sedimentologisk miljö i Ekoln. Del 1. Målsättning och analysmetodik. Univ Uppsala, UNGI Rap 11, 35 p

Axelsson V, Håkanson L (1972) Sambandet mellan kvicksilverförekomst och sedimentologisk miljö i Ekoln. Del 2. Sedimentens egenskaper och kvicksilverinnehåll. Univ Uppsala, UNGI Rap 14, 89 p

Axelsson V, Håkanson L (1975) Sambandet mellan kvicksilverförekomst och sedimentologisk miljö i Ekoln. Del 4. Deposition av sediment och kvicksilver 1971 och 1972. Univ Uppsala, UNGI Rap 35, 42 p

Axelsson V, Håkanson L (1978) A gravity corer with a simple valve system. J Sediment Petrol 48:630–633

Axelsson V, Händel SK (1972) X-radiography of unextruded sediment cores. Geogr Ann 54A: 34–37

Baccini P, Roberts PV (1976) Die Belastung der Gewässer durch Metalle. Beil Forsch Tech Neue Zuerich Z 18:57–58

Bågander L-E (1976) Sediment description. In: Dybern BI, Ackefors H, Elmgren R (eds) Recommendations on methods for marine biological studies in the Baltic Sea. Baltic Mar Biol, Publ 1 Stockholm, pp 35–50

Bagnold RA (1954) The physics of blown sand and desert dunes. Methuen, London, 265 p

Baker RA (1980) Contaminants and sediments, vol II. Analysis, chemistry, biology. Ann Arbor Sci, Michigan, 627 p

Barko JW, Smart RM (1979) Mobilization of sediment phosphorus by submersed freshwater macrophytes. Freshwater Biol 10:229–238

Barnes MA, Barnes WC (1978) Organic compounds in lake sediments. In: Lerman A (ed) Lakes: chemistry, geology, physics Springer, Berlin Heidelberg New York, pp 127–152

Beach Erosion Board (1972) Waves in inland reservoirs. Tech Mem 132. Beach Erosion Board Corps of Engineers, Washington DC

Beijer K, Bengtsson B-E, Jernelöv A, Laveskog A, Lithner G, Westermark T (1977) Svenska vattenkvalitetskriterier – metaller. Del I, II. IVL Rapp B 398, Stockholm

Benninger LK, Aller RC, Cochran JK, Turekian KK (1979) Effects of biological sediment mixing on the Pb-210 chronology and trace metal distribution in a Long Island Sound sediment core. Earth Planet Sci Lett 43:241–259

Berg K (1938) Studies on the bottom animals of Esrom Lake. K Dan Vidensk Selsk Naturvidesk Math Afd 8:1–255

Berger WH, Heath GR (1968) Vertical mixing in pelagic sediments. J Mar Res 26:134–143

Berner RA (1980) Early diagenesis. A theoretical approach. Princeton Ser Geochem, Princeton University Press, New Jersey, 241 p

Berner RA (1981) A new geochemical classification of sedimentary environments. J Sediment Petrol 51:359–365

Beyers RJ (1963) The metabolism of twelve aquatic laboratory microecosystems. Ecol Monogr 33:281–306

Bloesch J (1977) Sedimentation rates and sediment cores in two Swiss lakes of different trophic status. In: Golterman HL (ed) Interactions between sediments and freshwater. Junk, The Hague, pp 65–71

Bloesch J, Burns NM (1980) A critical review of sedimentation trap technique. Schweiz Z Hydrol 42:15–55

Blomqvist S (1982) Ekologiska bedömningsgrunder för muddring och muddertippning. Natl Swed Environ Prot Board, SNV PM 1613, Solna, 113 p

Blomqvist S, Håkanson L (1981) A review on sediment traps in aquatic environments. Arch Hydrobiol 91:101–132

Blomqvist S, Kofoed C (1981) Sediment trapping – a subaquatic in situ experiment. Limnol Oceanogr 26:585–590

Boström B (1982) Potential mobility of phosphorus in lake sediments of different character. Diss thesis, Univ Uppsala

Boström B, Petterson K (1982) Different patterns of phosphorus release from lake sediments in laboratory experiments. Hydrobiologia 92:415–429

Boström B, Jansson M, Forsberg C (1982) Phosphorus release from lake sediments. Arch Hydrobiol Beih Ergebn Limnol 18:5–59

Bouma AH (1969) Methods for the study of sedimentary structures. Wiley, New York, 458 p

Bowen HJM (1966) Trace elements in biochemistry. Academic Press, London New York, 241 p

Briggs D (1977) Sources and methods in geography – sediments. Butterworths, London, 192 p

Brinkhurst RO (1974) The benthos of lakes. Macmillan, London, 190 p

Brinkhurst RO, Cook DG (eds) (1980) Aquatic oligochaete biology. Plenum Press, New York, 529 p

Brinkhurst RO, Chua KE, Kaushik NK (1972) Interspecific interactions and selective feeding by tubificid oligochaetes. Limnol Oceanogr 17:122–123

Brinkman AG, Raaphorst W van, Lijklema L (1982) In situ sampling of interstitial water from lake sediments. Hydrobiologia 92:659–663

Broberg A (1980) Measurement of electron-transport-system activity in freshwater sediments. Proc 8th Nord Symp Sedimentol, Biol Inst, Odense Univ, Denmark, pp 172–193

Brock Neely W (1982) The definition and use of mixing zones. Environ Sci Technol 16:518–521

Brundin L (1949) Chironomiden und andere Bodentiere der Südschwedische Urgebirgsseen. Rep Inst Freshwater Res, Drottningholm 30:1–914

Brunskill GJ, Povoledo D, Graham BW, Stainton M (1971) Chemistry of surface sediments of sixteen lakes in the Experimental Lakes Area, northwestern Ontario. J Fish Res Board Can 28:277–294

Buikema AL, Rutherford CL, Cairns Jr (1980) Screening sediments for potential toxicity by in vitro enzyme inhibition. In: Baker RA (ed) Contaminants and sediments, vol I. Ann Arbor Sci Publ, Ann Arbor, pp 463–475

Cahill RA (1981) Geochemistry of recent Lake Michigan sediments. Ill State Geol Surv Circ 517: 94

Cairns J Jr (1981) Testing for effects of chemicals on ecosystems. Natl Acad Press, Washington, DC, 103 p

Callender E (1968) The postglacial sedimentology of Cevils Lake, North Dakota. Ph D thesis, Univ North Dakota

Carter L (1973) Surficial sediments of Barkley Sound and the adjacent continental shelf, West Coast Vancouver Island. Can J Earth Sci 10:441–459

Cato I (1977) Recent sedimentological and geochemical conditions and pollution problems in two marine areas in south western Sweden. STRIAE Univ Uppsala 6:1–158

Chase RRP (1979) Settling behavior of natural aquatic particulates. Limnol Oceanogr 24:417–426

Chau YK, Saitoh H (1973) Mercury in international Great Lakes. Proc 16th Conf Great Lakes Res, Ann Arbor, pp 221–232

Chester R, Hughes MJ (1967) A chemical technique for the separation of ferromanganese minerals, carbonate minerals and adsorbed trace elements from pelagic sediments. Chem Geol 2: 249–262

Chmelik FB (1967) Electro-osmotic core cutting. Mar Geol 5:321–325

Christiansen JP, Packard TT (1977) Sediment metabolism from the north-west African upwelling system. Deep-Sea Res 24:331–343

Cline JT, Upchurch SB (1973) Mode of heavy metal migration in the upper strata of lake sediments. Proc 16th Conf Great Lakes Res, Ann Arbor, pp 349–350

Coleman JM (1981) Deltas: processes of deposition and models for exploration. IHRDC Publ, 124 p

Cranwell PA (1974) Monocarboxylic acids in lake sediments: indicators, derived from terrestrial and aquatic biota, of paleoenvironmental trophic levels. Chem Geol 14:1–14

Cranwell PA (1976) Organic geochemistry of lake sediments. In: Nriagu JO (ed) Environmental biogeochemistry, vol I. Ann Arbor Sci Publ, Ann Arbor, pp 75–88

Csanady GT (1978) Water circulation and dispersal mechanisms. In: Lerman A (ed) Lakes: chemistry, geology, physics. Springer, Berlin Heidelberg New York, pp 21–64

Cummins KW (1973) Trophic relations of aquatic insects. Annu Rev Entomol 18:183–206

Damiani V, Thomas RL (1974) The surficial sediments of the Big Bay Section of the Bay of Quinte, Lake Ontario. Can J Earth Sci 11:1562–1576

Davis RB (1974) Stratigraphic effects of tubificids in profundal lake sediments. Limnol Oceanogr 19:466–488

Degens ET, Herzen RP von, Wong H-K, Deuser WG, Jannasch HW (1973) Lake Kivu: structure, chemistry and biology of an East African rift lake. Geol Rundsch 62:245–277

Dell CI (1973) A special mechanism for varve formation in a glacial lake. J Sediment Petrol 43: 838–840

Digerfeldt G (1972) The post-glacial development of Lake Trummen. Folia Limnol Scand 16: 1–104

Dominik J, Mangini A, Müller G (1981) Determination of recent deposition rates in Lake Constance with radioisotopic methods. Sedimentology 28:653–677

Dyer KR (ed) Estuarine hydrography and sedimentation. Cambridge Univ Press, Cambridge, 230 p

Edberg N, Hofsten B von (1973) Oxygen uptake of bottom sediments studied in situ and in the laboratory. Water Res 7:1285–1294

Edgington DN, Robbins JA (1976) Records of lead deposition in Lake Michigan sediments since 1800. Environ Sci Technol 10:266–274

Edgren M (1978) Tungmetaller i Mälarens och Östersjöns sediment. Natl Swed Environ Prot Board, SNV PM 1018, Solna, 110 p

Edmondson WT (1975) Microstratification of Lake Washington sediments. Verh Int Ver Limnol 19:770–775

Einsele W (1936) Über die Beziehungen des Eisenkreislaufs zum Phosphatkreislauf im eutrophen See. Arch Hydrobiol 29:664–686

Einsele W (1937) Physikalisch-chemische Betrachtung einiger Probleme des limnischen Mangan- und Eisenkreislaufs. Verh Int Ver Limnol 5:69–84

Einsele W (1938) Über chemische und kolloidchemische Vorgänge in Eisen-Phosphat-Systemen unter limnologischen und limnogeologischen Gesichtspunkten. Arch Hydrobiol 33:361–387

Enell M (1979) Critical analysis of a method for removal of interstitial water from lake sediment. In: Enell M, Gahnström G (eds) 7th Nord Symp Sediments, Univ Lund, pp 183–193

Ericsson B (1973) The cation content of Swedish post-glacial sediments as a criterion of paleosalinity. Geol Fören Stockholm Förh 94:5–21

Eugster HP, Hardie LA (1978) Saline lakes. In: Lerman A (ed) Lakes: chemistry, geology, physics. Springer, Berlin Heidelberg New York, pp 237–293

Evans RD, Rigler FH (1980) Measurement of whole lake sediment accumulation and phosphorus retention using lead-210 dating. Can J Earth Sci 37:817–822

Falkenmark M, Forsman A (1966) Vattnet i vår värld. Wahlström och Widstrand, Stockholm, 208 p

Fast AW, Wetzel RG (1974) A close-interval fractionator for sediment cores. Ecology 55:202–204

Fenchel T (1969) The ecology of marine microbenthos. Ophelia 6:1–182

Finch RH (1937) A tree-ring calender for dating volcanic events, Cinder Core, Lassen National Park, California. Am J Sci 33:140–146

Fischer HB, List EJ, Koh RCY, Imberger J, Brooks NH (1979) Mixing in inland and coastal waters. Academic Press, London New York, 483 p

Fisher JB, Lick WJ, McCall PL, Robbins JA (1980) Vertical mixing of lake sediments by tubificid oligochaetes. J Geophys Res 85:3997–4006

Fisher JS, Pickral J, Odum WE (1979) Organic detritus particles: Initiation of motion criteria. Limnol Oceanogr 24:529–532

Fleischer S (1972) Sugars in the sediments of Lake Trummen. Arch Hydrobiol 70:392–412

Förstner U (1977) Metal concentrations in freshwater sediments – natural background and natural effects. In: Golterman HL (ed) Interactions between sediments and freshwater. Junk, The Hague, pp 94–103

Förstner U, Müller G (1974) Schwermetalle in Flüssen und Seen. Springer, Berlin Heidelberg New York, 255 p

Förstner U, Prosi F (1979) Heavy metals pollution in freshwater ecosystems. In: Ravera O (ed) Biological aspects of freshwater pollution. Pergamon Press, Oxford, pp 129–161

Förstner U, Salomons W (1981) Trace metal analysis on polluted sediments. Delft Hydraulics Lab Publ No 248, pp 1–13

Förstner U, Wittmann GTW (1979) Metal pollution in the aquatic environment. Springer, Berlin Heidelberg New York, 486 p

Friedman GM, Sanders JE (1978) Principles of sedimentology. Wiley, New York, 792 p

Fukuda MK (1978) The entrainment of cohesive sediments in fresh water. Ph D Diss, Case Western Reserve Univ, Cleveland, Ohio

Fukuda MK, Lick W (1980) The entrainment of cohesive sediments in fresh water. J Geophys Res 85:2813–2824

Gardner WD (1980a) Sediment trap dynamics and calibration: a laboratory evaluation. J Mar Res 38:17–39

Gardner WD (1980b) Field assessment of sediment traps. J Mar Res 38:41–52

Gibbs RJ (1982) Floc stability during coulter-counter size analysis. J Sediment Petrol 52:657–660

Gilberg R (1975) Sedimentation in Lillooet Lake, British Columbia. Can J Earth Sci 12:1697–1711

Gjessing ET (1976) Physical and chemical characteristics of aquatic humus. Ann Arbor Sci, Ann Arbor, 120 p

Golterman HL (1975) Physiological limnology. Elsevier, Amsterdam, 489 p

Golterman HL (ed) (1977) Interactions between sediments and freshwater. Junk, The Hague, 473 p

Gorham E (1960) Chlorophyll derivates in surface muds from the English lakes. Limnol Oceanogr 5:29–33
Goudie A (ed) (1981) Geomorphological techniques. Allen and Unwin, London, 395 p
Graf WH, Mortimer CH (eds) (1979) Hydrodynamics of lakes. Elsevier, Amsterdam, 360 p
Granéli W (1979) The influence of *Chironomus plumosus* larvae on the exchange of dissolved substances between sediment and water. Hydrobiologia 66:149–159
Gruendling GK (1971) Ecology of the epipelic algal communities in Marion Lake, British Columbia. J Phycol 7:239–249
Guinasso NL, Schink DR (1975) Quantitative estimates of biological mixing rates in abyssal sediments. J Geophys Res 80:3032–3043
Haines TA (1981) Acidic precipitation and its consequences for aquatic ecosystems: a review. Trans Am Fish Soc 110:669–707
Haines DA, Bryson RA (1961) An empirical study of wind factor in Lake Mendota. Limnol Oceanogr 6:356–364
Håkanson L (1972) Sambandet mellan kvicksilverföremomst och sedimentologiskt miljö i Ekoln. Del 3. Transport och deposition av kvicksilver. Univ Uppsala, UNGI Rap 15:1–56
Håkanson L (1973) Kvicksilver i några svenska sjöars sediment – möjligheter till tillfriskning. IVL-publ, A 92, Stockholm, pp 171–186
Håkanson L (1974) A mathematical model for establishing numerical values of topographical roughness for lake bottoms. Geogr Ann 56A:183–200
Håkanson L (1975) Kvicksilver i Vänern – nuläge och prognos. Natl Swed Environ Prot Board, SNV PM 563, Uppsala, 121 p
Håkanson L (1976) A bottom sediment trap for recent sedimentary deposits. Limnol Oceanogr 21:170–174
Håkanson L (1977a) An empirical model for physical parameters of recent sedimentary deposits of Lake Ekoln and Lake Vänern. VATTEN 3:266–289
Håkanson L (1977b) The influence of wind, fetch, and water depth on the distribution of sediments in Lake Vänern, Sweden. Can J Earth Sci 14:397–412
Håkanson L (1977c) On lake form, lake volume and lake hypsographic survey. Geogr Ann 59A: 1–29
Håkanson L (1977d) Sediments as indicators of contamination – investigations in the four largest Swedish lakes. Natl Swed Environ Prot Board, SNV PM 839, Uppsala, 159 p
Håkanson L (1978) Kapitel 6. Bottnar och sediment. In: Vänern en naturresurs. Liber, Stockholm, pp 75–94
Håkanson L (1980a) An ecological risk index for aquatic pollution control – a sedimentological approach. Water Res 14:975–1101
Håkanson L (1980b) The quantitative impact of pH, bioproduction and Hg-contamination on the Hg-content in fish (pike). Environ Pollut 18:285–304
Håkanson L (1981a) A manual of lake morphometry. Springer, Berlin Heidelberg New York, 78 p
Håkanson L (1981b) Sjösedimenten i recipientkontrollen; principer, processer och praktiska exempel. Natl Swed Environ Prot Board, SNV PM 1398, Uppsala, 242 p
Håkanson L (1981c) Determination of characteristic values for physical and chemical lake sediment parameters. Water Resour Res 17:1625–1640
Håkanson L (1982a) A modified Ponar grab sampler for coarse and consolidated sediments. Preprint Natl Swed Environ Prot Board, Uppsala
Håkanson L (1982b) Bottom dynamics in lakes. In: Sly PG (ed) Sediment/freshwater interaction. Junk, The Hague, pp 1–22
Håkanson L (1982c) A new apparatus for in situ determination of sediment type. Preprint Natl Swed Environ Prot Board, Uppsala
Håkanson L (1982d) Lake bottom dynamics and morphometry – the dynamic ratio. Water Resour Res 18:1444–1450
Håkanson L (1982e) Bioturbation and lake sediment age. Manuscr Natl Swed Environ Prot Board, Uppsala
Håkanson L (1983a) Metals in fish and sediments – on sample representativity and dose-response relationships. Manuscr Natl Swed Environ Prot Board, Uppsala.

Håkanson L (1983b) Aquatic contamination and ecological risk. An attempt to a conceptual framework. Manuscr Natl Swed Environ Prot Board, Uppsala

Håkanson L (1983c) On the relationship between lake trophic level and lake sediments. Manuscr Natl Swed Envriron Prot Board, Uppsala

Håkanson L, Ahl T (1976) Vättern – recenta sediment och sedimentkemi. Natl Swed Environ Prot Board, SNV PM 740, Uppsala, 164 p

Håkanson L, Källström A (1978) An equation of state for biologically active lake sediments and its implications for interpretations of sediment data. Sedimentology 25:205–226

Håkanson L, Uhrberg R (1981) Undersökningar i Kolbäcksåns vattensystem. XIII. Metaller i fisk och sediment. Natl Swed Environ Prot Board, SNV PM 1408, Uppsala, 215 p

Hallberg R, Bågander L-E, Engvall A-G, Holm N (1978) Kemisk sedimentologi. Fysiks Riksplanering Nr 5, Stockholm, pp 104–165

Hamblin PF, Carmack EC (1978) River-induced currents in a fjord lake. J Geophys Res 83:885–899

Hansebo S (1957) A new approach to the determination of the stear strength of clays by the fall-cone test. R Swed Geol Inst Proc Nr 14:1–47

Hansen K (1956) The profundal bottom deposits of Gribsø. In: Berg K, Petersen IC (eds) Studies on the humic acid Lake Gribsø. Folia Limnol Scand, No 8, pp 16–24 and pp 250–252

Hansen K (1959a) The terms gyttja and dy. Hydrobiologia 13:309–315

Hansen K (1959b) Sediments from Danish lakes. J Sediment Petrol 29:38–46

Hansen K (1961) Lake types and lake sediments. Verh Int Ver Limnol 14:285–290

Hargrave BT (1970a) The utilization of benthic microflora by *Hyalella azteca* (Amphipoda). J Anim Ecol 39:427–437

Hargrave BT (1970b) Distribution, growth, and seasonal abundance of *Hyalella azteca* (Amphipoda) in relation to sediment microflora. J Fish Res Board Can 27:685–699

Hargrave BT (1970c) The effcct of a deposit-feeding amphipod on the metabolism of benthic microflora. Limnol Oceanogr 15:21–30

Harrison MJ, Pacha RW, Morita RY (1972) Solubilization of inorganic phosphate by bacteria isolated from Upper Klemath lake sediment. Limnol Oceanogr 17:50–57

Hart BT (1982) Uptake of trace metals by sediments and suspended particulates: a review. In: Sly PG (ed) Sediment/freshwater interaction. Junk, The Hague, pp 299–313

Hem JD (1977) Reactions of metal ions at surfaces of hydrous iron oxide. Geochim Cosmochim Acta 41:527–538

Hesslein RH (1976) An in situ sampler for close interval pore water studies. Limnol Oceanogr 21: 912–914

Hieltjes AHM, Lijklema L (1980) Fractionation of inorganic phosphates in calcareous sediments. J Environ Qual 9:405–407

Hillerdal E (1975) Limnologiska undersökningar i Norrviken, Edssjön och Oxundasjön. Med 18: Sedimentundersökningar i Edssjön och Oxundasjön 1972. Tech Rep Inst Limnol, Uppsala

Hjulström F (1935) Studies on the morphological activity of rivers as illustrated by the River Fyris. Bull Geol Inst Uppsala 25:221–527

Holdgate MW (1979) A perspective of environmental pollution. Cambridge Univ Press, Cambridge, 278 p

Holdren GC, Armstrong DE (1980) Factors affecting phosphorus release from intact lake sediment cores. Environ Sci Technol 14:79–87

Holdren GC, Armstrong DE, Harris RF (1977) Interstitial inorganic phosphorus concentrations in Lakes Mendota and Wingra. Water Res 11:1041–1047

Hopkins TL (1964) A survey of marine bottom samplers. In: Sears M (ed) Progress in oceanography, vol II. Pergamon-MacMillan, New York, pp 213–256

Horie S (1969) Asian lakes. In: Eutrophication: causes, consequences, correctives. Natl Acad Sci, Washington DC, pp 98–124

Hutchinson GE (1957) A treatise on limnology. I. Geography, physics, and chemistry. Wiley, New York, 1015 p

Hutchinson GE (1967) A treatise on limnology. II. Introduction to lake biology and the limnoplankton. Wiley, New York, 1115 p

Hutchinson GE (1973) Eutrophication. The scientific background of a contemporary practical problem. Am Sci 61:269–279

Hutchinson GE (1975) A treatise on limnology. III. Limnological botany. Wiley, New York, 660 p

Hutchinson GE, Löffler H (1956) The thermal classification of lakes. Proc Natl Acad Sci USA 42:84–86

Huttunen P, Meriläinen J (1978) New freezing device providing large unmixed sediment samples from lakes. Ann Bot Fenn 15:128–130

Hvorslev MJ (1949) Subsurface exploration and sampling of soils for civil engineering purposes. US Army Corps Eng Waterways Exp Stn, Vicksburg, Miss, 521 p

Imboden DM, Gächter R (1979) The impact of physical processes on the trophic state of a lake. In: Ravera O (ed) Biological aspects of freshwater pollution. Pergamon Press, Oxford, pp 93–110

Imboden DM, Lerman A (1979) Chemical models of lakes. In: Lerman A (ed) Lakes: chemistry, geology, physics. Springer, Berlin Heidelberg New York, pp 341–356

Igelman KR, Hamilton EL (1963) Bulk densities of mineral grains from Mohole samples (Guadalupe Side). J Sediment Petrol 33:474–478

Jackson KS, Janasson IR, Skippen GB (1978) The nature of metals – sediment-water interactions in freshwater bodies, with emphasis on the role of organic matter. Earth-Sci Rev 14:97–116

Jansson M (1977) Enzymatic release of phosphate in water from subarctic lakes in northern Sweden. Hydrobiologia 56:175–180

Jansson M (1978) Experimental lake fertilization in the Kuokkel area, northern Sweden: Budget calculations and fate of nutrients. Verh Int Ver Limnol 20:857–862

Jenne EA (1977) Trace elements sorption by sediments and soils – sites and processes. In: Chappell W, Peterson K (eds) Molybdenum in the environment. Marcel-Dekker, New York, pp 425–553

Johansson K (1980) Tungmetaller i små skogssjöar. Natl Swed Environ Prot Board, SNV PM 1359, Solna, 83 p

Johnson DW (1919) Shore processes and shoreline development. Wiley, New York, 584 p

Johnson MA (1964) Turbidity currents. Oceanogr Mar Biol Annu Rev 2:31–43

Johnson TC (1980) Sediment redistribution by waves in lakes, reservoirs and embayments. Proc Symp Surface Water Impoundments, ASCE, June 2–5, 1980, Minneapolis, Minnesota, Pap No 7–9, pp 1307–1317

Jonasson A (1977) New devices for sediment and water sampling. Mar Geol 24:1413–1421

Jones BF, Bowser CJ (1978) The mineralogy and related chemistry of lake sediments. In: Lerman A (ed) Lakes: chemistry, geology, physics. Springer, Berlin Heidelberg New York, pp 179–235

Jones JG (1979) Microbial activity in lake sediments with particular reference to electrode potential gradients. J Gen Microbiol 115:19–26

Jones JG (1982) Activities of aerobic and anaerobic bacteria in lake sediments and their effect on the water column. In: Nedwell DB, Brown CM (eds) Sediment microbiology. Academic Press, London New York, pp 107–145

Jones JG, Simon BM (1980) Decomposition processes in the profundal region of Blelham Tarn and the Lund Tubes. J Ecol 68:493–512

Jones JG, Simon BM (1981) Differences in microbial decomposition processes in profundal and littoral lake sediments, with particular reference to the nictrogen cycle. J Gen Microbiol 123: 297–312

Jopling AV (1960) An experimental study on the mechanics of bedding. Doct Diss, Harvard Univ, Cambridge, Mass

Jopling AV (1963) Hydraulic studies on the origin of bedding. Sedimentology 2:115–121

Källström A, Håkanson L (1982) Physics of cone penetration in recent sediments. Manuscr Natl Swed Environ Prot Board, Uppsala

Kamp-Nielsen L (1974) Mud-water exchange of phosphate and other ions in undisturbed sediment cores and factors affecting the exchange rates. Arch Hydrobiol 73:218–237

Keller GH, Bennett RH (1970) Variations in the mass physical properties of selected submarine sediments. Mar Geol 9:215–223

Kemp ALW (1971) Organic carbon and nitrogen in the surface sediments of lakes Ontario, Erie and Huron. J Sediment Petrol 41:537–548

Kemp ALW, Mudrochova A (1973) The distribution and nature of amino acids and other nitrogen-containing compounds in Lake Ontario surface sediments. Geochim Cosmochim Acta 37: 2191–2206

Kemp ALW, Thomas RL, Dell CI, Jaquet J-M (1976) Cultural impact on the geochemistry of sediments in Lake Erie. J Fish Res Board Can Spec Issue 33:440–462

Kenner RA, Ahmed SI (1975) Measurements of electron transport activities in marine phytoplankton. Mar Biol 33:119–127

Kögler FC, Larsen B (1979) The West Bornholm basin in the Baltic Sea: geological structure and Quaternary sediments. Boreas 8:1–22

Kohnke H (1968) Soil physics. McGraw-Hill, New York, 224 p

Komar PD (1976) Beach processes and sedimentation. Prentice-Hall, Englewood Cliffs, NJ, 429 p

Komar PD, Miller MC (1973) The threshold of sediment movement under oscillatory water waves. J Sediment Petrol 43:1101–1110

Komar PD, Miller MC (1975) On the comparison between the threshold of sediment motion under waves and unidirectional currents with a discussion on the practical evaluation of the threshold. J Sediment Petrol 45:362–367

Komar PD, Neudeck RH, Kulm LD (1972) Observations and significance of oscillatory ripple marks on the Oregon Continental Shelf. In: Swift DJP, Duane DB, Pilkey OH (eds) Shelf sediment transport. Dowden Hutchinson and Ross, Straudsberg, PA, pp 601–619

Kranck K (1973) Flocculation of suspended sediment in the sea. Nature 246:348–350

Kranck K (1979) Dynamics and distribution of suspended particulate matter in the St. Lawrence Estuary. Nat Can 106:163–179

Kranck K (1979) Particulate matter grain-size characteristics and flocculation in a partially mixed estuary. Sedimentology 28:107–114

Krenkel PA (ed) (1975) Heavy metals in the aquatic environment. Pergamon Press, Oxford, 352 p

Krishnaswami S, Lal D (1978) Radionuclide limnochronology. In: Lerman A (ed) Lakes: chemistry, geology, physics. Springer, Berlin Heidelberg New York, pp 153–177

Krone RB (1978) Aggregation of suspended particles in estuaries. In: Kjerve B (ed) Estuarine transport processes. Univ SC Press, South Carolina, pp 177–190

Krumbein WC, Pettijohn FJ (1938) Manual of sedimentary petrography. Appelton-Century, New York, 549 p

Krumbein WC, Sloss LL (1963) Stratigraphy and sedimentation. Freeman, San Francisco, 660 p

Ku WC, Diliano FA, Feng TH (1978) Factors affecting phosphate adsorption equilibria in lake sediments. Water Res 12:1069–1074

Kukal Z (1971) Geology of recent sediments. Academic Press, London New York, 490 p

Lastein E (1976) Recent sedimentation and resuspension of organic matter in eutrophic Lake Esrom, Denmark. OIKOS 27:44–49

Lee GF (1970) Factors affecting the transfer of materials between water and sediments. Univ Wisconsin, Water Res Center, Eutroph Inf Program, 50 p

Leeder MR (1982) Sedimentology. Allen and Unwin, London, 384 p

Lenz F (1925) Chironomiden und Seetypenlehre. Naturwissenschaften 13:5–10

Lerman A (ed) (1978) Lakes: chemistry, geology, physics. Springer, Berlin Heidelberg New York, 363 p

Lindell T (1975) Lake Vänern (in Swedish with English summary). In: SNV (Natl Swed Environ Prot Board). Vänern, Vättern, Mälaren, Hjälmaren, en översikt. Liber, Stockholm, pp 21–35

Lindell T (1980) Hydrographic characteristics. In: Welch EB, Ecological effects of waste water. Cambridge Univ Press, Cambridge, pp 17–47

Lindeström L (1979) Bioturbation i sediment i olika vattenmiljöer och dess betydelse för tungmetallers omsättning – en litteraturstudie. Inst Vatten Luftv Forskn, IVL B482, 39 p

Ludlam SD (1974) Fayetteville Green Lake, New York. 6. The role of turbidity currents in lake sedimentation. Limnol Oceanogr 19:656–664

Ludlam SD (1981) Sedimentation rates in Fayetteville Green Lake, New York, U.S.A. Sedimentology 28:85–96

Lund JWG (1954) The seasonal cycle of the plankton diatom, *Melosira italica* (Ehr.) Kütz. subsp. *subarctica* O. Müll. J Ecol 42:151–179

Lund JWG (1955) Further observations on the seasonal cycle of *Melosira italica* (Ehr.) Kütz. subsp. *subarctica* O. Müll. J Ecol 43:90–102

Lundbeck J (1936) Untersuchungen über die Bodenbesiedlung der Alpenrandseen. Arch Hydrobiol Suppl 10:208–358

Lundin L-C, Håkanson L (1982) Fortia-Pharmacia och vattenmiljön. Univ Uppsala, UNGI Rap 55, 99 p

Lundqvist G (1938) Sjösediment från Bergslagen (Kolbäcksåns vattenområde). Sver Geol Unders Ser C 420:1–186

Lundqvist G (1942) Sjösediment och deras bildningsmiljö. Sver Geol Unders Ser C 444:1–126

Lüthi S (1980) Some new aspects of two-dimensional turbidity currents. Sedimentology 28: 97–105

Mackereth FJJ (1966) Some chemical observations on post-glacial lake sediments. Philos Trans R Soc London 250:167–213

Mare MF (1942) Paper published in: J Mar Biol Assoc UK 25:517–554 (quoted in Rhoads 1974)

Marzolf GR (1965) Substrate relations of the burrowing amphipod *Pontoporeia affinis* in Lake Michigan. Ecology 46:579–592

Mason B (1966) Principles of geochemistry. Wiley, New York, 329 p

Mayer LM (1976) Chemical water sampling in lakes and sediments with dialysis bags. Limnol Oceanogr 21:909–912

McCall PL, Fisher JB (1980) Effects of tubificid oligochaetes on physical and chemical properties of Lake Erie sediments. In: Brinkhurst RO, Cook DG (eds) Aquatic oligochate biology. Plenum Press, New York, pp 253–317

Middleton GV (1966) Experiments on density (sic) and turbidity currents. 1. Motion of the head. Can J Earth Sci 3:523–546

Milbrink G (1973) On the vertical distribution of oligochaetes in lake sediments. Inst Freshwater Res, Drottningholm 53:34–50

Mortimer CH (1941) The exchange of dissolved substances between mud and water in lakes. I. J Ecol 29:280–329

Mortimer CH (1942) The exchange of dissolved substances between mud and water in lakes. II. J Ecol 30:147–201

Muir Wood AM (1969) Coastal hydraulics. McMillan, London, 187 p

Müller G (1964) Methoden der Sedimentuntersuchungen. Schweizerbart, Stuttgart, 303 p

Müller G (1967) Sedimentary petrology. Part I: Methods of sedimentary petrology. Hafner Press, New York, 283 p

Müller G (1971) Aragonite: inorganic precipitation in a freshwater lake. Nature Phys Sci 229:18

Müller G (1979) Schwermetalle in den Sedimenten des Rheins – Veränderungen seit 1971. Umsch Wiss Tech 79:778–783

Müller G, Förstner U (1968a) Sedimenttransport im Mündungsgebiet des Alpenrheins. Geol Rundsch 58:229–259

Müller G, Förstner U (1968b) General relationship between suspended sediment concentration and water discharge in the Alpenrhein and some other rivers. Nature 217:244–245

Naumann E (1931) Limnologische Terminologie. Handb Biol Arbeitsmethod, Abt IX, Teil 8. Urban & Schwarzenberg, Berlin, 776 p

Nauwerck A (1981) Studien über die Bodenfauna des Latnjajaure (Schwedisch Lappland). Ber Nat Med Ver Innsbruck 68:79–98

Neame PA (1977) Phosphorus flux across the sediment-water interface. In: Golterman HL (ed) Interactions between sediments and freshwater. Junk, The Hague, pp 307–312

Nielsen J, Nielsen N (1978) Kustmorfologi. Geografförlaget, Brenderup, 185 p

Niemistö L (1974) A gravity corer for studies of soft sediments. Havsforskn Inst Skriftser 238: 33–38

Nikaido M (1978) Ratios of organic carbon to nitrogen in the core sediments from Lake Kojima. Jpn J Limnol 39:15–21

Nissenbaum A, Presley BJ, Kaplan IR (1972) Early diagenesis in a reducing fjord, Saanich Inlet, British Columbia. I. Chemical and isotopic changes in major components of interstitial water. Geochim Cosmochim Acta 36:1007–1027

Norrman JO (1964) Lake Vättern. Investigations on shore and bottom morphology. Geogr Ann 1–2:1–238

Norrman JO, Königsson L-K (1972) The sediment distribution in Lake Vättern and some analyses of cores from its southern basin. Geol Fören Stockholm Förh 94:489–513

Ohle W (1937) Kolloidgele als Nährstoffregulanten der Gewässer. Naturwissenschaften 25:471–474

Ohle W (1958) Die Stoffwechseldynamik der Seen in Abhängigkeit von der Gasausscheidung ihres Schlammes. Vom Wasser 25:127–149

Ohle W (1964) Interstitiallösung der Sedimente, Nährstoffgehalt des Wassers und Primärproduktion des Phytoplanktons in Seen. Helgol Wiss Meeresunters 10:411–429

Ohle W (1978) Ebullition of gases from sediment, conditions, and relationship to primary production of lakes. Verh Int Ver Limnol 20:957–962

Olausson E, Cato I (eds) (1980) Chemistry and biogeochemistry of estuaries. Wiley, Chichester, 452 p

Olsén P (1980) Population development of introduced *Mysis relicta* and impact on char and brown trout. Acta Univ Uppsala Abstr 570

Ottow JCG, Munch JC (1978) Mechanisms of reductive transformations in the anaerobic microenvironment of hydromorphic soils. In: Krumbein WE (ed) Environmental biogeochemistry and geomicrobiology. Ann Arbor Sci, Michigan, pp 483–491

Overrein LN, Seip HM, Tollan A (1980) Acid precipitation – effects on forest and fish. Norw Min Environ Res Rep 19:1–175

Pamatmat MM, Bhagwat AM (1973) Anaerobic metabolism in Lake Washington sediments. Limnol Oceanogr 18:611–627

Partheniades E (1972) Results of recent investigations on erosion and deposition of cohesive sediments. In: Shen HW (ed) Sedimentation. Symposium to honor Prof HA Einstein. Colorado State Univ, Fort Collins, pp 20-1 to 20-39

Pearson TH, Rosenberg R (1976) A comparative study on the effects on the marine environment of wastes from cellulose industries in Scotland and Sweden. Ambio 5:77–79

Pennington W, Cambray RS, Fisher EM (1973) Observations on lake sediments using fallout Cs-137 as a tracer. Nature 242:324–326

Persson G, Holmgren SK, Jansson M, Lundgren A, Nyman B, Solander D, Ånell C (1977) Phosphorus and nitrogen and the regulation of lake ecosystems: Experimental approaches in subarctic Sweden. In: Proc Circump Conf North Ecol, Sept 1975, III. Natl Res Counc Can, Ottawa, pp 1–19

Pharo CH, Carmack EC (1979) Sedimentation processes in a short residence-time intermontane lake, Kamploops Lake, British Columbia. Sedimentology 26:523–541

Post H v (1862) Studier öfver nutidens koprogena jordbildningar, gyttja, torf, mylla. Sver Vetensk Akad Handl 4

Postma H (1967) Sediment transport and sedimentation in the estuarine environment. In: Lauff GH (ed) Estuaries. Am Assoc Adv Sci, Washington DC, pp 158–179

Postma H (ed) (1981) Sediment and pollution interchange in shallow seas. Rapports et procès-verbaux des réunion, vol 181. Conf Int Explor Mer, Copenhague, 122 p

Potonié H (1908) Die rezente Kaustobiologie. Abh Preuss Geol Landesanst NF Heft 55

Povoledo D (1972) Some model experiments on detritus formation and on some possible functions of suspended and deposited lake organic matter. In: Tonolli L (ed) Detritus and its role in aquatic systems. Mem Ist Ital Idrobiol Suppl 33:485–524

Premazzi G, Ravera O (1977) Chemical characteristics of Lake Lugano sediments. In: Golterman HL (ed) Interactions between sediments and freshwater. Junk, The Hague, pp 121–124

Prentki RT, Miller MC, Barsdate RJ, Alexander U, Kelley J, Coyne P (1980) Chemistry. In: Hobbie JE (ed) Limnology of Tundra Ponds. US/IBP Synthesis Ser 13. Dowden, Hutchinson and Ross, Stroudsberg, pp 76–179

Ravera O, Parise G (1978) Eutrophication of Lake Lugano "read" by means of planktonic remains in the sediment. Schweiz Z Hydrol 40:40–50

Rawson DS (1955) Morphometry as a dominant factor in the productivity of large lakes. Verh Int Ver Limnol 12:164–175

Reeves CC Jr (1968) Introduction to paleolimnology. Elsevier, Amsterdam, 228 p

Reineck H-E, Singh IB (1975) Depositional sedimentary environments. Springer, Berlin Heidelberg New York, 439 p

Renberg I (1981a) Improved methods for sampling, photographing and varvecounting of varved lake sediments. Boreas 10:255–258

Renberg I (1981b) Formation, structure and visual appearance of iron-rich, varved lake sediments. Verh Int Ver Limnol 21:94–101

Rhoads DC (1963) Rates of sediment reworking by Yoldia limatula in Buzzards Bay, Massachusetts and Long Island Sound. J Sediment Petrol 33:723–727

Rhoads DC (1967) Biogenic reworking of intertidal and subtidal sediments in Barnstable Harbor and Buzzards Bay, Massachusetts. J Geol 75:461–476

Rhoads DC (1974) Organism-sediment relations on the muddy seafloor. Oceanogr Mar Biol Annu Rev 12:263–300

Richard DT (1975) Kinetics and mechanism of pyrite formation at low temperatures. Am J Sci 275:636–652

Richards AF, Hirst TJ, Parks JM (1974) Bulk density-water content relationship in marine silts and clays. J Sediment Petrol 44:1004–1009

Richards FA (1965) Anoxic basins and fjords. In: Riley JP, Skirrow G (eds) Chemical oceanography, vol I. Academic Press, London New York, pp 611–695

Ripl W, Lindmark G (1979) The impact of algae and nutrient composition on sediment exchange dynamics. Arch Hydrobiol 86:45–65

Rippey B (1977) The behaviour of phosphorus and silicon in undisturbed cores of Lough Neagh sediments. In: Golterman HL (ed) Interactions between sediments and freshwater. Junk, The Hague, pp 348–353

Robbins JA (1978) Geochemical and geophysical applications of radio-active lead. In: Nriagu JO (ed) Biogeochemistry of lead in the environment. Elsevier, Amsterdam, pp 285–393

Robbins JA, Gustinis I (1976) A squeezer for efficient extraction of pore water from small volumes of anoxid sediments. Limnol Oceanogr 21:905–909

Roberts DJ, Lindell T, Kvarnäs H (1982) Environmental factors governing regional lake water quality differences. Natl Swed Environ Prot Board, SNV PM 1621, Uppsala, 32 p

Roberts HH (1972) X-ray radiography of recent carbonate sediments. J Sediment Petrol 42:690–693

Rodhe W (1958) Primärproduktion und Seetypen. Verh Int Ver Limnol 13:121–141

Rodhe W (1969) Crystallization of eutrophication concepts in Northern Europe. In Rohlich GA (ed) Eutrophication: causes, consequences, corrective. Natl Acad Sci, Washington DC, pp 50–64

Rohlich GA (ed) (1969) Eutrophication: causes, consequences, correctives. Natl Acad Sci, Washington DC, 661 p

Round FE (1957) Studies on bottom-living algae in some lakes of the English Lake District. Part I. Some chemical features of the sediments related to algal productivities. J Ecol 45:133–148

Round FE (1960) Studies on bottom-living algae in some lakes of the English Lake District. Part IV. The seasonal cycles of the Bacillariophyceae. J Ecol 48:529–547

Round FE (1961) Studies on bottom-living algae in some lakes of the English Lake District. Part VI. The effect of depth on the epipelic algal community. J Ecol 49:245–254

Round FE, Eaton JW (1966) Persistent, vertical migration rhythms in benthic microflora. III. The rhythm of epipelic algae in a freshwater pond. J Ecol 54:609–615

Round FE, Happey CM (1965) Persistent, vertical-migration rhythms in benthic microflora. IV. A Diurnal rhythm of the epipelic diatom association on non-tidal flowing water. Br Phycol Bull 2:463–471

Ruttner F (1931) Hydrographische und hydrochemische Beobachtungen auf Java, Sumatra und Bali. Arch Hydrobiol Suppl 8:197–454

Ryding S-O, Borg H (1973) Sedimentkemiska studier i Lilla Ullevifjärden. Natl Swed Environ Prot Board, NLU Rap 58, Uppsala, 77 p

Ryding S-O, Forsberg C (1977) Sediments as a nutrient source in shallow polluted lakes. In: Golterman HC (ed) Interactions between sediments and freshwater. Junk, The Hague, pp 227–234

Saarnisto M, Huttunen P, Tolonen K (1977) Annual lamination of sediments in Lake Lovojärvi, southern Finland, during the past 600 years. Ann Bot Fenn 14:35–45

Saether OA (1979) Chirnomid communities as water quality indicators. Holarct Ecol 2:65–74

Salomons W, Förstner U (1970) Trace metal analysis on polluted sediments. Part II: Evaluation of environmental impact. Environ Tech Lett 1:506–517

Schindler DW, Hesslein R, Kipphut G (1977) Interactions between sediments and overlying waters in an experimentally eutrophied Precambrian Shield Lake. In: Golterman HL (ed) Interactions between sediments and freshwater. Junk, The Hague, pp 235–243

Schnitzer M (1971) Metal-organic matter interactions in soils and waters. In: Faust SJ, Hunter JV (eds) Organic compounds in aquatic environment. Marcel Dekker, New York, pp 297–315

Scott RF (1963) Principles of soil mechanics. Addison-Wesley, New York, 550 p

Seibold E, Berger WH (1982) The sea floor. An introduciton to marine geology. Springer, Berlin Heidelberg New York, 288 p

Serruya C, Edelstein M, Pollingher U, Serruya S (1974) Lake Kinneret sediments: Nutrient composition of pore water and mud water exchange. Limnol Oceanogr 19:489–508

Shapiro J (1948) The core-freezer, a new sampler for lake sediments. Ecology 39:748

Sheng YP, Lick W (1979) The transport and resuspension of sediments in a shallow lake. J Geophys Res 84:1809–1825

Shepard FP (1954) Nomenclature based on sand-silt-clay ratios. J Sediment Petrol 24:151–158

Simons TJ (1980) Circulation models of lakes and inland seas. Can Bull Fish Aquat Sci 203:1–146

Skougstad MW, Fishman MJ, Friedman LC, Erdmann DE, Duncan SS (eds) (1979) Methods for determination of inorganic substances in water and fluvial sediments. US Gov Print Off, Washington DC, 626 p

Sly PG (1969) Bottom sediment sampling. Proc 12th Conf Great Lakes Res, Ann Arbor, pp 883–898

Sly PG (1975) Statistical evaluation of recent sediment geochemical sampling. Proc 9th Int Congr Sedimentol, Nice

Sly PG (1977) Sedimentary environments in the Great Lakes. In: Golterman HL (ed) Interactions between sediments and freshwater. Junk, The Hague, pp 76–82

Sly PG (1978) Sedimentary processes in lakes. In: Lerman A (ed) Lakes: chemistry, geology, physics. Springer, Berlin Heidelberg New York, pp 65–89

Sly PG (ed) (1982) Sediment/freshwater interaction. Developments in hydrobiology, vol IX. Junk, The Hague, 701 p

Smith IR (1975) Turbulence in lakes and rivers. Freshwater Biol Assoc Sci Publ.No 29, 79 p

Smith IR, Sinclair IJ (1972) Deep water waves in lakes. Freshwater Biol 2:387–399

SNV (1976) Om metaller – Statens naturvårdsverk (SNV). Liber, Stockholm, 262 p

SNV (1978) Lake Vänern (in Swedish with English summary). Statens naturvårdsverk (SNV = Natl Swed Environ Prot Board). Liber, Stockholm, 372 p

SNV (1980a) Riktvärden för vattenkvalitet. Zink, förekomst, miljöeffekter, riktvärden för sjöar och vattendrag. In: Wiederholm T. Natl Swed Environ Prot Board, (mimeo) Uppsala

SNV (1980b) Riktvärden för vattenkvalitet. Kadmium, förekomst, miljöeffekter, riktvärden för sjöar vattendrag. In: Wiederholm T. Natl Swed Environ Prot Board, (mimeo) Uppsala

Sørensen I (1982) Reduction of ferric iron in anaerobic marine sediment and interaction with reduction of nitrate and sulfate. Appl Environ Microbiol 43:319–324

Spiegel MR (1972) Statistics. McGraw-Hill, New York, 359 p

Sternberg RW, Larson LH (1975) Threshold of sediment movement by open ocean waves: observations. Deep-Sea Res 22:299–309

Stockner JG (1972) Paleolimnology as a means of assessing eutrophication. Verh Int Ver Limnol 18:1018–1030

Stokes GG (1851) Collected papers, vol III. Cambridge Trans, vol IX. [see, e.g., Lamb H (1945) Hydrodynamics, 6th edn. Dover Publ, New York, 450 p]

Stumm W, Leckie JO (1971) Phosphate exchange with sediments; its role in the productivity of surface waters. Eidg Tech Hochsch Duebendorf 406:1–16

Stumm W, Morgan JJ (1970) Aquatic chemistry. Wiley-Interscience, New York, 583 p

Sturm M (1975) Depositional and erosional sedimentary features in a turbidity current controlled basin (Lake Brienz). Proc IXth Int Congr Sedimentol, Nice

Sturm M (1979) Origin and composition of clastic varves. In: Schlüchter Ch (ed) Moraines and varves. Proc INQUA Symp Genosis Lithol Quatern Deposits, Zuerich. Balkema, Rotterdam, pp 281–285

Sturm M, Matter A (1972) The electro-osmotic guillotine, a new devide for core cutting. J Sediment Petrol 42:987–989

Sturm M, Matter A (1978) Turbidities and varves in Lake Brienz (Switzerland): deposition of clastic detritus by density currents. Spec Publ Int Assoc Sedimentol 2:147–168

Sundborg Å (1956) The River Klarälven. A study of fluvical processes. Geogr Ann 38:127–316

Swain FM (1970) Non-marine organic geochemistry. Cambridge Univ Press, Cambridge, 445 p

Sweeney RE, Kaplan IR (1973) Pyrite framboid formation. Laboratory synthesis and marine sediments. Econ Geol 68:618–634

Swift DJP, Duane DB, Pilkey OH (eds) (1972) Shelf sediment transport: Processes and pattern. Dowden, Hutchinson and Ross, Stroudsburg, 656 p

Syers JK, Harris RF, Armstrong DE (1973) Phosphate chemistry in lake sediments. J Environ Qual 2:1–14

Tadajewski A (1966) Bottom sediments in different limnetic zones of a eutrophic lake. Ecol Pol 15:321–341

Taylor D (1975) Mercury as an environmental pollutant. A bibliography, 5th edn. Imp Chem Ind Ltd Brisham Lab. Freshwater Quarry, Brixham, UK, 269 p

Terwindt JHJ (1977) Deposition, transportation and erosion of mud. In: Golterman HL (ed) Interactions between sediments and freshwater. Junk, The Hague, pp 19–24

Tessenow U (1975) Lösungs-, Diffusions- und Sorptionsprozesse in der Oberschicht von Seesedimenten. Arch Hydrobiol Suppl 47:325–412

Tessier A, Campbell PGC, Bisson M (1979) Sequential extraction procedure for the speciation of particulate trace metals. Anal Chem 51:844–851

Tessier A, Campbell PGC, Bisson M (1980) Trace metal speciation in the Yamaska and St. Francois Rivers (Quebec). Can J Earth Sci 9:636–651

Theis TL, McCabe PJ (1978) Phosphorus dynamics in hypereutrophic lake sediments. Water Res 12:677–685

Thienemann A (1925) Die Binnengewässer Mitteleuropas. Eine limnologische Einführung. Binnengewässer 1:1–225

Thienemann A (1931) Der Produktionsbegriff in der Biologie. Arch Hydrobiol 22:616–622

Thomas RL (1972) The distribution of mercury in the sediment of Lake Ontario. Can J Earth Sci 9:636–651

Thomas RL, Kemp ALW, Lewis CFM (1972) Distribution, composition and characteristics of the surficial sediments of Lake Ontario. J Sediment Petrol 42:66–84

Thomas RL, Jaquet J-M, Kemp ALW, Lewis CFM (1976) Surficial sediments in Lake Erie. J Fish Res Board Can 33:385–403

Tirén T (1977) Denitrification in sediment-water systems of various types of lakes. In: Golterman HL (ed) Interactions between sediments and freshwater. Junk, The Hague, pp 363–369

Tirén T, Thorin J, Nömmik H (1976) Denitrification measurements in lakes. Acta Agric Scand 26:175–184

Turekian KK, Wedepohl KH (1961) Distribution of the elements in some major units of the Earth's crust. Bull Geol Soc Am 72:175–192

Ulén B (1977) Seston and sediment in Lake Norrviken. III. Nutrient release from sediment. Scr Limnol Upsaliensia 448

US Army Coastal Engineering Research Center (1977) Shore protection manual, vol I, 3rd edn. US Gov Print Off, Washington DC, 714 p

Valle KJ (1927) Ökologisch-limnologische Untersuchungen über die Boden- und Tiefenfauna in einigen Seen nördlich vom Ladoga-See. Acta Zool Fenn 4:1–231

Vollenweider RA (1968) Scientific fundamentals of the eutrophication of lakes and flowing waters, with particular reference to nitrogen and phosphorus as factors in eutrophication. Rep OECD, DAS/SCI/68.27, Paris, 192 p

Welch EB, Perkins MA, Lynch D, Hufschmidt P (1979) Internal phosphorus related to rooted macrophytes in a shallow lake. In: Breck JE, Prentki RT, Loucks OL (eds) Aquatic plants, lake management and ecosystem consequences of lake harvesting. Conf Proc, Madison, Wis, pp 81–99

Wesenberg-Lund C (1901) Studier øver Søkal, Bønnemalm og Søgytje i danske indsøer. Medd Dan Geol For 7

Wetzel RG (1970) Recent and postglacial production rates of a marl lake. Limnol Oceanogr 15: 491–503

Wetzel RG (1972) The role of carbon in hard-water marl lakes. In: Likens GE (ed) Nutrients and eutrophication: the limiting-nutrient controversy. Spec Symp Soc Limnol Oceanogr 1:83–91

Wetzel RG (1975) Limnology. Saunders, Philadelphia, 743 p

Wetzel RG, Rich PH, Miller MC, Allen HL (1972) Metabolism of dissolved and particulate detrital carbon in a temperate hard-water lake. Mem Ist Ital Idrobiol Suppl 29:185–243

Wiederholm T (1978) Vänerns vattenkvalitet. Bottenfauna i Vänern. In: SNV 1978, Vänern en naturresurs. Liber, Stockholm, pp 168–184

Wiederholm T (1979) Use of benthic communities in lake monitoring. In: The use of ecological variables in environmental monitoring. Natl Swed Environ Prot Board, SNV PM 1151, Uppsala, pp 196–211

Wiederholm T (1980) Use of benthos in lake monitoring. J Water Pollut Control Fed 52:537–547

Williams JDH, Mayer T (1972) Effects of sediment diagenesis and regeneration of phosphorus with special reference to lakes Erie and Ontario. In: Allen HE, Kramer JR (eds) Nutrients in natural waters. Wiley, New York, pp 281–315

Williams JDH, Syers JK, Harris RF, Armstrong DE (1971a) Fractionation of inorganic phosphate in calcareous lake sediments. Soil Sci Soc Am Proc 45:250–255

Williams JDH, Syers JK, Shukla SS, Harris RF (1971b) Levels of inorganic and total phosphorus in lake sediments as related to other sediment parameters. Environ Sci Technol 5:1113–1120

Williams JDH, Jaquet J-M, Thomas RL (1976) Forms of phosphorus in the surficial sediments of Lake Erie. J Fish Res Board Can 33:413–429

Wintle AG, Huntley DJ (1980) Thermoluminescence dating of ocean sediments. Can J Earth Sci 17:348–360

Wise S (1980) Cs-137 and Pb-210: a review of the techniques and some applications in geomorphology. In: Lewin J, Cullingford R, Davidson D (eds) Timescales in geomorphology. Wiley, New York, pp 109–127

Wright HE, Cushing EJ, Livingstone DA (1965) Coring devices for lake sediments. In: Kummel B, Raup D (eds) Handbook of paleontological techniques. Freeman, San Francisco, pp 494–520

Wright RF, Matter A, Schweingruber M, Siegenthaler U (1980) Sedimentation in Lake Biel, an eutrophic, hard-water lake in north-western Switzerland. Schweiz Z Hydrol 42:101–126

Young DK, Rhoads DC (1971) Animal-sediment relations in Cape Cod Bay, Massachusetts, I.A. Transect study. Mar. Biol 11:242–254

Zink-Nielson I (1975) Interkalibrering av sedimentkemiska analysmetoder. II. Nordforsk Miljövårdssekr Publ 1975:6

Subject Index

L. Håkanson

A Manual of Lake Morphometry

1981. 49 figures. IX, 78 pages
ISBN 3-540-10480-1

Contents: Introduction. - Echosoundings: Introduction. Instrumentation. Preparations. Fieldwork. - Bathymetric map construction. - Morphometry: Optimization of lake hydrography surveys - the information value of bathymetric maps. The intensity of the survey. Practical use of the optimization model-manuals. Morphometrical parameters. - Acknowledgements. - Appendix. - References.

A prerequisite for almost all types of lake studies is the establishment of basic morphometric facts, among them lake area and volume, shoreline length, mean depth, and lake bottom contour.

A Manual of Lake Morphometry, written by a renowned scientist who has made many outstanding contributions to the field, is the first modern introduction to the design, execution and evaluation of such studies. In it, Dr. Håkanson describes both the instrumentation necessary for successful soundings and measurements and the definition of morphometrical parameters, their determination and interpretation.

This book will prove an indispensable guide for both researchers and students in hydrology, limnology and physical geography, as well as for governmental agencies and consulting firms involved with water issues.

"... His [the author's] scientific contributions have distinguished him on the North American continent as well as in Europe. His experience and impressive publication record related to lake morphology and sedimentary geochemistry have served as the basis for this new and valuable contribution... The strength of the volume lies in the fact that is does exactly what it purports to do, i. e., provide the reader with **a clear and lucid explanation** of the fundamentals of lake morphometry... The order that Håkanson brings to the chaotic array of terminology that was lake morphometry is both refreshing and worth the effort to re-learn. **This slim volume is the authoritative stuff of which definitive classics are made... in less than a year** Håkanson's manual has become **an indispensable volume,** belonging at the edge of the desk of every limnologist, hydrologist, and physical geographer, handy for ready reference." *Geo-Journal*

"... this brief treatment of lake morphometry is well done and provides a useful compilation of the terms and methods one needs to determine morphological parameters of lakes. It should be in the libraries of all limnological groups and agencies concerned with freshwater use and management." *Earth Science Reviews*

Springer-Verlag
Berlin
Heidelberg
New York
Tokyo